Fiber Optic Telecommunicatic
Lit Fiber Services Agre

Introduction

Fiber optic telecommunications networks are no longer the exclusive province of the major telecommunications carriers. In recent years there has been a great evolution and proliferation of privately owned fiber optic telecommunications networks. Much of this has been accomplished either by incumbent telephone companies or by competitive telecommunications carriers. Private users of telecommunications services, "enterprise customers," have also developed their own private fiber optic networks designed specifically to serve their unique needs.

The services delivered by a provider over its network may consist of a measured volume of bandwidth or "lit fiber" services. The following is a schedule of common measurements of lit fiber capacity.

Bandwidth Schedule:

DS-0	64 kilobits per second. This is a single voice circuit.
DS-1 (a.k.a. "T-1")	24 DS-0s 1.544 megabits per second
DS-3	28 DS-1s 45 Mbps
OC-1	54 Mbps

Designed to carry a DS-3

OC-3	3 OC-1s 155 Mbps
OC-12	12 OC-1s 622 Mbps
OC-192	192 OC-1s

This treatise is intended to explain the principal clauses of typical lit fiber services agreements. Very often a lit services agreement consists of three essential components. First, there is a Master Services Agreement, or MSA, that contains general contract terms that are intended to apply to all services offered by the vendor.

Second, there is a service description. This is a more technical explanation of the services that are offered by the vendor. The service description would set out specific service features and terms and conditions that for a particular service.

Sometimes the three elements are merged into a single document. However, where they are not, as in the sample form that accompanies this treatise, each contract element should be signed by the parties and specifically identified with a common Master Services Agreement.

The chapters of this book follow the organization of a typical lit fiber services agreement and related exhibits and agreements. However, by adopting this approach I do not mean to imply that there is a single, universally accepted standard for such agreements. Lit fiber service agreements are as varied as the businesses of the companies utilizing them. With that in mind, I have offered numerous alternate clauses to cover the same topic of many articles, as well as commentary to provide rationale for different selections.

James F. Booth
JFBTelecom@gmail.com

Fiber Optic Telecommunications Networks: Lit Fiber Services Agreements

Table of Contents

INTRODUCTION 1

MASTER SERVICES AGREEMENT 5

PART 1 - THE MASTER SERVICES AGREEMENT - INTRODUCTION 12

Part 1, Chapter 1 - The Term 13

Part 1, Chapter 2 - Testing and Acceptance of a Service 14

Part 1, Chapter 3 - Billing, Payment, and Billing Disputes 16

Part 1, Chapter 4 - Early Termination Charges 24

Part 1, Chapter 5 - Limitation of Liability 37

Part 1, Chapter 6 - Late Delivery 49

Part 1, Chapter 7 - System Maintenance 56

Part 1, Chapter 8 - Default 64

PART 2 SERVICE DESCRIPTIONS AND SLA 72

Part 2, Chapter 1 - Optical Wavelength Service 73

Part 2, Chapter 2 - Service Interruptions 86

Part 2, Chapter 3 - Chronic Outage 127

Part 2, Chapter 4 - Low Latency Service 135

Part 2, Chapter 5 - Latency Service Level Agreement 145

Part 2, Chapter 6 - Diversity 153

Part 2, Chapter 7 - Undersea Networks 167

PART 3 - THE SERVICE ORDER FORM 173

PART 4 - OTHER FORMS AND CLAUSES 176

Part 4, Chapter 1 - Portability 177

Part 4, Chapter 2 - On-Net, Off-Net, and Near-Net Services 190

Part 4, Chapter 3 - Most Favored Treatment 200

Part 4, Chapter 4 - Minimum Purchase Obligation 209

Part 4, Chapter 5 - Preferred Provider 224

Part 4, Chapter 6 - Acceptable Use Policy 231

Part 4, Chapter 7 - Termination Agreement 255

Part 4, Chapter 8 - Third Party Vendor 261

Part 4, Chapter 9 - Intellectual Property 266

Part 4, Chapter 10 - Bilateral Carrier Services Agreement 280

Part 4, Chapter 11 - Reseller Agreement 302

Part 4, Chapter 12 - Independent Referral Agreement 325

Part 4, Chapter 13 - Bandwidth Trading 339

Master Services Agreement

This Master Services Agreement ("**MSA**" or "**Agreement**") is made and entered into between _______________________, a ______________________________ ("**Seller**") and ____, a ________________________________ ("**Customer**"), with offices at

____, and is effective as of the date when signed by both Parties (the "**Effective Date**"). The terms and conditions of this MSA, any exhibits, attachments, and any service order form(s) ("**Service Order Forms**" or "**SOFs**") will govern the provision of data and communications services ("**Services**") from Seller to Customer.

The term of this MSA shall commence on the Effective Date, and shall terminate on the later of: (a) twenty four (24) months following the Effective Date; or (b) the expiration of the last effective SOF, unless terminated earlier as provided in this MSA. The term of each Service (the "**Service Term**") shall commence on the Service Activation Date and shall continue for the Service Term specified in the applicable SOF. Each Service Term shall automatically renew at the end of the Service Term on a month-to-month basis, on the same terms until terminated by either Party on not less than thirty (30) days' notice to the other Party.

1. Service Order Procedure:

(a) To order a Service, Customer must execute an SOF provided by Seller. Customer may order additional Services from time to time by executing additional SOFs. Upon receipt of an executed SOF, Seller will email an order acknowledgment letter to Customer. Within five (5) business days following the issuance of the order acknowledgement letter, Seller will either: (a) accept the SOF by emailing a Firm Order Commitment ("**FOC**") date to Customer; (b) request clarification of information on the SOF; or (c) reject the SOF. Seller shall be under no obligation to accept an SOF.

(b) After installing a Service, Seller will email an order completion notification to Customer. If Customer does not notify Seller in writing within seventy-two (72) hours following receipt of the order completion notification that the Services do not conform to Seller's specifications (with evidence of such non-conformance included in the notice), or if Seller has not performed the acceptance testing by the FOC date due to Customer's failure to satisfy any of its obligations under this MSA relating to installation, or if Customer begins using the Service for any purpose other than testing, the Service shall be deemed accepted, and such date shall constitute the "**Service Activation Date**."

(c) If the Service Activation Date for a Service has not occurred within thirty (30) days following the FOC date, except in the case of a delay caused by Customer or a Force Majeure Event, and except in the case of an Off-Net Service, Customer may cancel the Service without liability, provided that Customer notifies Seller in writing of cancellation before Seller has completed the acceptance testing.

(d) Seller's obligation to provide Services to Customer pursuant to this Agreement is subject to approval by Seller of the applicable Service Order Form and Customer's credit status. At any time during the Term, Seller may require Customer to provide a security deposit or an increased security deposit, as the case may be (the "**Security Deposit**"); provided, however, that in no event shall the amount of the Security Deposit exceed the greater of two (2) months' estimated or actual usage charges, or twice the MRCs and other amounts payable by Customer to Seller hereunder. Any such Security Deposit shall be maintained as security for Customer's performance of its obligations pursuant to this Agreement. In its sole discretion, Seller may offset any amounts past due from Customer to Seller against the Security Deposit without waiving any additional rights or remedies or making an election of remedies. Within thirty (30) days following Customer's fulfillment of its obligations under this Agreement, Seller shall return to Customer the balance of the Security Deposit.

2. Billing and Payment: Beginning on the Service Activation Date, Seller will invoice Customer monthly in advance, and all amounts shall be due and payable no later than thirty (30) days following the invoice date. Pricing on each SOF includes all applicable discounts. A Monthly Recurring Charge for any period that is less than a full month shall be a prorated portion of the Monthly Recurring Charge based on the actual number of days in such month.

At any time during the term, if Customer has failed to make two (2) payments when due in accordance with this Section 2, Seller may require Customer to remit all payments by electronic funds transfer via Automated Clearinghouse ("**ACH**") debits initiated by Seller. Customer agrees to execute from time to time all appropriate documentation reasonably requested by Seller to give effect to this Section.

(a) Past due balances are subject to an interest charge calculated from the date thirty-one (31) calendar days from the invoice date through the date of receipt of payment at the lesser rate of one and one-half percent (1.5%) per month or the maximum lawful rate allowable under applicable state law. If Customer disputes any charges billed hereunder, Customer must submit a documented claim regarding the billed amount within ninety (90) days following receipt of the invoice on which the disputed charges appear. All claims regarding disputed charges not submitted to Seller within such time frame will be deemed waived.

(b) All stated charges herein do not include, and Customer agrees to pay, any and all applicable foreign, federal, state and local taxes (other than taxes on Seller's net income), including without limitation, all sales, use, value-added, excise, franchise, property, commercial, gross receipts, license, privilege and other taxes, levies, surcharges, duties, fees, and tax-related charges (including the Universal Service Fund surcharge) and those charges resulting from Regulatory Activity, whether charged to or against Seller or Customer with respect to the sale or use of the Services or the facilities provided by Seller. Customer shall provide Seller with appropriate tax exemption certificates demonstrating that it maintains tax-exempt status from collection of all or part of these types of charges.

(c) Customer agrees to keep its billing address and contact information current, and shall be responsible for paying all reasonable collection costs incurred by Seller (including without limitation, reasonable attorneys' fees) related to unpaid invoices.

3. Termination:

(a) Either Party may terminate a Service for Default by the other Party. In the event of a Default by Customer, Seller shall have the right to any or all of the following remedies: (1) suspend Service(s) to Customer; (2) cease accepting or processing orders for Service(s); and (3) terminate this Agreement and any applicable SOF. If Seller terminates this MSA due to a Default by Customer, Customer shall immediately pay to Seller the early termination charge as described below. If the Service provided under any Service order hereunder has been suspended or terminated by Seller for Default and thereafter Seller agrees to restore such Service, Customer first must pay all past due charges, plus a re-connect charge equal to one (1) month's MRC, plus a deposit equal to two (2) months' MRCs.

(b) Except as otherwise provided in Subsection 1(c), if Customer terminates a Service or SOF before the expiration of its applicable Service Term for any reason other than a Default by Seller, in addition to any charges incurred for Service provided herein, Customer shall incur an early termination charge. Notice of early termination must be delivered to Seller in writing, and will be effective thirty (30) days after receipt. The early termination charge shall be an amount equal to: (1) 100% of the MRC for each terminated Service multiplied by the number of months remaining in the first year of the Service Term, if any; plus (2) 50% of the MRC for each terminated Service multiplied by the number of months remaining in the second and succeeding years of the Service Term; plus (3) all third party charges incurred by Seller which are directly related to the installation or termination of the Service; plus (4) all supplemental charges and NRC charges (including all nonrecurring charges that were waived by Seller at the time that Customer entered into the Service Order Form), if not already paid; plus (5) in the case of collocation space, the costs incurred by Seller in returning the collocation space to a condition suitable for use by other parties. The Parties agree that if a Service is terminated, the actual anticipated loss that Seller would likely suffer would be difficult or impossible to determine and the charges set forth herein constitute a reasonable estimate of the anticipated loss from such termination and do not constitute a penalty.

(c) Either Party may terminate this Agreement, including all attachments, exhibits and SOFs, without liability if the other Party: (1) becomes or is declared insolvent or bankrupt; (2) is the subject of any proceeding related to its liquidation, insolvency or for the appointment of a receiver or similar officer for it; (3) makes an assignment for the benefit of its creditors; (4) enters into an agreement for the composition, extension, or readjustment of all or substantially all of its obligations.

(d) Seller may suspend or terminate this MSA or a Service without liability if: (1) Seller reasonably determines that Customer or an End User has failed to comply with any foreign, federal, state or local law or regulation related to the Service, or Seller has a reasonable belief that Customer or an End User has committed any illegal act relating to the Service, including but not limited to, use of the Services for illegal purposes; or (2) a regulatory body, governmental authority, or a court of competent jurisdiction, restricts or prohibits Seller from providing a Service on the same terms and conditions as agreed herein; or (3) Seller is unable to obtain or maintain, on acceptable terms and pricing, any access right, permit or right of way necessary to provision the Services; or (4) if such suspension or termination is necessary to protect the technical integrity of the Seller network due to actions by Customer or an End User. Any termination pursuant to subsection (1) or (4) shall constitute a Default by Customer without notice to Customer.

4. <u>Customer's Use of Services</u>: Customer is solely responsible for obtaining and maintaining all licenses, approvals, and regulatory authority for its operations and the use of the Services.

5. <u>Customer Premises Equipment and Interconnection</u>: Seller will provide Service between the demarcation points listed on the SOF, or if it does not so specify, then between Seller's fiber distribution panels at the specified locations. Customer is responsible for all costs incurred on Customer's side of the demarcation points, including but not limited to, costs for customer equipment, interconnections, cross connections, hand-offs, installation charges, and any costs incurred at Customer's request.

(a) If Seller installs fiber optic cable termination or related equipment ("**<u>Seller Equipment</u>**") in Customer's owned or controlled premises ("**<u>Premises</u>**"), Seller will notify Customer and provide Customer applicable environmental specifications, power, HVAC and fire suppression requirements. Customer shall comply with these specifications, keep the Seller Equipment secure, and supply properly sized and protected power, HVAC and fire suppression systems. Customer shall be responsible to Seller for its acts or omissions that result in damage the Seller Equipment, and shall pay Seller the replacement cost if it is lost, damaged or destroyed. Customer shall notify Seller in advance of any repairs or maintenance to the Premises that may affect the Seller Equipment, and shall not interfere with its operation. Seller Equipment will remain the personal property of Seller, notwithstanding the fact that it may be affixed or attached to the Premises, and it will belong to and be removable by Seller during the term of this Agreement or thereafter.

(b) Customer is responsible for obtaining and maintaining access rights to Premises and the building where the Premises are located during the applicable Service Term so that Seller may install, repair, maintain, inspect, operate and remove Service components. If possible, Customer shall provide Seller personnel access to Service components for maintenance activities twenty-four (24) hours per day, seven (7) days per week, on two (2) hours' notice. If this access is not possible, Customer must provide commercially reasonable access, and any interruption in Service due to Seller's inability to gain access to Seller Equipment will be excluded from calculations of service credits.

6. <u>Disclaimers and Limitation of Liability</u>: (a) SELLER MAKES NO EXPRESS OR IMPLIED WARRANTY AS TO ANY SERVICE PROVISIONED HEREUNDER. SELLER SPECIFICALLY DISCLAIMS ALL IMPLIED WARRANTIES, INCLUDING, WITHOUT LIMITATION, IMPLIED WARRANTIES OF MERCHANTABILITY, FITNESS FOR A PARTICULAR PURPOSE, TITLE, NON-INFRINGEMENT OF THIRD PARTY RIGHTS, AND PERFORMANCE OR INTEROPERABILITY OF THE SERVICE WITH ANY CUSTOMER OR END USER PROVIDED EQUIPMENT.

(b) NEITHER PARTY SHALL BE LIABLE TO THE OTHER FOR ANY INDIRECT, CONSEQUENTIAL, EXEMPLARY, SPECIAL, INCIDENTAL, COVER-TYPE OR PUNITIVE DAMAGES, INCLUDING, WITHOUT LIMITATION, LOST REVENUE OR PROFITS, LOSS OF USE OR LOST BUSINESS, OR GOODWILL, ARISING IN CONNECTION WITH THIS MSA OR SELLER'S PROVISIONING OF THE SERVICES (INCLUDING BUT NOT LIMITED TO: (A) ANY SERVICE IMPLEMENTATION DELAYS OR FAILURES; (B) LOST, DELAYED OR ALTERED MESSAGES OR TRANSMISSIONS; OR (C) UNAUTHORIZED ACCESS TO OR THEFT OF CUSTOMER'S TRANSMITTED DATA), BASED ON ANY THEORY, CAUSE OF ACTION OR CLAIM, INCLUDING TORT, CONTRACT, WARRANTY, STRICT LIABILITY OR NEGLIGENCE, EVEN IF THE PARTY HAS BEEN ADVISED, KNEW OR SHOULD HAVE KNOWN OF THE POSSIBILITY OF SUCH DAMAGES.

(c) THE TOTAL LIABILITY OF SELLER TO CUSTOMER IN CONNECTION WITH THIS AGREEMENT SHALL BE LIMITED TO THE LESSER OF DIRECT DAMAGES OR THE PREVIOUS MONTH'S MRC ASSOCIATED WITH THE AFFECTED SERVICE. THE FOREGOING LIMITATION APPLIES TO ALL CAUSES OF ACTIONS AND CLAIMS, INCLUDING WITHOUT LIMITATION, BREACH OF CONTRACT, BREACH OF WARRANTY, NEGLIGENCE, STRICT LIABILITY, MISREPRESENTATION AND OTHER TORTS. CUSTOMER ACKNOWLEDGES AND ACCEPTS THE REASONABLENESS OF THE FOREGOING DISCLAIMER AND LIMITATIONS OF LIABILITY. NEITHER PARTY MAY ASSERT ANY CAUSE OF ACTION AGAINST THE OTHER PARTY UNDER ANY THEORY WHICH ACCRUES MORE THAN ONE (1) YEAR PRIOR TO THE INSTITUTION OF A LEGAL PROCEEDING ALLEGING SUCH CAUSE OF ACTION. For purposes of this Section 6, all references to Seller and Customer include their respective officers, directors, shareholders, members, managers, and employees, Affiliates, End Users, agents, lessors and providers of service to Seller.

(d) The foregoing notwithstanding, the waivers of claims and limitations of liability set forth in this Agreement shall not apply to termination charges described in Section 3, and shall not apply to a claim arising from a breach of the restrictions on transfer or assignment or use of the Services described in Subsections 8(a) and 8(b), and shall not apply to the indemnity obligations described in Subsection 8(b).

7. Service Credits: Seller may offer service credits related to installation intervals, Service availability, latency, and time to restore Service, which shall be set forth in the applicable Service Order Forms. These credits are the Customer's sole and exclusive remedy for Service related claims. To qualify for a service credit, Customer must not have any invoices that are past due, and must notify Seller that a trouble ticket should be opened to document the event. In no event shall the total amount of all service credits in a month exceed the total MRC (or NRC if applicable) for the affected Service for such month. If Seller specifies in an applicable SOF that the Service will be provided through the use of a third-party vendor, and if there is a delay in installation or interruption of such service obtained from such third party vendor, Customer shall be entitled to remedies for such delay or interruption of service only if and to the extent of the service credit to which Seller is entitled under its agreement with such third-party vendor.

8. Assignment:

(a) Customer may not transfer or assign all or any part of its right to receive the Services or bandwidth or capacity derived therefrom except to Customer's Affiliates, nor may Customer make the Services available in any manner to any third party, other than to Affiliates of Customer, in whole or in part, without the consent of Seller, which consent may be given or withheld in Seller's sole discretion.

(b) Nothing in this Agreement shall preclude Customer from reselling or providing to third party end users services derived from Customer's use of the Services, such as information services, provided that such services do not include the provision of bandwidth or lit fiber services, and provided that any such use of such Services is subject to the terms and conditions of this Agreement. Further, Customer agrees to indemnify, defend and hold harmless Seller, its Affiliates, and their employees, agents, officers and directors, from and against, and assumes all liability for, all suits, actions, damages or claims made against Seller relating to such resale or third-party use, including, but not limited to, claims based on failure, breakdown, interruption or deterioration of a Service or claims that the use of a Service, including that the placement of traffic on the Service, or any application used in connection with a Service is illegal or infringes on any intellectual property right.

(c) Customer may not assign its interest in this Agreement without the prior written consent of Seller, which consent may not be unreasonably withheld, conditioned or delayed, except that Customer may assign all, but not less than all, of its interest in this Agreement without the consent of Seller, but with prior notice to Seller (1) to any parent, Affiliate or subsidiary of Customer; (2) pursuant to a merger, acquisition, reorganization, sale or transfer of all or substantially all of the assets of Customer; or (3) as a collateral assignment for purposes of financing of Customer.

(d) Subject to the foregoing, this MSA shall be binding on Customer and its respective Affiliates, successors, and permitted assigns.

9. General:

(a) Force Majeure: Neither Party shall be liable to the other for any delay or failure in performance of any part of this Agreement to the extent that a Force Majeure Event causes such delay or failure. Further, Seller shall not be liable for any delay or failure in performance to the extent caused by Customer's failure to perform any of its obligations under this Agreement.

(b) Governing Law: This MSA is governed by and shall be construed in accordance with the laws of the State of New York without regard to its choice of law principles, except and to the extent that the Communications Act of 1934, as amended by the Telecommunications Act of 1996, and as interpreted by the FCC, applies to this Agreement.

(c) No Waiver. The failure of a Party hereto to enforce any provision of this Agreement, or the waiver thereof in any instance, shall not be construed as a general or continuing waiver or a waiver of any subsequent breach of covenant or other matter occurring, or a waiver of any other provision hereunder. No waiver by Seller or Customer of any violation or breach of any of the terms, provisions and covenants contained in this Agreement shall be deemed or construed to constitute a waiver of any other violation or breach of any of the terms, provisions and covenants contained in this Agreement. Seller's acceptance of the payment of rental or other payments after the occurrence of a Default shall not be construed as a waiver of such default, unless Seller so notifies Customer in writing. Forbearance by Seller enforcing one or more of the remedies provided in this Agreement upon a default shall not be deemed or construed to constitute a waiver of such Default or of Seller's right to enforce any such remedies with respect to such Default or any subsequent Default.

(d) Authority: Customer represents and warrants that the full legal name of the legal entity intended to receive the benefits and Services hereunder is accurately set forth herein. Each Party represents and warrants that: (1) the person signing this MSA has been duly authorized to execute on its behalf; and (2) the execution hereof is not in conflict with law, the terms of any charter or bylaw, or any agreement to which such Party is bound or affected.

(e) Headings: The headings used in this MSA are for convenience only and do not in any way limit or otherwise affect the meaning of any terms herein.

(f) Third Party Beneficiaries: Except as expressly set forth in Section 6 and subsection 9(p), the representations, warranties, covenants and agreements of the Parties set forth herein are not intended for, nor shall they be for the benefit of or enforceable by, any third party or person not a party hereto, including without limitation, Customer's End Users.

(g) Relationship: Neither Party shall have the authority to bind the other by contract or otherwise or make any representations or guarantees on behalf of the other. Both Parties acknowledge and agree that the relationship arising from this MSA is one of independent contractor, and does not constitute an agency, joint venture, partnership, employee relationship or franchise.

(h) Severability: If any provision of this MSA is held to be invalid or unenforceable, the remainder of the MSA terms and conditions will remain in full force and effect, unless such survival would be inconsistent with any express termination right provided for herein. If any such provision may be made enforceable by a limitation of its scope or time period, such provision will be deemed to be amended to the minimum extent necessary to render it enforceable.

(i) Integration: This MSA, the exhibits and the applicable SOF(s) set forth the entire agreement of the Parties with respect to the subject matter hereof, and supersede and merge all prior agreements and understandings whether written or oral. No amendment or waiver of any provision of this MSA shall be effective unless it is in writing and signed by the Party granting such amendment or waiver. The attachments, exhibits and SOFs applicable to this MSA are hereby incorporated by reference as though fully set forth herein. If there is a conflict or inconsistency between this MSA and an exhibit or SOF, the order of precedence shall be as follows: (1) the applicable exhibits (including the Service descriptions set forth therein); (2) this MSA; (3) the applicable tariff or schedule of terms and conditions published by Seller, if any; (4) the applicable Service Order Form. The foregoing notwithstanding, if an SOF or Exhibit specifically cites a provision of this MSA or an exhibit that is to be modified, superseded, or changed, the SOF shall control and take precedence.

(j) Survival: The expiration or termination of this MSA shall not relieve either Party of those obligations that by their nature are intended to survive.

(k) Remedies; Arbitration; Jurisdiction: Any dispute arising between the Parties in connection with this MSA shall be resolved by binding arbitration in the location of the Party not initiating the action, as set forth in this MSA, in accordance with the Commercial Arbitration Rules of the American Arbitration Association. In addition to such Rules, the arbitration shall be conducted in accordance with the Federal Rules of Civil Procedure, including, without limitation, the applicable rules therein with respect to discovery and the introduction of evidence. The arbitration shall be conducted by a panel of three arbitrators. Each Party shall select one arbitrator. The two chosen arbitrators shall then select the third arbitrator. The arbitrators shall have experience in telecommunications matters. Such award shall be final when rendered. The Parties shall not file any lawsuit or seek judicial review unless in accordance with this Section. Judgment on any award rendered by the arbitrators under this Section may be entered in any court having jurisdiction thereof. Any court having jurisdiction shall enforce as a binding and final arbitral award any interim measures ordered by the arbitral tribunal. Where a dispute involves a monetary claim, each Party acknowledges and agrees that it shall be required to place all disputed sums in an arbitrator approved escrow account during the pendency of the arbitration proceeding.

(l) Planned System Maintenance: Seller usually conducts Planned System Maintenance outside of normal working hours, on weekdays between 8:00 p.m. and 3:00 a.m. Eastern time, and on weekends after 5:00 p.m. Eastern time on Friday and before 5:00 p.m. Eastern time on Sunday. Seller will use reasonable efforts to minimize any Service interruptions that might occur as a result of Planned System Maintenance.

(m) Notice Information: All notices and other communications (including invoices) required or permitted under this Agreement shall be in writing and shall be given by first class mail (or its equivalent), postage prepaid, registered or certified, return receipt requested, or transmitted by electronic mail, or by hand delivery (including by means of a professional messenger service or overnight mail). Any notice or other communication shall be deemed given when received or refused and shall be sent: (1) in the case of Customer, to the Customer information contact address set forth in the SOF, and in the case of Seller, to the address below:

To Seller: ______________________________
____
____
____
Attn: ______________________________
Email Address: ______________________________

With a copy to: ______________________________
____
____
____
Attn: ______________________________
Email Address: ______________________________

Either Party may, by similar notice given, change the notice address to which future notices or other communications shall be sent.

Delivery of this MSA or an SOF as signed by a Party in pdf format, or by electronic mail shall have the same force and effect as delivery of original signatures, and that Party may use such electronic or facsimile signature as evidence of the execution and delivery of the MSA or SOF by both Parties to the same extent that an original signature could be used. Both Parties agree to exchange original executed versions as soon as is reasonably practicable after exchanging pdf versions.

(n) Confidential Information: Commencing on the Effective Date and continuing for a period of one (1) year following the expiration or other termination of this MSA, each Party shall protect as confidential and not disclose to any third party any Confidential Information received from the disclosing Party or otherwise discovered by the receiving Party during the term of this MSA, including, but not limited to, the pricing and terms of this MSA, and any information relating to the disclosing Party's technology, business affairs, and marketing or sales plans, provided that such Confidential Information is marked as confidential or, given the nature of the information or the circumstances surrounding its disclosure, such information reasonably should be considered as confidential (collectively the "**Confidential Information**"). The foregoing restrictions on disclosure of Confidential Information do not apply to information that: (1) is in the possession of the receiving Party at the time of its disclosure and is not otherwise subject to obligations of confidentiality; (2) is or becomes known through no wrongful act or omission of the receiving Party; (3) is received without restriction from a third party free to disclose it without obligation to the disclosing Party; (4) is developed independently by the receiving Party without reference to the Confidential Information of the other Party, or (5) is required to be disclosed by law, regulation, court or governmental order, administrative agency, or arbitration proceeding; provided, however, that the receiving Party shall provide prompt notice of such court order or requirement to the disclosing Party to enable the disclosing Party to seek an appropriate protective order or otherwise prevent or restrict such disclosure.

(o) Public Disclosures. All media releases, public announcements, and public disclosures relating to this Agreement or the subject matter of this Agreement, including promotional or marketing material, but not including announcements intended solely for internal distribution or disclosures to the extent required to meet legal or regulatory requirements beyond the reasonable control of the disclosing Party, shall be coordinated with and shall be subject to approval by both Parties prior to release.

(p) No Personal Liability. Every action or claim against any Party arising under or relating to this Agreement or a Service Order Form shall be made only against such Party as a corporation or company, and any liability relating thereto shall be enforceable only against the corporate or company assets of such Party. No Party shall seek to pierce the corporate veil or otherwise seek to impose any liability relating to, or arising out of, this Agreement or a Service Order Form against any shareholder, employee, officer, director, member, agent or representative of the other Party. Each of such persons is an intended beneficiary of the mutual promises set forth in this Section and shall be entitled to enforce the obligations of this Section.

10. Definitions: (a) "**Affiliate**" means an entity controlling, controlled by, or under common control with a Party. The term "control" and its correlative meanings, "controlling," "controlled by" and "under common control with," means the legal, beneficial or equitable ownership, directly or indirectly, of more than fifty percent (50%) of the aggregate of all voting equity interests in such entity; (b) "**Circuit**" means the Services provided between designated end points for the specified Service Term; (c) "**Default**" means: (1) in the case of a failure to pay any amount when due under this MSA or any SOF, if Customer fails to pay such amount within ten (10) days following notice specifying such failure; or (2) if Customer is in material breach of Section 4 of this Agreement; or

(3) in the case of any other material breach of this MSA or a Service Order Form by either Party, a Party fails to cure such breach within thirty (30) days after notice specifying such breach; (d) "**Direct Damages**" means those damages that follow immediately upon the act done and which arise naturally or ordinarily from breach of contract, but as used herein shall expressly exclude any cover-type damages; (e) "**End User(s)**" mean Customer's end-users or customers that use a Service; (f) **Excused Outage**: Any outage, unavailability, delay or other degradation of Service related to, associated with or caused by Planned System Maintenance, Customer actions or inactions, Customer provided power or equipment, any third party other than a third party directly involved in the operation and maintenance of the Seller network, including, without limitation, Customer's end users, third party network providers, traffic exchange points controlled by third parties, or any power, equipment or services provided by third parties, or a Force Majeure Event; (g) "**FCC**" means the Federal Communications Commission; (h) "**Force Majeure Event**" means an event (other than a failure to comply with payment obligations) caused by any of the following conditions: act of God; fire; flood; labor strike; sabotage; fiber cut; material shortages or unavailability or other delay in delivery not resulting from the responsible Party's failure to timely place orders therefor; power blackouts; lack of or delay in transportation; government codes, ordinances, laws, rules, regulations, permits or restrictions; failure of a governmental entity or other party to grant or recognize a right of way, war or civil disorder; or any other cause beyond the reasonable control of such Party; (i) **MRC**" means monthly recurring charge; (j) "**NRC**" means nonrecurring charge: (k) "**Off-Net**" means a service that originates from or terminates to a location that is not on the Seller network; (l) "**On-Net**" means service that originates from and terminates to locations that are both on the Seller network; (m) "**Party**" means either Seller or Customer, and "**Parties**" means collectively Seller and Customer; (n) **"Planned System Maintenance"** means maintenance on a network facility that is related to service delivery, either directly (maintenance of transmission equipment, fiber cable, etc.) or indirectly (maintenance of power, environmental systems, etc.) for which Seller has provided Customer with at least twenty-four (24) hours' notice; (o) "**Regulatory Activity**" means any regulation or ruling (including modifications thereto) by any governmental or quasi-governmental authority, regulatory agency, or court of competent jurisdiction; (p) "**Services**" means the Seller services provisioned pursuant to this MSA; (q) **"Service Activation Date"** is defined in Section 1.

IN WITNESS WHEREOF, the Parties have executed this Agreement as of the day and year written.

Customer:

By: ______________________________
Print Name: ________________________
Title: ____________________________
Date: ____________________________

Seller:

By: ______________________________
Print Name: ________________________
Title: ____________________________
Date: ____________________________

Part 1 - The Master Services Agreement - Introduction

A Master Services Agreement is one of the three principal elements of a contract for the provision of lit fiber services. An MSA may be used for multiple service orders and for many different types of services.

A Seller and a Customer will enter into a single MSA. Pursuant to the MSA the parties may thereafter enter into many service orders. (See MSA Section 1(a).) The contract terms and conditions that would be common to all service orders and all types of services would be found in the MSA.

The two other elements of a service contract are the service description and the service order form. A service description will be prepared for each type of service that is offered by the Seller, and each service description will describe with specificity the unique features of that service.

Finally, a service order form will be prepared for each service order. A complete service contract will consist of the combined elements of the MSA, the service description and the service order form. The service order form will integrate by reference the other elements of the contract and will add the commercial provisions and other terms that are specific to that service order. Each service order will constitute a new service contract.

Part 1, Chapter 1 - The Term

The term of a Master Services Agreement is defined by reference to the services ordered under the MSA. The parties may enter into a service agreement at any time during the term. Each service ordered will have its own term, and will incorporate the terms and conditions of the MSA. Therefore, whatever the stated term of the MSA, it cannot expire before the expiration of all services ordered under the MSA.

> The term of this MSA shall commence on the Effective Date, and shall terminate on the later of: (a) twenty four (24) months after the Effective Date; or (b) the expiration of the last effective SOF, unless terminated earlier as provided in this MSA. The term of each Service (the "**Service Term**") shall commence on the Service Activation Date and shall continue for the Service Term specified in the applicable SOF. Each Service Term shall automatically renew at the end of the Service Term on a month-to-month basis, until terminated by either Party on not less than thirty (30) days' notice to the other Party.

The Service Order Form is discussed at length detail later in this treatise.

One variation on this clause would be to foreclose the ability to enter into new service orders after the stated term of the MSA. The term of the agreement would continue so long as services are being provided, but no additional service orders could be submitted. In the sample form that accompanies this treatise all requests for service are subject to approval of Seller. The Seller may accept or reject any request for services. Therefore, the question of when service orders may no longer be submitted is irrelevant. In those instances where a customer has a right to require a seller to accept requests for service, a clause such as the following may be useful.

Example Clause:

> This Agreement shall begin on the Effective Date and shall automatically expire on the later of (a) three (3) years after the Effective Date or (b) the expiration of the last effective Service Term (defined below), unless earlier terminated as provided in this Agreement (the "**Term**"). The foregoing notwithstanding, Customer's ability to order Services under this Agreement shall automatically terminate on the third anniversary of the Effective Date.

Part 1, Chapter 2 - Testing and Acceptance of a Service

Before a service term begins and a customer is required to begin paying for the service, the service must conform to agreed service specifications. In this form it is presumed that the service does conform to the specifications. If the customer believes that the service does not conform, the customer must present evidence of nonconformity.

1(b) After installing a Service, Carrier will email an order completion notification to Customer. If Customer does not notify Carrier in writing within seventy-two (72) hours following receipt of the order completion notification that the Services do not conform to Carrier's specifications (with evidence of such non-conformance included in the notice), or if Carrier has not performed the acceptance testing by the FOC date due to Customer's failure to satisfy any of its obligations under this MSA related to installation, or if Customer begins using the Service for any purpose other than testing, the Service shall be deemed accepted, and such date shall constitute the "**Service Activation Date**."

(c) If the Service Activation Date for a Service has not occurred within thirty (30) days following the FOC date, except in the case of a delay caused by Customer or a Force Majeure Event, and except in the case of an Off-Net Service, Customer may cancel the Service without penalty, provided that Customer notifies Carrier in writing of cancellation before Carrier has completed the acceptance testing.

In some agreements delivery of a service is not complete until the customer indicates that the service is accepted. The clause below is one example of this approach. It calls for a joint acceptance procedure. While the two clauses are similar in many respects, the approach described below would give the customer somewhat more control of the manner and timing of acceptance. It differs from the above clause in several important respects. First, it requires the Seller to give the Customer prior notice of a time for testing and an opportunity to be present to observe the testing. Second, acceptance is not complete until the Customer executes a service acceptance statement. There is not simply a presumption that the service conforms to specifications. Third, it expressly allows the Customer to terminate the Service if the Seller fails to bring the Service within specifications within a limited period of time.

Example Clause:

Acceptance Testing

Seller shall notify the appropriate Customer field coordination contact by facsimile, electronic mail or telephone (at the numbers specified on the applicable Service Order) that the Service is ready for testing, the place where the test or retest will take place and ***the date and time of testing or retesting***, which date and time shall be not less than two (2) business days after notification. Seller shall perform such testing pursuant to the Specifications. Following completion of each test or retest, Seller shall deliver a completed copy of the Service Acceptance Notice in the form attached hereto as ***Exhibit B*** (the "**Service Acceptance Notice**") and a copy of all test results to the Customer field representative present at the testing or, if an Customer field representative is not present, Seller shall deliver same by overnight delivery service, facsimile or electronic mail to the Customer field coordination contact identified in the Service Order.

If the testing results meet the Specifications and the Service conforms to the Service Order, the ***Customer field representative shall verify Service acceptance by signing the Service Acceptance Notice***. If a Customer representative was not present at the testing, Customer shall execute the Service Acceptance Notice and provide a copy to Seller within ten (10) days after receipt of the completed notice and test results. In either event, Seller shall provide a copy of the fully-executed Service Acceptance Notice (by facsimile) to the representative placing the order for Service as indicated on each Service Order. If Customer believes that the test results do not meet the Specifications, the Customer field representative present at the testing shall so notify Seller's representatives, or if no Customer representative is present, Customer shall so notify Seller within ten (10) days after receipt of the completed notice and test results, in either event specifying the inadequacy in the test results. Notwithstanding anything to the contrary contained herein, if Customer does not verify Service acceptance in writing (by executing a Service Acceptance Notice and providing a copy to Seller) or notify Seller of its rejection of the test results within such ten (10) day period, Seller shall submit to the Customer representative placing the order for Service as indicated on the applicable Service Order, a Service Acceptance Reminder Notice in the form attached hereto as ***Exhibit C*** (the "**Service Acceptance Reminder Notice**"). If Seller does not receive a Service Acceptance Notice or notice of rejection of test results within ten (10) days after receipt by Customer of Seller's Service Acceptance Reminder Notice, Customer's rejection of the Service shall be deemed to have occurred.

If Seller determines that its test results fail to meet the applicable Specifications, or if Customer notifies Seller, in accordance with Section ___ above, that the test results fail to meet the applicable Specifications, Seller shall use its best efforts to cure all Service defects within a maximum of ten (10) days ("**Acceptance Cure Period**"). If the Service does not meet the applicable Specifications after two (2) service acceptance retests, Customer may either: (1) ***cancel the Service Order*** without liability; or (2) accept the Service with amended Specifications. If Customer elects to cancel such Service Order, Customer shall provide written notice to Seller of such election and shall remove any traffic or test signal within ten (10) days after the applicable Acceptance Cure Period. Seller shall reimburse Customer for all interconnection expenses incurred in preparation for commencement of Service within thirty (30) days after receipt of Customer's invoice. If Customer elects to accept the Service with amended Specifications, the Specifications for that particular Service shall be amended to reflect the actual test results for such Service.

Part 1, Chapter 3 - Billing, Payment, and Billing Disputes

Billing begins once a service has been accepted by the customer.

This form provides for electronic payment under certain circumstances. If a Customer fails to make two payments when due, the Seller may require the Customer to thereafter make payments electronically. A provision for electronic payments may be useful as a matter of course or only as a protective measure in the event of late payment or a credit risk.

1(d) Seller's obligation to provide Services to Customer pursuant to this Agreement is subject to approval by Seller of the applicable Service Order Form and Customer's credit status. At any time during the Term, Seller may require Customer to provide a security deposit or an increased security deposit, as the case may be (the "Security Deposit"); provided, however, that in no event shall the amount of the Security Deposit exceed the greater of two (2) months' estimated or actual usage charges, or twice the MRCs and other amounts payable by Customer to Seller hereunder. Any such Security Deposit shall be maintained as security for Customer's performance of its obligations pursuant to this Agreement. In its sole discretion, Seller may offset any amounts past due from Customer to Seller against the Security Deposit without waiving any additional rights or remedies or making an election of remedies. Within thirty (30) days following Customer's fulfillment of its obligations under this Agreement, Seller shall return to Customer the balance of the Security Deposit.

2. Billing and Payment: Beginning on the Service Activation Date, Seller will invoice Customer monthly in advance, and all amounts shall be due no later than thirty (30) days following the invoice date. Pricing on each SOF includes all applicable discounts.

At any time during the term, if Customer has failed to make two (2) payments when due in accordance with this Section 2, Seller may require Customer to remit all payments by electronic funds transfer via Automated Clearinghouse ("ACH") debits initiated by Seller. Customer agrees to execute from time to time all appropriate documentation reasonably requested by Seller to give effect to this Section.

(d) Past due balances are subject to an interest charge calculated from the date thirty-one (31) calendar days from the invoice date through the date of receipt of payment at the lesser rate of one and one-half percent (1.5%) per month or the maximum lawful rate allowable under applicable state law. If Customer disputes any charges billed hereunder, Customer must submit a documented claim regarding the billed amount within ***ninety (90) days following receipt of the invoice*** on which the disputed charges appear. All claims regarding disputed charges not submitted to Seller within such time frame will be deemed waived.

(e) All stated charges herein do not include, and Customer agrees to pay, any and all applicable foreign, federal, state and local taxes (other than taxes on Seller's net income), including without limitation, all sales, use, value-added, surcharges, excise, franchise, property, commercial, gross receipts, license, privilege and other taxes, levies, surcharges, duties, fees, or other tax-related surcharges (including the Universal Service Fund surcharge) and those charges resulting from Regulatory Activity, whether charged to or against Seller or Customer with respect to the sale or use of the Services or the facilities provided by Seller. Customer shall provide Seller with appropriate tax exemption certificates demonstrating that it maintains tax-exempt status from collection of all or part of these types of charges.

(f) Customer agrees to keep its billing address and contact information current, and shall be responsible for paying all reasonable collection costs incurred by Seller (including without limitation, reasonable attorneys' fees) related to unpaid invoices.

Late payments bear interest at one and one-half percent per month. This is a fairly common provision. One and one-half percent interest on late payments is not a punitive provision; it merely represents an estimate of the time value of a late payment.

A security deposit is not always required, or is not required initially at the time a contract is entered into. Sometimes a security deposit will be required only under certain circumstances, such as a failure to make payments in a timely manner or discovery of a negative credit rating.

Example Clause #1:

Seller's obligation to provide Services to Customer pursuant to this Agreement is subject to approval by Seller of each applicable Service Order Form and Customer's credit status. At any time during the Term, if: (1) Customer ***fails to make a payment when due*** in accordance with ***Section 2***; or (2) ***Seller receives a credit assessment*** from a credit rating agency that indicates an ***unacceptable*** financial stress of the Customer, then Seller may require a security deposit or, if Customer has previously provided security, Seller may require additional security (the "**Security Deposit**"). Customer's failure to provide the requested Security Deposit within five (5) days following Seller's request shall constitute a default; provided, however, that in no event shall the amount of the Security Deposit exceed the greater ***of two (2) months'*** estimated or actual usage charges, the MRCs, and other amounts payable by Customer to Seller under all Service Order Forms. Any such Security Deposit shall be maintained as security for Customer's performance of its obligations under this Agreement. In its sole discretion, Seller may offset any amounts past due from Customer to Seller against the Security Deposit without waiving any additional rights or remedies or making an election of remedies. At the expiration of the Service Term the amount of the Security Deposit will be credited to Customer's account, and any remaining credit balance will be refunded to Customer within thirty (30) days thereafter.

The following three clauses are very similar and reflect a simpler approach to security deposits. However, the amount of the maximum security deposit is not limited. The seller may require a security deposit if it merely decides that one is necessary.

Example Clause #2:

Credit Approval and Deposits. Customer will provide Seller with credit information as requested, and delivery of Service is subject to credit approval. Seller may require Customer to make a deposit as a condition to Seller's acceptance of any Customer Order, or as a condition to Seller's continuation of Service. The deposit will be held by Seller as security for payment of Customer's charges. When Service to Customer is terminated, the amount of the deposit will be credited to Customer's account and any remaining credit balance will be refunded.

Example Clause #3:

Charges, Billing and Payment: Provision of Service is subject to Seller's approval of Customer's credit standing. Seller may require a deposit prior to the provision of Service or as a condition to the continued provision of Service, if Customer's credit standing or payment record so indicates. Billing for Services begins on the Service Date and will not be delayed due to Customer premises equipment or Customer's readiness to accept or use Service.

Example Clause #4:

Seller's obligation to provide Services to Customer pursuant to this Agreement is subject to approval by Seller of the applicable Service Order Form and Customer's credit status. At any time during the Term, (defined below) Seller may require Customer to provide a security deposit or an increased security deposit, as the case may be (the "Security Deposit"). Any such Security Deposit shall be maintained as security for Customer's performance of its obligations pursuant to this Agreement. In its sole discretion, Seller may offset any amounts due from Customer to Seller against the Security Deposit without waiving any additional rights or remedies or making an election of remedies. Seller shall return the balance of the Security Deposit to Customer within thirty (30) days of Customer's fulfillment of its obligations under this Agreement.

In contrast to simple approaches above, the following clauses reflect a more systematic approach to credit reviews and security deposits. In this first clause the seller reserves the right to terminate the service agreement entirely if the customer becomes "financially unstable." Then, of course, "financially unstable" must be defined.

Example Clause #5:

Deposits. Using its Deposit standards, Seller has assessed and Customer shall pay the Initial Deposit amount specified on the Cover Sheet before Services are provided. Seller may require Customer, during the term of this Agreement, to tender a deposit in an amount to be determined by Seller in its reasonable discretion. Seller will rely upon commercially reasonable factors to determine the need for and amount of any deposit. These factors may include, but are not limited to, payment history, number of years in business, history of service with Seller, bankruptcy history, current account treatment status, financial statement analysis, and commercial credit bureau rating, as well as commitment levels and anticipated monthly charges. Any deposit will be held by Seller as a guarantee for the payment of charges. A deposit does not relieve Customer of the responsibility for the prompt payment of bills. Interest (at the rate of 6% per year or such other rate as is applicable by law) will be paid to Customer for any period that a cash deposit is held by Seller.

Either Party may elect to terminate this Agreement if the other Party becomes Financially Unstable. *"**Financially Unstable**"* means an entity (including any entity controlling it, where "control" means direct or indirect (i) ownership of at least forty percent (40%) of the equity, (ii) ability to direct at least forty percent (40%) of voting power, or (iii) ability otherwise to direct management policies) that (a) generally fails to pay (or admits in writing its inability to pay) material debts when they become due, or (b) becomes subject to any writ, judgment, warrant of attachment, execution or similar process against a substantial part of its property, assets or business. The Financially Unstable Party must notify the other Party of the occurrence of either of the foregoing events within three (3) calendar days of its receipt of notice or other knowledge thereof.

The following clause would allow the seller to require a security deposit only if the customer fails to make a payment when due.

Example Clause #6:

Security Deposit. ***If Customer fails to make a monthly payment*** as required under an applicable Network Order, Seller reserves the right during the Term (as defined in Section 11) to require Customer, within ten (10) days of Seller's demand, to pay a security deposit to Seller as set forth in the Fee Schedule ("Security Deposit"), post an irrevocable letter of credit or provide other adequate assurances, in an amount determined at Seller's reasonable discretion, that all payments due hereunder will be made, if Customer's financial position, taking into consideration Customer's payment history, number of years in business, history with Seller, bankruptcy history, current account treatment status, financial analysis and/or commercial credit rating becomes unacceptable to Seller. Each Security Deposit (which shall not bear interest to Customer unless required to do so by any provision of law) shall be considered as security for the payment and performance by Customer of all of Customer's obligations, covenants, conditions and promises under this Agreement or any other Agreement Customer may have with Seller. In the event that Seller draws down any portion of the Security Deposit, Customer shall replenish all drawn down amounts within ten (10) days of Seller's demand therefor. Promptly upon expiration of the Term or any renewal or extension thereof, and in no event later than ninety days (90) after the expiration of the Term or any renewal or extension thereof, Seller shall return such Security Deposit to Customer, provided Customer is not in Default under the terms hereof less any reasonable amounts appropriated and properly documented by Seller to make good on Customer's obligations hereunder.

The following clause reflects a much stricter approach to credit approval and requirements for a security deposit. The criteria of credit approval are not defined. What would constitute an "adverse change in Customer's financial condition" is not defined. Thus the seller would have a great deal of discretion in extending credit and requiring financial safeguards. Furthermore, the customer would be required to respond on very short notice to additional conditions of service imposed by the seller at any time during the term.

Example Clause #7:

> Credit Approval and Deposits. All Services hereunder are ***subject to credit approval***. Customer shall provide Supplier with credit information as requested. Supplier shall establish a credit limit ("Credit Limit"), which will be specified in a rider that shall be incorporated into this Agreement by reference (the "Credit Approval Rider"). The Credit Approval Rider shall be issued to Customer prior to the provision of Services. In the event the Credit Limit is exceeded, at the request of Supplier Customer shall within one (1) business day provide the amount of MRC and/or unbilled usage charges exceeding the Credit Limit or, at Supplier's option, a deposit in accordance with Section ___, Additional Assurance, below. In the event Customer does not respond to Supplier's request, Supplier reserves the right to suspend Services without further notice until Supplier's requirements are met. The Credit Limit is subject to periodic review by Supplier and, as a result of such review, may be adjusted upon written notice to Customer; such adjustment may be made without an amendment hereto.
>
> Additional Assurances. If at any time during the Term of this Agreement ***there is a material and adverse change in Customer's financial condition***, business prospects or payment history, which shall be determined by Supplier in its sole and reasonable discretion, Supplier may demand that Customer provide Supplier with a Security Deposit or increase the amount of the Security Deposit, as the case may be, as security for the Full and faithful performance of Customer of the Terms, conditions and covenants of this Agreement; in no event shall the amount of the Security Deposit ever exceed two (2) months estimated or actual usage charges, MRC and/or other amount payable by Customer to Supplier hereunder. A Security Deposit also may be required in the Credit Approval Rider or prior to Supplier's acceptance of any SOF.

The following clause has one additional feature that has not appeared in other clauses. This seller requires a security interest in its customer's equipment.

Example Clause #8:

> Credit Approval.
>
> A. At Seller's request, Customer must provide Seller with financial statements or other indications of Customer's financial and business circumstances so that Seller may review Customer's credit standing.
>
> B. Seller may require security (or additional security, if previously provided) if:

(1) during the Term, Customer's financial or business circumstances or payment history are unacceptable to Seller;
(2) Customer fails to make payment(s) when due under Section ___;
(3) Customer exceeds its credit line or limit; or
(4) Seller has reason to believe that Customer is or may become insolvent.

C. Customer's failure to provide the requested security within five business days after Seller's request is a material breach of this Agreement.

D. At Customer's request, Seller will provide Customer written information regarding its credit line or limit for purchasing Services under this Agreement.

E. If, at any time, Customer exceeds its credit line or limit by more than 20%, the excess amount is due immediately at Seller's request. If Customer fails to provide payment within five business days of Seller's request, Customer will have materially breached this Agreement.

F. ***Customer grants Seller a purchase money security interest in equipment purchased under this Agreement***, together with all replacements, parts, additions, repairs and accessories incorporated in or affixed to the equipment, and all proceeds of the sales of the equipment, until all charges in this Agreement (including interest, if any) are paid in full. Customer will execute and deliver any documents that Seller reasonably requests for filing or recording, as may be necessary to perfect the security interest created under this Agreement. The parties agree that the equipment will remain personal property, not a part of the land or building, regardless of the manner of affixation.

Example Clause #9:

Credit Approval. Customer will promptly provide Seller with such financial and credit information regarding Customer and/or its ability to pay for Services hereunder as Seller may at any time and from time to time request. Delivery of Service hereunder is subject in all respects to Seller's satisfaction that Customer meets Seller's then-prevailing creditworthiness requirements. Seller may require a letter of credit, prepayment, security deposit, parental guaranty(ees), or other form of payment security as a condition to Seller's providing or continued provision of any Service. Any payment security tendered to Seller hereunder will be held by Seller as security for payment of Customer's obligations. Upon the termination of this Agreement, any Customer payment security held by Seller will be applied against amounts due or that may become due by Customer to Seller under Customer's account with Seller. After Customer's account has been finally settled, the balance, if any, of any cash payment security of Customer will be refunded to Customer. Seller reserves the right at any time to withdraw credit approval for Customer and limit the size of Customer's account. Customer hereby authorizes (i) Seller, directly or indirectly, to collect, obtain, or exchange any information that may be required relative to the creditworthiness of Customer from any source, including Customer's lenders or financial institutions and trade suppliers and (ii) each such source to provide such information to Seller. If Customer is a sole proprietorship or partnership,

Customer hereby consents to allowing Seller to obtain a consumer credit report for the purpose of evaluating Customer's creditworthiness.

Example Clause #10:

(a) Customer's initial monthly credit limit hereunder shall be $50,000 (the "Monthly Credit Limit"). If Customer's charges for the Services are projected to exceed (based on Seller's measurement of Customer's daily usage run rate), or do exceed, its Monthly Credit Limit, Seller may at any time: (1) require additional security of its choice from Customer in an amount equal to Customer's projected monthly run rate or Customer's highest Invoice over the prior six month period (or such lesser period if this Agreement has not been in effect for six months) as a condition to continuing to provide the Services; and (2) demand immediate payment in the amount of such actual or projected excess, notwithstanding the date payment would ordinarily be due for that billing cycle. In addition, if (1) Customer is delinquent in payment of an Invoice, or (2) Customer's overall financial condition changes adversely during the term hereof (in Seller's reasonable business judgment), and Seller does not have security from Customer in an amount equal to Customer's highest Invoice over the prior six month period (or such lesser period if this Agreement has not been in effect for six months), Seller may require additional security of its choice from Customer at two times such amount. Any such additional security shall be provided by Customer to Seller within forty-eight (48) hours if the security is to be other than a letter of credit and within ten (10) Business Days if the security is to be a letter of credit, from its receipt of Seller's written request for additional security.

Billing Disputes:

Comment: Subparagraph (a) of Section 2 above imposes a limitation on Customer's right to dispute invoices. If a Customer questions a charge, the matter must be raised within 90 days following receipt of the pertinent invoice. Customer would be barred from asserting a claim after that time. This is sometimes referred to as a "sunset clause."

A related point that is not discussed in these passages is a customer's right to withhold disputed amounts. Although a customer would have a right to question an invoice at any time within 90 days, none of the clauses discussed above would allow a customer to withhold disputed amounts. This is sometimes a difficult issue for both parties and there are important considerations on both sides of the question. To a seller that has provided service and incurred expenses, the prospect of a customer withholding payment is very undesirable. A right to withhold payment of disputed amounts can sometimes be manipulated unfairly. On the other hand, a customer that believes it is entitled to a service credit is sometimes reluctant to trust that the seller will resolve a claim promptly and fairly.

The clause below would allow a customer to withhold disputed amounts, but would impose parameters around the right, thus affording limited protection of the seller against abuse. A customer would have a right to withhold payment only if and to the extent it could satisfy the requirements of this clause. In addition to a sunset clause as discussed above:

1. Customer may withhold only for amounts billed currently, only for amounts disputed on a current invoice. Customer may not withhold for amounts disputed on previous invoices that have been paid. The Customer may still raise the issue and demand a refund for amounts previously paid, but Customer cannot withhold the disputed amount.

2. The Customer must furnish "Complete Documentation" of any disputed amount.

Example Clause #11:

Withholding. If Customer has a Bona Fide Dispute with respect to a charge that appears on an invoice, Customer may, on or before the due date for payment of the invoice, notify Seller that it is withholding a disputed sum. For any such notice to be valid, Customer must submit Complete Documentation with such notification. Promptly following such notice the parties shall exercise reasonable good faith efforts to resolve the dispute. If the parties fail to resolve such dispute within forty-five (45) ***days following the date that payment of the invoice is due, Customer shall deposit all withheld amounts into escrow pending resolution by arbitration.***

The foregoing notwithstanding, Customer may withhold only for a Bona Fide Dispute pertaining to a charge appearing on the ***current invoice***, or for a sum that by the terms of this Agreement would be credited to such current invoice; and Customer may not withhold or set-off against sums previously paid, but Customer may claim a refund within ninety (90) days following the invoice date for amounts previously paid. Customer waives the right to dispute any charges not disputed within such 90-day period.

For purposes of this Agreement: (1) "**Bona Fide Dispute**" means a good faith assertion of a right or claim to which Customer believes it is entitled under the applicable terms of Service. A Bona Fide Dispute shall not include (and Customer may not withhold any amounts invoiced for) actual usage of Customer, Customer's End Users or unauthorized third parties (e.g., fraudulent usage); and (2) "**Complete Documentation**" means documents and other detailed written support which identify with specificity the charges disputed by Customer, which explain the reasons for such dispute and support Customer's position and justify the withholding for the Service interruption credit or other credit to which Customer believes itself entitled; and states the amounts being withheld by Customer pending resolution of such dispute.

Part 1, Chapter 4 - Early Termination Charges

A service may be terminated for many reasons before the normal expiration of its term. A seller may terminate a service in the event of a default by the customer. The customer may choose to terminate a service because its business needs change.

Section 3(b) would allow a customer to terminate a service for any reason, a termination for convenience. Some agreements would not permit a customer to terminate a service for convenience without paying the full contract price that was agreed to be paid for the full term. (See Example Clause #1 below.) Other agreements would allow a customer to pay a lesser amount as a termination charge. Section 3(b) of the MSA sets forth a termination charge of:

(a) 100% of the first year of the term of the Service, plus
(b) 50% of amounts owed for subsequent years.

Note that in this clause the early termination charges are calculated by reference to the term of the applicable service, not the remainder of the term following a decision to terminate the service. That is, if the term of a service is three years, and the customer terminates the service in the second year of the service term, the termination charge would simply be 50% of the remaining charges that had been agreed.

3. Termination:

(e) Either Party may terminate a Service for Default by the other Party. In the event of a Default by Customer, Carrier shall have the right to any or all of the following remedies: (1) suspend Service(s) to Customer; (2) cease accepting or processing orders for Service(s); and (3) terminate this Agreement and any applicable SOF. If Carrier terminates this MSA due to a Default by Customer, Customer shall immediately pay to Carrier the early termination charge as described below or in the applicable Service Order Form. If the Service provided under any Service Order hereunder has been suspended or terminated by Carrier for Default and thereafter Carrier agrees to restore such Service, Customer first must pay all past due charges, plus a re-connect charge equal to one (1) month's MRC, plus a deposit equal to two (2) months' MRCs.

(f) Except as otherwise provided in Subsection 1(c), if Customer terminates a Service or SOF before the expiration of its applicable Service Term for any reason other than a Default by Carrier, in addition to any charges incurred for Service provided herein, Customer shall incur an early termination charge. The Parties acknowledge and agree that the early termination charge is a reasonable estimate of the likely loss and damage suffered by Carrier and is not a penalty. Notice of early termination must be delivered to Carrier in writing, and will be effective thirty (30) days after receipt. The early termination charge shall be an amount equal to: (1) ***100% of the MRC for each terminated Service multiplied by the number of months remaining in the first year of the Service Term***, if any; plus (2) ***50% of the MRC for each terminated Service multiplied by the number of months remaining in the second and succeeding years of the Service Term***; plus (3) all third party charges incurred by Carrier which are directly related to the installation or termination of the Service; plus (4) all supplemental charges and NRC charges (including all nonrecurring charges that were waived by Carrier at the time that Customer entered into the Service Order Form), if not already paid; plus (5) in the case of collocation space, the costs incurred by Carrier in returning the collocation space to a condition suitable for use by other parties. The Parties agree that if a Service is terminated, the actual anticipated loss that Carrier would likely suffer would be difficult or impossible to determine and the charges set forth herein constitute a reasonable estimate of the anticipated loss from such termination and do not constitute a penalty.

Either Party may terminate this Agreement, including all attachments, exhibits and SOFs, without penalty if the other Party: (1) becomes or is declared insolvent or bankrupt; (2) is the subject of any proceedings related to its liquidation, insolvency or for the appointment of a receiver or similar officer for it; (3) makes an assignment for the benefit of its creditors; (4) enters into an agreement for the composition, extension, or readjustment of all or substantially all of its obligations.

Example Clause #1:

If this Agreement or any Service is terminated prior to the expiration of the initial or any renewal term of any Service (except if properly terminated by Customer for Seller's breach under Section ___), then Customer shall pay to Seller upon demand an early termination fee in an amount equal to the aggregate sum of each existing Service's MRC times the number of months remaining of the applicable term and all third party termination liability for Customer's termination. Customer agrees that such a termination fee is based on an agreed revenue expectation based on actual Customer Service data and is not a penalty.

Note that a default by Customer would also give rise to an obligation to pay termination charges. (Section 3(a).)

Here are two more examples in which the termination charges are scaled according to the remaining term of the agreement. As more of the term of the service passes, the charge for terminating the service would be decline.

Example Clause #2:

Customer may also terminate the Agreement, or any Service provided hereunder, at its convenience, upon thirty (30) days written notice; however, the Customer shall pay in addition to any charges for Services used, an early termination fee equal to: (a) ***100% of the*** monthly recurring charge ("MRC") for the terminated Service multiplied by the number of remaining months in the ***first year*** of the Service Term; plus (b) ***75% of the MRC*** for the terminated Service multiplied by the number of remaining months in the ***second year*** of the Service Term, if applicable; plus (c) **50%** of the MRC for the terminated Service multiplied by the number of remaining months in the ***third year*** of the Service Term, if applicable; plus (d) ***25%*** of the MRC for the terminated Service multiplied by the number of remaining months in the ***fourth year*** of the Service Term;

plus (e) ***10%*** of the MRC for the terminated Service multiplied by the number of remaining months in any additional year of the Service Term, if applicable.

<u>Example Clause #3</u>:

Early Termination. If a Product Order is terminated prior to expiration by reason of: (i) Customer termination, if such termination is not an exercise of Customer's rights or remedies under the Agreement, or (ii) a Customer Event of Default, then in addition to all other sums due and owing, Customer agrees to immediately pay early termination charge (as liquidated damages, and not a penalty) equal to the following:

1st year termination – 100% of all unpaid monthly MRCs for the first year, plus 65% of all unpaid monthly MRCs for the second year, plus 45% of all unpaid monthly MRCs for the third year, plus 30% of all unpaid monthly MRCs for the fourth year, plus 20% of all unpaid monthly MRCs for the fifth year, plus 10% of all unpaid monthly MRCs for the sixth through tenth years.

2nd year termination – 65% of all unpaid monthly MRCs for the second year, plus 45% of all unpaid monthly MRCs for the third year, plus 30% of all unpaid monthly MRCs or the fourth year, plus 20% of all unpaid monthly MRCs for the fifth year, plus 10% of all unpaid monthly MRCs for the sixth through tenth years.

3rd year termination – 55% of all unpaid monthly MRCs for the third year, plus 35% of all unpaid monthly MRCs for the fourth year, plus 20% of all unpaid monthly MRCs for the fifth year, plus 10% of all unpaid monthly MRCs for the sixth through tenth years.

4th year termination – 20% of all unpaid monthly MRCs for the fourth year, plus 15% of all unpaid monthly MRCs for the fifth year, plus 10% of all unpaid monthly MRCs for the sixth through tenth years.

5th year termination – 10% of all unpaid monthly MRCs for the fifth year, plus 5% of all unpaid monthly MRCs for the sixth through tenth years.

6th through 20th year termination – 5% of all unpaid monthly MRCs for the sixth through tenth years, and no Early Termination Charge for 11th through 20th year termination.

The following several examples take a similar approach but would call for a reduced termination charge if customer terminates before the service activation date. The first clauses allows for a cancellation window, before a service is turned up, during which the customer would be incur a lesser termination charge the service is terminated before the projected service activation date, and a termination charge of three months' MRC if the service is terminated later than five days before the service activation date.

Example Clause #4: The following clause takes a similar approach.

Early Termination. Licensee may terminate any Licensed Fiber Addendum on notice to Seller. If Licensee terminates a Licensed Fiber Addendum before its expiration by reason of: (1) Licensee termination, if such termination is not an exercise of Licensee's rights or remedies under the Agreement or a Licensed Fiber Addendum, or (2) a Licensee default, then in addition to all other sums due and owing, Licensee agrees to immediately pay early termination charge (as liquidated damages, and not a penalty) equal to the following:

1. 1st year termination - 100% of all unpaid monthly License Fees for the first year, plus 65% of all unpaid monthly License Fees for the second year, plus 40% of all unpaid monthly License Fees for the third year, plus 20% of all unpaid monthly License Fees for the fourth year, plus 10% of all unpaid monthly License Fees for the fifth year through the final year of the License Term.

2. 2nd year termination - 65% of all unpaid monthly License Fees for the second year, plus 40% of all unpaid monthly License Fees for the third year, plus 20% of all unpaid monthly License Fees for the fourth year, plus 10% of all unpaid monthly License Fees for the fifth year through the final year of the License Term.

3. 3rd year termination - 40% of all unpaid monthly License Fees for the third year, plus 20% of all unpaid monthly License Fees for the fourth year, plus 10% of all unpaid monthly License Fees for the fifth year through the final year of the License Term.

4. 4th year termination - 20% of all unpaid monthly License Fees for the fourth year, plus 10% of all unpaid monthly License Fees for the fifth year through the final year of the License Term.

5. 5th year termination - 10% of all remaining unpaid monthly License Fees.

The foregoing notwithstanding, Licensee may terminate a Licensed Fiber Addendum on sixty (60) days prior written notice without payment of any early termination charge by purchasing other services from Seller at a price that is equal to or greater than the monthly License Fee of the Licensed Fiber Addendum.

Example Clause #5:

Termination Charges. (A) Customer may cancel a Customer Order following Seller's acceptance of the same and ***prior to the Customer Commit Date*** [1]on notice to Seller. If Customer does so, or if delivery of such Service is terminated by Seller prior to delivery of a Connection Notice due to a failure of Customer to comply with these Terms, Customer shall pay Seller a cancellation charge equal to the sum of, in the case of

[1] In the form from which this term was borrowed, the term "Customer Commit Date" means the date that the seller has agreed that the service will be in operation.

Collocation Space, the costs incurred by Seller in returning the Collocation Space to a condition suitable for use by third parties, plus the sum of the following:

(a) Any third party cancellation/termination charges related to the installation and/or cancellation of Service;

(b) The non-recurring charges (including any nonrecurring charges that were waived by Seller at the time of the Customer Order) for the cancelled Service; and

(c) As the case may be, (1) ***one (1) month's monthly recurring charges*** for the cancelled Service if written notice of cancellation is received by Seller ***more than five (5) business days prior to the Customer Commit Date***, or (2) ***three (3) month's monthly recurring charges*** for the cancelled Service if written notice of cancellation is received by Seller ***five (5) business days or less prior to the Customer Commit Date***.

Customer's right to cancel any particular Service under this Section 3.8(A) shall automatically expire and shall no longer apply upon Seller's delivery to Customer of a Connection Notice for such Service.

(B) In addition to Customer' right of cancellation under Section ___(A) above, Customer may terminate Service prior to the end of the Service Term upon thirty (30) days' prior written notice to Seller, subject to the following termination charges. In the event that after either the Customer Commit Date or Customer's receipt of the Connection Notice for a particular Service (whichever occurs first) and prior to the end of the Service Term, Customer terminates Service or in the event that the delivery of Service is terminated due to a failure of Customer to comply with these Terms, Customer shall pay Seller a termination charge equal to the sum of (1) in the case of Colocation Space, the costs incurred by Seller in returning the Colocation Space to a condition suitable for use by third parties, plus (2):

(a) Any third party cancellation/termination charges related to the installation and/or termination of Service;

(b) The non-recurring charges (including any nonrecurring charges that were waived by Seller at the time of the Customer Order) for the cancelled Service, if not already paid; and

(c) The percentage of the monthly recurring charges for the terminated Service calculated from the effective date of termination as (1) 100% of the remaining monthly recurring charges that would have been incurred for the Service for months 1-12 of the Service Term, plus (2) 50% of the remaining monthly recurring charges that would have been incurred for the Service for months 13 through the end of the Service Term.

Example Clause #6:

If Customer terminates a Service or Service Order ***after the Cancellation Window***, which is defined as after Service Acceptance and before the expiration of its applicable Service Term for any reason other than pursuant to sections ___ or ____ or as otherwise provided hereunder, or if Provider terminates for Customer default that is not cured after thirty (30) days prior written notice, or for any of the reasons specified in paragraph ____ or ____, in addition to any charges incurred for Service provided to the date of termination, Customer will incur an early termination charge, which will be Provider's sole and exclusive remedy. The parties acknowledge and agree that this early termination charge is a reasonable estimate of the likely loss and damage suffered by Provider and is not a penalty. Notice of early termination must be delivered to Provider in writing, and will be effective thirty (30) days after receipt. The early termination charge shall be an amount equal to: (a) 100% of the MRC for each terminated Service multiplied by the number of remaining months in the first year of the Service Term, plus (b) 75% of the MRC for each terminated Service multiplied by the number of remaining months in the second year of the Service Term, if applicable, plus (c) 50% of the MRC for each terminated Service multiplied by the number of remaining months in any additional year of the Service Term, if applicable, plus (d) any termination liability associated with any third party provided service, plus (e) any special construction charges and NRC) listed on the Service Order if not already paid. To the extent possible, Provider and Customer will work to mitigate such charges. Notwithstanding anything to the contrary contained above, in no event shall the termination charge exceed 100% of the MRC for the remaining term of a terminated Service.

If Customer terminates a Service or Service Order before Service Acceptance, but after the Firm Order Commitment ("FOC") date has been sent to Customer, then Customer will incur an early termination charge that will be equal to the amount necessary to reimburse Provider for all actual and documented third party charges incurred in preparing to deliver the Service.

Example Clause #7:

Subject to the following, Customer may terminate a Service ***prior to the applicable Service Term start date*** by providing Seller notice at least fifteen (15) days prior to the start date set forth in the applicable Service Order Form. In the event Customer so terminates a Service, Customer shall pay Seller all applicable charges incurred to that date, including any third party costs. In the event Customer so terminates a Service by notifying Seller less than fifteen (15) days but at least ten (10) days prior to the start date set forth in the applicable Service Order Form, then, in addition to paying Seller all applicable charges incurred to that date, including any third party costs, Customer shall upon such termination pay Seller a termination fee equal to any and all of the Fees that would have become due and payable during the first two (2) months of the Service Term. Further, in the event Customer terminates a Service by notifying Seller less than ten (10) days prior to the specified start date, then, in addition to paying Seller all applicable charges incurred to that date, including any third party costs, Customer shall upon such termination pay Seller a termination fee equal to any and all of the Fees that would have become due and payable during the first four (4) months of the Service Term. The Parties acknowledge and agree that if Customer so terminates a Service, the termination fee set forth in this Section 5.3 is a genuine pre-estimation of the loss and damage likely to be suffered by Seller and not a penalty.

Subject to the following, Customer may terminate a Service prior to the end of the applicable Service Term. In the event Customer terminates a Service prior to the end of its Service Term, then, in addition to paying Seller all applicable charges incurred to that date, Customer shall upon such termination pay Seller a termination fee equal to one-hundred percent (100%) of any and all of the remaining Fees and other amounts payable with respect to the first year of the Service Term and fifty percent (50%) of the remaining Fees and other amounts payable for all years (if any) after the first year with respect to the Service Term for such Service. The Parties acknowledge and agree that if Customer so terminates a Service, the termination fee set forth in this Section 5.4 is a genuine pre-estimation of the loss and damage likely to be suffered by Seller and not a penalty.

Example Clause #8:

Service Commitment Period.

Cancellation. The term applicable to the Service ("Service Commitment Period") will be set forth in the ASR. There will be no cancellation charge for cancellation of a Service prior to the issuance of FiberNet's design layout record ("DLR"). If Service is canceled after the DLR is issued but prior to the start of Service, Customer shall pay one month's recurring charges, plus any applicable service ordering and installation charges. If Service has been activated, then Customer may terminate a particular Service prior to the expiration of the Service Commitment Period by providing FiberNet thirty days' prior written notice and paying, on or before the effective date of such termination, an early cancellation charge equal to:

(a) 100% of the monthly recurring charges that would have been incurred for the Service for months one through six of the agreed Service Commitment Period; plus
(b) 75% of the monthly recurring charge that would have been incurred for the Service for the months six through twelve of the agreed Service Commitment Period; plus
(c) 50% of the monthly recurring charge that would have been incurred for the Service after month twelve through the end of the Service Commitment Period.

Portability:

Comment: Services agreements frequently allow a customer to terminate a service without incurring an early termination charge provided that the customer replaces the service with another service offered by the seller. That is, a customer would have a right to terminate a service if it is replaced by another service offered by the seller. This feature is sometimes offered for all services available under the MSA, or only for certain services. Where this feature is offered, it can be subject to many conditions, such as minimum service term and bandwidth.

Example Clause #9:

Customer may terminate any Service on _____ days' notice and substitute a new service as a replacement service without incurring the early termination charged described in Section ___, provided that Customer simultaneously replaces the existing Circuit with another Circuit order, and provided that the following conditions are met:

1. The Circuit must have been installed for at least twelve (12) months;

2. The Circuit must be replaced simultaneously by a new Circuit with a term that is equal to or greater than the term remaining on the Circuit being terminated, and otherwise on the terms and conditions of this Agreement.

3. The term of the replacement Circuit must be the longer of: (a) the remainder of the term of the Circuit being replaced, or (b) one year.

4. The MRC and NRC of the new Circuit are equal to or greater than those of the Circuit that is being replaced.

5. Customer may terminate no more than of ten percent (10%) of Customer's total active Circuits ordered pursuant to this Service Exhibit, or ***two (2)*** total circuits, whichever is greater, in any ***during any 12-month period***.

6. The replacement Circuit must be of equal or higher bandwidth.

7. The Circuit being terminated must be On-Net and the replacement Circuit must be On-Net where Seller has capacity available.

8. There must not be any amount past due for any Service.

In addition, if Customer exercises this portability option, Customer agrees to pay for:

(1) All nonrecurring disconnect and installation charges due to Seller for the terminated Service, including all termination charges for local access services or any other third party services or provided facilities that are affected by the disconnection and replacement of the circuit;

(2) All other third party charges, including but not limited to cross-connect charges, early termination charges, cancellation charges, one-time charges, and the new monthly recurring charges, if applicable.

"On-Net Services" are Services that are provided between existing demarcation points and connected to Seller's network by fiber optic cable, without the necessity to incur any construction or other costs.

The following two clauses are very similar except that the exercise of the right of portability would not require the customer to substitute another service immediately. The substitute order may be placed at some time after the initial service is terminated.

Example Clause #10:

Portability: On thirty (30) days' notice, Customer may change one or both endpoints of an existing Circuit without incurring early disconnection fees if the following conditions are met:

1. The Circuit must have been installed for at least twelve (12) months;

2. The replacement Circuit must be ordered by Customer within thirty (30) days of Customer's cancellation of the original Circuit, and Customer must request a planned delivery interval that is within Seller's standard installation intervals set forth in this MSA. Such replacement Circuit shall have a term that is equal to or greater than one year.

3. The new Circuit's MRC must be equal to or greater than that associated with the Circuit being replaced.

4. Customer's right to exercise this portability option shall be limited to the greater of (i) three (3) Circuits during a calendar year, or (ii) fifteen percent (15%) of all active Circuits under this MSA during a calendar year.

5. The Circuit being replaced must have been an On-Net Circuit and the new Circuit must be On-Net, and sufficient capacity must be available.

In addition, if Customer exercises this portability option, Customer agrees to pay:

1. Any disconnect and installation charges for the terminated Circuit, as determined by Seller and agreed to by Customer.

2. All Third Party Provider charges, including but not limited to cross-connect charges, early termination charges, cancellation charges, one-time charges, and the new monthly recurring charges, if applicable.

Example Clause #11:

On thirty (30) days prior written notice, Customer may terminate an existing On-Net circuit during the initial Service Order period (a "Terminated Circuit") without incurring any early termination charges so long as: (a) Customer replaces the Terminated Circuit with another On-Net circuit having an equal (or greater) monthly recurring charge as the Terminated Circuit (the "Replacement Circuit"); (b) ***the Replacement Circuit is ordered*** by Customer under a binding Service Order ***within thirty (30) calendar days*** following the notice of intended termination of the Terminated Circuit and the initial Service Order period of the Terminated Circuit continues with the Replacement Circuit, provided the remaining Service Order Period is at least (12) twelve months, and if it is less than (12) twelve months, it will be deemed to be (12) twelve months; and (c) there is sufficient network capacity available for the Replacement Circuit in Provider's sole discretion. After Customer submits its request for a Replacement Circuit that is acceptable to Provider, Provider will send Customer a new Service Order that will list the new endpoint locations and new MRC and NRC. If the new Service Order is acceptable to Customer, Customer will execute and return to Provider and provisioning of the Replacement Service will commence. If no Replacement Service is ordered pursuant to this paragraph, Customer shall pay all applicable early termination charges for the Terminated Circuit as set forth herein.

The following form would allow portability only for services that are operating at lower bandwidth capacity than the new replacement service. Portability would not be permitted for larger bandwidth services.

Example Clause #12:

Portability. Customer may cancel an existing ***On-Net Circuit with up to OC-12 capacity*** (a "Cancelled Circuit") without incurring any cancellation or early termination fees or penalties, so long as:

(a) Cancelled Circuit has been installed and in use for at least sixty (60) calendar days;

(b) Customer replaces the Cancelled Circuit with another On-Net circuit having equal (or greater): (i) IXC MRCs, and (ii) term commitment (the " Replacement Circuit");

(c) The Replacement Circuit is ordered by Customer before the effective date of the cancellation of the Cancelled Circuit (the effective date of cancellation being the "Cancellation Date");

(d) The total number of Cancelled Circuits does not exceed ten percent (10%) of their total active Circuits in any annual period; and

(e) Customer is not otherwise in breach of this Agreement.

If the number of Cancelled Circuits exceeds ten percent (10%) of their total active Circuits in any annual period, or if Customer fails to meet any of the other above criterion, Customer shall pay all applicable shortfall charges and early termination fees for each additional Cancelled Circuit.

Mergers and Acquisitions:

Comment: Another circumstance under which a customer might be excused from paying early termination charges is a merger between two companies that are both customers of the seller. Below is an example of such a clause. Under this clause, termination without termination charge would not be available if the customer has made a minimum revenue commitment to the seller which would be adversely affected by an early termination.

Example Clause #13:

> Acquired Entity. If Customer or Customer's Affiliate acquires through purchase, merger, or other means of acquisition an entity as an Affiliate during the term of this Agreement, Customer and Affiliate may elect one of the following: (1) Affiliate may be deemed to be a joint party with Customer under this Agreement; or (2) if the Affiliate is also contracting with Seller for the same or similar services as provided for in this Agreement, then Customer and Affiliate may elect, provided the election will not adversely affect any volume or revenue commitments made in the aggregate by the Customer and the Affiliate, which agreement will govern the parties going forward. If alternative (2) is selected, Customer and Affiliate may terminate, without cost or liability, including any early termination liability or liquidated damages, one of the existing agreements and become joint parties under the remaining agreement ("Surviving Agreement"). Customer shall provide written notice to Seller that Customer and Affiliate are exercising such election, identifying which agreement will be terminated and which agreement shall remain effective among Seller, Customer, and Affiliate. For the purposes of this section, "Affiliate" shall mean any entity which controls, is controlled by, or is under common control with Customer; and "control" shall mean at least fifty percent (50%) ownership or at least fifty percent (50%) of the voting interests of the entity. The parties acknowledge and agree that the Surviving Agreement shall be amended to correctly reflect the parties' names, but in all other respects shall have identical rates, charges, terms and conditions as contained in the Surviving Agreement prior to the election of the Customer and Affiliate. The effective date of the Surviving Agreement shall be thirty (30) days from the date of the written notice to Seller unless the parties agree to an earlier effective date and the Surviving Agreement shall renew, expire or terminate in accordance with its original terms.

Business Downturn:

Comment: A Customer sometimes request a right to terminate its agreement in the event that they experience adverse business conditions. A clause such as this is less common, but is sometimes accepted by a seller for more significant customers.

Such clauses raise difficult issues for a seller because the circumstances that give rise to a cancellation right are not within the control of seller, are impossible to predict, and often are not clearly defined. Therefore, sellers will usually agree to a clause such as these only reluctantly.

Example Clause #14:

> Business Downturn: For purposes of this agreement, the term "Business Downturn" is hereby defined to mean an unplanned, measurable change in business conditions

affecting Customer's business that is outside of Customer's control and that materially and negatively affects Customer's need for the level of Services provided hereunder. In the event Customer, after the first year of this Agreement, is unable to fulfill its obligations for the Annual Commitment due to a Business Downturn, and not due to a transfer of services to another provider shall not be assessed cost for the remainder of term.

Example Clause #15:

Business Downturn. Except for Services described in ________, in the event of a business downturn beyond the control of Customer that significantly reduces the volume of Services needed by Customer with the result that Customer is unable to meet its revenue and/or volume commitments, if any, under this Agreement or any Service order, Seller will waive any shortfall obligation in an aggregate amount not to exceed 30% of the revenue and/or volume commitment caused directly by such business downturn. In the event that such shortfall exceeds 30% of the revenue and/or volume commitment, Seller shall make appropriate and commercially reasonable changes to the applicable Addendum, which may include adjustments to price, term, annual commitment, or a combination thereof. Further, in the event Customer's business activity is utilizing Seller's equipment or software licenses provided under this Agreement and such business activity is divested by sale or spin-off into a separate entity, such resulting separate business entity is entitled to receive transfer of such equipment and software licenses without any transfer fees (unless Seller is charged a fee by its vendor, in which case Seller will pass through such fee), provided it enters into a subsequent agreement with Seller containing terms and pricing as contained herein and assumes full financial responsibility for payment for use of the equipment and software license.

Example Clause #16:

BUSINESS DOWNTURN. In the event of a business downturn beyond Customer's control or a divestiture of an Affiliate of Customer that significantly reduces the volume of network services required by Customer with the result that Customer will be unable to meet its revenue and/or volume commitments under this Agreement (notwithstanding Customer's best efforts to avoid such a shortfall), Seller and Customer will cooperate in efforts to develop a mutually agreeable alternative proposal that will satisfy the concerns of both parties and comply with all applicable legal and regulatory requirements. By way of example and not limitation, such alternative proposals may include changes in rates, nonrecurring charges, revenue and/or volume commitments, discounts, the multi-year services period and other provisions. Subject to all applicable legal and regulatory requirements, including the requirements of the Federal Communications Commission and the Communications Act of 1934 (as revised and amended), Seller will prepare and file any tariff revisions necessary to implement such mutually agreeable alternative proposal. This Section shall not apply to change resulting from a decision by Customer to: (i) reduce its overall use of telecommunications; or (ii) transfer portions of its traffic or projected growth to carriers other than Seller. Customer must give Seller sixty (60) days' prior written notice of the conditions it believes will require the application of this

Section. This Section does not constitute a waiver of any charges, including shortfall charges, incurred by Customer prior to the time the parties mutually agree to amend or replace this Agreement.

Part 1, Chapter 5 - Limitation of Liability

There are several ways in which a seller may be exposed to potential liability to a customer under a lit services agreement. A seller may fail to make the service available by the date that was agreed. Or once a service has been activated, there may be an interruption of service or service outage.

Typically, a seller will employ several strategies to limit its liability to its customers. There are several such clauses set forth in Section 6 of the sample MSA that accompanies this chapter. (Section 6 of the MSA.)

A customer's recourse in the event of a failure of seller to perform under a lit services agreement is typically narrowly circumscribed. The principal strategies are to expressly limit a customer's recourse to service credits, or to disclaim warranties, require a waiver of claims for consequential and other indirect damages, and to impose specific dollar limitations on all other claims.

Notably absent from the liability clauses that are part of this form is Seller's responsibility for claims for interruptions of service, even those caused by the fault of Seller. It is Seller's objective to limit Customer's recourse for interruptions of service to agreed service credits. Seller will typically not indemnify a Customer for losses suffered by Customer from interruptions of service, regardless of the cause.

Service credits are discussed in more detail in Part 2.

There are some common exceptions to limitation of claims against the Carrier or Seller. The most common are indemnity for claims for bodily injury and property damage (Example Clause 4 below), and claims for infringement of intellectual property rights (Example Clause 5 below). These exceptions are discuss below.

Note that the duty to indemnify might be limited to a fixed maximum monetary amount of liability and subject to the waivers of certain kinds of claims described elsewhere in the limitation of liability clauses (Example Clause #3). Or it might not be subject to these limitations (Example Clause #1, Section 9.3).

The several clauses below are substantially similar to the clause that appears in the sample MSA herein, although the respective carriers phrase the limitations in different ways.

In addition, a limitations of liability clause is important to define the liability of customers. The customer is liable for charges for services. The customer might also be liable for early termination charges and for breaches of its covenants under this agreement. If the Customer were to breach the restrictions on assignment, the damages to the Carrier might be very substantial. Therefore, a carrier will frequently exclude breaches of the restrictions on assignment from the limitations of liability. (MSA Section 6(d) below.)

MSA Section 6:

Disclaimers and Limitation of Liability: (a) SELLER MAKES NO EXPRESS OR IMPLIED WARRANTY AS TO ANY SERVICE PROVISIONED HEREUNDER. SELLER SPECIFICALLY DISCLAIMS ALL IMPLIED WARRANTIES, INCLUDING, WITHOUT LIMITATION, IMPLIED WARRANTIES OF MERCHANTABILITY, FITNESS FOR A PARTICULAR PURPOSE, TITLE, NON-INFRINGEMENT OF THIRD PARTY RIGHTS, AND PERFORMANCE OR INTEROPERABILITY OF THE SERVICE WITH ANY CUSTOMER OR END USER PROVIDED EQUIPMENT.

(b) NEITHER PARTY SHALL BE LIABLE TO THE OTHER FOR ANY INDIRECT, CONSEQUENTIAL, EXEMPLARY, SPECIAL, INCIDENTAL, COVER-TYPE OR PUNITIVE DAMAGES, INCLUDING, WITHOUT LIMITATION, LOST REVENUE OR PROFITS, LOSS OF USE OR LOST BUSINESS, OR GOODWILL, ARISING IN CONNECTION WITH THIS MSA OR SELLER'S PROVISIONING OF THE SERVICES (INCLUDING BUT NOT LIMITED TO: (A) ANY SERVICE IMPLEMENTATION DELAYS OR FAILURES; (B) LOST, DELAYED OR ALTERED MESSAGES OR TRANSMISSIONS; OR (C) UNAUTHORIZED ACCESS TO OR THEFT OF CUSTOMER'S TRANSMITTED DATA), BASED ON ANY THEORY, CAUSE OF ACTION OR CLAIM, INCLUDING TORT, CONTRACT, WARRANTY, STRICT LIABILITY OR NEGLIGENCE, EVEN IF THE PARTY HAS BEEN ADVISED, KNEW OR SHOULD HAVE KNOWN OF THE POSSIBILITY OF SUCH DAMAGES.

(e) THE TOTAL LIABILITY OF SELLER TO CUSTOMER IN CONNECTION WITH THIS AGREEMENT SHALL BE LIMITED TO THE LESSER OF DIRECT DAMAGES OR THE PREVIOUS MONTH'S MRC ASSOCIATED WITH THE AFFECTED SERVICE. THE FOREGOING LIMITATION APPLIES TO ALL CAUSES OF ACTIONS AND CLAIMS, INCLUDING WITHOUT LIMITATION, BREACH OF CONTRACT, BREACH OF WARRANTY, NEGLIGENCE, STRICT LIABILITY, MISREPRESENTATION AND OTHER TORTS. CUSTOMER ACKNOWLEDGES AND ACCEPTS THE REASONABLENESS OF THE FOREGOING DISCLAIMER AND LIMITATIONS OF LIABILITY. NEITHER PARTY MAY ASSERT ANY CAUSE OF ACTION AGAINST THE OTHER PARTY UNDER ANY THEORY WHICH ACCRUES MORE THAN ONE (1) YEAR PRIOR TO THE INSTITUTION OF A LEGAL PROCEEDING ALLEGING SUCH CAUSE OF ACTION. For purposes of this Section 6, all references to Seller and Customer include their respective officers, directors, shareholders, members, managers, and employees, Affiliates, End Users, agents, lessors and providers of service to Seller.

(f) The foregoing notwithstanding, the waivers of claims and limitations of liability set forth in this Agreement shall not apply to termination charges described in Section 3, and shall not apply to a claim arising from a breach of the restrictions on transfer or assignment or use of the Services described in Subsections 8(a) and 8(b), and shall not apply to the indemnity obligations described in Subsection 8(b).

Example Clause #1:

LIMITATIONS OF LIABILITY

9.1. Unauthorized Access to Customer or End User Facilities. Seller will not be liable in contract or in tort to Customer or its End Users for unauthorized access to transmission facilities or premise equipment, or for unauthorized access to or alteration, theft, or destruction of data files, programs, procedure, or information through accident, wrongful means or devices, or any other method.

9.2. Disclaimers. Seller will not be liable for claims or damages resulting from or caused by:

A. Customer's fault, negligence or failure to perform Customer's responsibilities;
B. Claims against Customer by any other party (except for claims indemnified under Section ___;
C. Any act or omission of any other party, including End Users; or
D. Equipment or services furnished by a third party, including End Users.

9.3. Limitation of Liability. Seller's entire liability for its failure to perform any of its obligations under this Agreement will not exceed an amount equal to the lesser of $_______ or the monthly Services' charges net of any discounts or credits for the affected Services during the preceding three months. Seller will not be liable for any unavoidable damage to Customer's premises. ***This limitation of liability will not apply to claims arising from the parties' indemnification obligations under the Agreement.***

9.4. Consequential Damages. NEITHER PARTY WILL BE LIABLE FOR ANY CONSEQUENTIAL, INCIDENTAL, OR INDIRECT DAMAGES FOR ANY CAUSE OF ACTION, WHETHER IN CONTRACT OR TORT. CONSEQUENTIAL, INCIDENTAL, AND INDIRECT DAMAGES INCLUDE, BUT ARE NOT LIMITED TO, LOST PROFITS, LOST REVENUES, AND LOSS OF BUSINESS OPPORTUNITY, AND INCIDENTAL DAMAGE TO CUSTOMER'S PREMISES FOR SERVICE INSTALLATION, WHETHER OR NOT THE OTHER PARTY WAS AWARE OR SHOULD HAVE BEEN AWARE OF THE POSSIBILITY OF THESE DAMAGES. THIS LIMITATION OF LIABILITY DOES NOT APPLY TO CLAIMS FOR INDEMNIFICATION UNDER SECTION 10 OF THE AGREEMENT.

9.5. Disclaimer of Warranties. Seller disclaims all warranties, express or implied, that are not explicitly stated in this Agreement, and in particular disclaims all warranties of merchantability and fitness for a particular purpose and warranties related to third party equipment, material, services, or software. The services and any materials or equipment are supplied "as is" to the full extent permitted by law.

9.6. Liability for Content. Seller is not responsible for the content of any information transmitted or received through the Services.

Example Clause #2:

DISCLAIMER OF WARRANTIES. SELLER MAKES NO REPRESENTATIONS OR WARRANTIES, WHETHER EXPRESS, IMPLIED OR STATUTORY, REGARDING THE SERVICES, SYSTEM EQUIPMENT OR SELLER-OWNED OR PROVIDED EQUIPMENT USED BY THE CUSTOMER, INCLUDING ANY EQUIPMENT WITH RESPECT TO WHICH TITLE MAY TRANSFER TO CUSTOMER (EXCEPT TO THE EXTENT SET FORTH IN A SEPARATE SALE TRANSFER DOCUMENT). THIS INCLUDES, BUT IS NOT LIMITED TO, ANY IMPLIED WARRANTIES OF MERCHANTABILITY, FITNESS OF THE SERVICE OR EQUIPMENT FOR A PARTICULAR PURPOSE AND NON-INFRINGEMENT OF ANY THIRD PARTY RIGHTS. Additional Warranty limitations related to specific products may be found at (___________.com).

LIMITATION OF LIABILITY. 1. WITH RESPECT TO CLAIMS OR SUITS BY CUSTOMER, OR ANY OTHERS, FOR DAMAGES RELATING TO OR ARISING OUT OF ACTS OR OMISSIONS UNDER THIS AGREEMENT AND/OR SERVICES PROVIDED HEREUNDER, SELLER's LIABILITY FOR SERVICE INTERRUPTIONS OR PROBLEMS, IF ANY, SHALL BE LIMITED TO CREDIT ALLOWANCES AS EXPRESSLY PROVIDED IN APPLICABLE TARIFFS OR AS OTHERWISE SET FORTH IN THE TERMS AND CONDITIONS.

2. SELLER SHALL NOT BE LIABLE FOR ANY LOSSES OR DAMAGES RESULTING FROM: (A) THE DELIVERY, INSTALLATION, MAINTENANCE, OPERATION, USE OR MISUSE OF AN ACCOUNT, EQUIPMENT, OR SERVICE; (B) ANY ACT OR OMISSION OF CUSTOMER, OR ITS END USERS OR AGENTS, OR ANY OTHER ENTITY FURNISHING EQUIPMENT, PRODUCTS OR SERVICES TO CUSTOMER; OR (C) ANY PERSONAL OR PROPERTY DAMAGES DUE TO THE LOSS OF STORED, TRANSMITTED OR RECORDED DATA RESULTING FROM THE SERVICE OR THE EQUIPMENT, EVEN IF SELLER HAS BEEN ADVISED OF THE POSSIBILITY OF SUCH DAMAGES. THE ONLY EXCEPTION SHALL BE TO THE EXTENT PROPERTY DAMAGE TO CUSTOMER'S PREMISES IS CAUSED DUE TO SELLER'S SOLE GROSS NEGLIGENCE OR WILLFUL MISCONDUCT, PROVIDED, HOWEVER, IN NO EVENT SHALL SELLER 'S LIABILITY FOR DIRECT DAMAGES BE GREATER THAN THE SUM TOTAL OF PAYMENTS MADE BY CUSTOMER TO SELLER DURING THE THREE MONTHS IMMEDIATELY PRECEDING THE EVENT FOR WHICH DAMAGES ARE CLAIMED, BUT IN NO EVENT TO EXCEED $10,000.

3. IN NO EVENT SHALL EITHER PARTY BE LIABLE FOR ANY INDIRECT, INCIDENTAL, EXEMPLARY, PUNITIVE OR OTHER CONSEQUENTIAL DAMAGES, WHETHER OR NOT FORESEEABLE, INCLUDING, BUT NOT

LIMITED TO, DAMAGES FOR THE LOSS OF DATA, GOODWILL OR PROFITS, SAVINGS OR REVENUE, OR HARM TO BUSINESS, WHETHER UNDER CONTRACT, TORT (INCLUDING NEGLIGENCE), STRICT LIABILITY OR ANY CAUSE WHATSOEVER.

Example Clause #3:

LIMITATION OF LIABILITY; DISCLAIMER OF WARRANTIES

A. In no event will Seller be liable, either in contract or in tort, for unauthorized access to Customer's transmission facilities or Customer premise equipment; or for unauthorized access to or alteration, theft, or destruction of Customer's data files, programs, procedure, or information through accident, fraudulent means or devices, or any other method.

B. Except to the extent caused by the negligence of Seller, Seller will not be liable for claims or damages resulting from or caused by: (i) Customer's fault, negligence or failure to perform Customer's responsibilities; (ii) claims against Customer by any other party (except for claims of copyright or patent infringement as specified herein); (iii) any act or omission of any other party; or (iv) equipment or services furnished by a third party.

C. Seller's entire liability for its failure to perform any of its obligations under this Agreement will not exceed an amount equal to the lesser of $_______ or the monthly charges paid for the effected Services during the preceding 12 months. Seller will not be liable for any unavoidable damage to Customer's premises.

D. Seller will not be liable for any consequential, special, incidental, indirect, exemplary or punitive damages for any cause of action, whether in contract or tort, arising out of this Agreement or in any way related to the relationship between the Parties. Consequential and indirect damages include, but are not limited to, lost profits, lost revenues or loss of business opportunity, whether or not Seller was aware or should have been aware of the possibility of such damages.

E. With respect to the Services, materials and equipment provided hereunder, SELLER HEREBY DISCLAIMS ALL WARRANTIES, EXPRESS OR IMPLIED, NOT EXPLICITLY STATED IN THE AGREEMENT, AND IN PARTICULAR DISCLAIMS ALL WARRANTIES OF MERCHANTABILITY AND FITNESS FOR A PARTICULAR PURPOSE.

Comment: Another way that claims may arise against a Seller is as a consequence of Customer reselling the service to its customers, either directly or bundled with other services of Customer. A customer of Customer might suffer an interruption of service with the result that a claim is made against Seller. A Seller in all cases will strive to limit its potential liability in the manner described above. Therefore, a Seller will often require its Customer to protect it from any such claims.

8(b) Nothing in this Agreement shall preclude Customer from reselling or providing to third party end users services derived from the Services, provided that any such use of such Services is subject to the terms and conditions of this Agreement and provided that Customer agrees to indemnify, defend and hold harmless Seller, its Affiliates, and their employees, agents, officers and directors, from and against, and assumes all liability for, all suits, actions, damages or claims made against Seller related to such resale or third-party use, including, but not limited to, claims based on failure, breakdown, interruption or deterioration of a Service or claims that the use of a Service, including that the placement of traffic on the Service, or any application used in connection with a Service is illegal or infringes on any intellectual property right.

Comment: Sellers will sometimes accept certain narrow exceptions to the limitations of liability described above. The clause below would require the Seller to indemnify the Customer for claims for property damage and bodily injury that are caused by the fault of the Seller. Such a passage does not appear in the standard form, but is frequently accepted by a seller. This is not a general indemnification for all claims resulting from the fault of the parties. The indemnification is limited to claims for bodily injury or property damage.

Example Clause #4:

INDEMNITY. For the purposes of this Section, "Losses" means all losses, liabilities, damages and costs (including Taxes) and all related costs and expenses (including reasonable attorney's fees and disbursements and costs of investigation, litigation and settlement). The indemnity provisions shall be subject to limitations set forth in Section ___.

Indemnification by Seller. Subject to Section ____ below, Seller shall indemnify, defend and hold Customer and its officers, directors, employees, agents, successors and assigns harmless from and against any and all Losses arising out of or relating to:

(a) A breach by Seller or its Affiliates of any material representation or warranty provided under this Agreement;

(b) A breach by Seller of the confidentiality provisions of this Agreement; and/or

(c) The death or injury of or damage to any person, or real or personal, tangible or intangible personal property to the extent such injury or damage is proximately caused by Seller's negligence or willful misconduct or that of its Affiliates.

By Customer. Subject to Section ____ below, Customer shall indemnify, defend and hold Seller, its respective officers, directors, employees, agents, affiliates, successors and assigns, harmless from and against any and all Losses arising out of or relating to:

(a) a breach by Customer or its Affiliates of any material representation of warranty provided under this Agreement;

(b) A breach by Customer of the confidentiality provisions of this Agreement; and/or

(c) The death or injury of or damage to any person, or real or personal, tangible or intangible personal property to the extent such injury or damage is proximately caused by Customer's negligence or willful misconduct; or

(d) A claim by a third party that the content, use and/or publication of information and communications transmitted by Customer or its customers or authorized end-users using the

Services, or accessible to third parties through the use by Customer or its customers or authorized end-users of the Services ("Content") infringes upon the rights of such third party, regardless of the form of action, whether in contract, tort, warranty, or strict liability and whether in respect of copyright infringement or any manner of intellectual property claims, defamation claims, claims of publication of obscene, indecent, offensive, racist, unreasonably violent, threatening, intimidating or harassing material, or claims of infringement of data protection legislation.

Indemnification Procedures. If any claim in respect of Losses is asserted or any civil, criminal, administrative or investigative action or proceeding (any such claim, action or proceeding, a "Claim") is threatened or commenced, in each case against any party seeking indemnification under this Section (an "Indemnified Party"), the Indemnified Party will promptly notify the indemnifying Party (the "Indemnifying Party") in writing thereof. Any failure or delay by the Indemnified Party in giving such written notice shall not constitute a breach of this Agreement and shall not excuse the Indemnifying Party's obligation under this Section, except to the extent (if any) that the Indemnifying Party is prejudiced by such failure or delay. If the Indemnifying Party acknowledges in writing an indemnification obligation under this Section, it will be entitled to elect, within thirty (30) days after its receipt of such notice, to assume sole control over the investigation, defense and settlement of such Claim at its own cost, risk and expense. Neither the Indemnifying Party nor the Indemnified Party shall enter into a settlement of a Claim without the prior written consent of the other, which consent shall not be unreasonably withheld. After notice of a Claim by the Indemnified Party, if the Indemnifying Party does not elect to assume sole control of the defense of such Claim, the Indemnified Party will have the right to defend such Claim in such reasonable manner as it may deem appropriate, at the cost, risk and expense of the Indemnifying Party. The Indemnifying Party will have the right to participate in such defense at its own cost and expense. Each party, at its own cost and expense, agrees to provide reasonable cooperation and assistance to the other party in the investigation, defense and settlement of any Claim, including but not limited to providing access to relevant information and employees.

In the event that a court of competent jurisdiction (without possibility of appeal) makes any award of damages against Seller (and/or its officers, employees, Affiliates or agents) as the ultimate provider of the Network or Services, regardless of the form of action, whether in contract, tort, warranty, or strict liability, in respect of copyright infringement or any manner of intellectual property claims, defamation claims, claims of publication of obscene, indecent, offensive, unreasonably violent, threatening, intimidating or harassing material, or claims of infringement of data protection legislation, based upon: (a) the content of any information transmitted by Customer or by any customer or authorized end user of Customer, (b) the use and/or publication of any and all communications or information transmitted by Customer or by any customer or authorized end user of Customer, or (c) the misuse of Service(s) by Customer or by any customer or authorized end user of Customer, then Customer agrees to indemnify and hold Seller and its officers, employees, Affiliates and agents harmless from and against any such damages, and any reasonable costs or expenses incurred in connection therewith.

A Seller will sometimes accept responsibility for indemnifying Customer against claims for infringement of intellectual property rights. This can be a difficult area for Seller for several reasons. An IP claim might arise from electronic equipment that the Seller has incorporated into its network but which it did not manufacture. Rather, the Seller will most likely have purchased the equipment from an electronics manufacturer. The equipment vendor will most likely have given only a highly qualified and limited indemnity that offers very little protection to Seller. Therefore, if a claim arises, the Seller may be asked to indemnify for a loss that it did not create and for which the Seller has little recourse. Under these circumstances it is difficult or impossible for a Seller to assess its risk. The following is a typical indemnity for infringement of intellectual property rights from an equipment vendor.

Manufacturer Indemnity

Patents and Copyrights

If a third party claims that Manufacturer Hardware or Software provided to Customer under this Agreement infringes that party's patent or copyright, Manufacturer will defend Customer against that claim at Manufacturer's expense and pay all costs and damages that a court finally awards or are agreed in settlement, provided that Customer (a) promptly notifies Manufacturer in writing of the claim and (b) allows Manufacturer to control, and cooperates with Manufacturer in, the defense and any related settlement negotiations. If such a claim is made or appears likely to be made, Manufacturer agrees to secure the right for Customer to continue to use the Hardware or Software, or to modify it, or to replace it with one that is equivalent. If Manufacturer determines that none of these alternatives is reasonably available, Customer agrees to return the Hardware or Software to Manufacturer on Supplier's written request. Manufacturer will then give Customer a credit equal to Customer's net book value for the Hardware or Software provided Customer has followed generally-accepted accounting principles. Any such claims against the Customer or liability for infringement arising from use of the Hardware or Software following a request for return by Manufacturer are the sole responsibility of Customer. This represents Manufacturer' entire obligation to Customer regarding any claim of infringement. Manufacturer has no obligation regarding any claim based on any of the following: (a) anything Customer provides which is incorporated into the Hardware or Software; (b) functionality provided by Manufacturer at the instruction of Customer; (c) Customer's modification of Hardware or Software; (d) Customer's modification of Hardware or Software; (e) the combination, operation, or use of Hardware or Software with other products not provided by Manufacturer as a system, or the combination, operation, or use of Hardware or Software with any product, data, or apparatus that Manufacturer did not provide; or (f) infringement by a Third Party Vendor Item alone, as opposed to its combination with Products Manufacturer provides to Customer as a system.

Limitation of Liability

In no event shall Manufacturer or its agents or suppliers be liable to Customer for more than the amount of any actual direct damages up to the greater of U.S. $_______ (or equivalent in local currency) or the charges for the Product or Service that are the subject of the claim, regardless of the cause and whether arising in contract, tort (including negligence) or otherwise. This limitation will not apply to claims for damages for bodily injury (including death) …

Comment: It is easy to understand that if a carrier's recourse against its equipment supplier is highly qualified and limited, it is difficult for a carrier to extend greater protection to its customers. Nevertheless, it is not uncommon to find an indemnification for claims of infringement of intellectual property rights in a carrier service contract such as the following.

Note that the duty to indemnify may or may not be subject to a fixed monetary maximum liability and waiver of certain types of claims described elsewhere in the limitation of liability clauses.

Example Clause #5:

INDEMNIFICATION

10.1. Property & Personal Injury Indemnity. Each party agrees to indemnify and defend the other party, its directors, officers, employees, agents and their successors from and against all third party claims for damages, losses, or liabilities, including reasonable attorneys' fees, arising directly from performance of this Agreement and relating to ***personal injury, death, or damage to personal property*** that is alleged to have resulted from the negligent or willful acts or omissions of the indemnifying party or its subcontractors, directors, officers, employees, or agents.

10.2. Third party Claims. Customer will indemnify and defend Seller from and against all third party claims for damages arising from or related to the use or misuse of the Services.

10.3. Customer Proprietary Rights Indemnity. If promptly notified of any claim asserted against Customer based on a claim that the Services as provided by Seller ***infringe on a United States patent or copyright***, Seller will defend the action at its expense and will pay any fees, costs, or damages that may be awarded in the action or resulting settlement. If a final injunction is obtained against Customer prohibiting use of Services by reason of infringement of a United States patent or copyright, Seller will, at its option and expense, either: (a) procure the right for Customer to continue using the Services; (b) procure alternative Services that furnish equivalent functionality; or (c) direct Customer to return the Services to Seller, and the Service Order relating to the returned Services will terminate.

10.4. Proper Notification. The party seeking indemnification under this Section must (a) promptly notify the other party in writing of any claim; (b) give the indemnifying party full and complete authority to resolve the matter; and (c) provide information and assistance for the claim's defense. The indemnifying party will retain the right, at its option, to settle or defend the claim, at its own expense and with its own counsel. The indemnified party will have the right, at its option, to participate in the settlement or defense of the claim, with its own counsel and at its own expense, but the indemnifying party will retain sole control of the claim's settlement or defense. To be indemnified under this Section, the party seeking indemnification must not by any act, including any admission or acknowledgement, materially impair or compromise a claim's defense.

Comment: Because an IP indemnity is a difficult issue, it often is qualified and limited. The following are two examples in which the carrier has agreed to extend an indemnification for IP infringement, but qualified the indemnity obligation. A carrier will frequently attempt to limit its obligations to what is sometimes referred to in common parlance as "repair, replace or refund."

Example Clause #6:

Supplier agrees to indemnify, defend and hold Customer, its officers, directors, employees, agents and contractors harmless from and against all loss, damage, liability, cost and expense (including reasonable attorney's fees and expenses) by reason of any claims or actions by third parties for bodily injury or death, and damage, loss or destruction of any real or tangible personal property, to the extent caused by Supplier's negligence, gross negligence, or willful misconduct.

In addition, Supplier agrees to indemnify, defend and hold Customer, its officers, directors, employees, agents and contractors harmless from and against all loss, damage, liability, cost and expense (including reasonable attorney's fees and expenses) by reason of any claims or actions by third parties that ***use of the Leased Fibers as permitted hereby infringes***, violates or otherwise misappropriates the intellectual property rights of any third party (an "**IP Claim**"), except to the extent the IP Claim arises as a result of (1) the combination of the Leased Fibers with other technology if such claim would have been avoided absent such combination, or (2) modification of the Leased Fibers by Customer or (3) the use of the Leased Fibers other than in accordance with this Agreement.

Example Clause #7:

Supplier has no indemnity obligations under this Section ____ to the extent arising out of Customer's use of the IRUs or Deliverables in combination with equipment not provided by Supplier, unless the combination was done at the direction of Supplier or as authorized by Supplier's documentation, where such indemnity obligation would not have arisen but for such combination.

Comment: In the following clause the Carrier has limited its indemnity to elements of its service that are created by or uniquely for the Carrier. That is, this indemnity would not extend to components of the service that are generally available to other carriers that provide similar services in the telecommunications industry. The indemnity would extend only to infringement claims relating to "Carrier's Intellectual Property."

Example Clause #8:

Carrier, at its expense, will indemnify, defend and hold harmless Customer Indemnified Parties from all liabilities, costs, losses, damages and expenses (including reasonable attorneys' and experts' fees and expenses as well as interparty damages caused by Carrier or third parties) and will reimburse such fees and expenses as they are incurred, including in connection with any claim or action threatened or brought against the Customer Indemnified Parties: (a) arising out of or relating to any claim that the provision or utilization of any Services or any portion thereof that constitutes Carrier's Intellectual Property results in a claim of infringement, violation, trespass, contravention or breach of any patent, copyright, trademark, license or other Intellectual Property or proprietary right of any third party, or constitutes the unauthorized use or misappropriation of any trade secret of any third party. "**Carrier's Intellectual Property**" means I***ntellectual Property that Carrier or its***

Affiliates, in whole or in part, created, developed, owns, or holds an exclusive license. Carrier's Intellectual Property does not include a non-exclusive licenses to IP that is owned by a third party.

If a claim is brought against a Service provided by Carrier, or is enjoined or is facing the realistic threat of being enjoined, based on a claim that Carrier's Intellectual Property infringes on a third-party's rights, Carrier shall at its sole expense take one of the following actions: either procure the right for Customer to continue using the affected Service, or replace or modify the affected Service so that it is non-infringing. If neither of these alternatives is available, Carrier will notify Customer and Customer may terminate the affected Service without termination liability (except for the avoidance of doubt, Carrier shall remain liable for any obligations as to amounts previously due for services provided or credits owed or amounts that may be due as set forth elsewhere in this Agreement). In such circumstance, Carrier shall credit or refund (at Customer's option) any amounts pre-paid by Customer related to the affected Services and shall take reasonable steps to facilitate Customer's migration to a similar or substitute service, if available.

Noninfringement. Carrier represents to Customer, that Carrier has not received any notice (orally or in writing) from any third party asserting an adverse claim that any use of or provision of the equipment or software, or the use of any of the Services as contemplated under this Agreement, do or will infringe or constitute the unauthorized use or misappropriation of any Intellectual Property of any third party (an "**Infringement Claim**"). Carrier further represents that to the best of its knowledge the vendors of the third party equipment and software used by Carrier have not received any such notice.

If Carrier or Customer receives a notice of an Infringement Claim, Carrier or Customer shall promptly notify the other Party, and Customer shall have the right to terminate this Agreement or the applicable SOF without liability or further obligation on notice to Carrier. In circumstances where Customer terminates pursuant to this Agreement, Carrier shall remain liable for any obligations as to amounts previously due for services provided or credits owed or amounts that may be due as set forth elsewhere in this Agreement and shall credit or refund (at Customer's option) any amounts pre-paid by Customer related to the affected Services and shall take reasonable steps to facilitate Customer's migration to a similar or substitute service, if available.

(a) "**Intellectual Property**" means all (1) patents, patent applications, patent disclosures and inventions (whether patentable or not); (2) trademarks, service marks, trade dress, trade names, logos, corporate names, Internet domain names, and registrations and applications for the registration thereof together with all of the goodwill associated therewith; (3) copyrights and copyrightable works (including computer programs and mask works) and registrations and applications thereof; (4) trade secrets, know-how and other confidential information; (5) waivable or assignable rights of publicity, waivable or assignable moral rights; and (6) all other forms of intellectual property, such as data and databases.

(b) This Section ___ (Indemnification) sets forth the sole and exclusive remedy of Customer against Carrier for any Infringement Claim.

Part 1, Chapter 6 - Late Delivery

Service credits may be available to a customer for many reasons. A customer may be entitled to a service credit if a service is not turned up according to schedule. After a service has been turned up, there might be an interruption of service or the service might not perform in accordance with specifications. This chapter addresses service credits and other remedies for late delivery of services. A discussion of service credits for interruption of services and other service related remedies are discussed under the chapter entitled "Service Interruption."

In some agreements a particular level of service credit is applied uniformly to all services. Other agreements allow for different levels of service or premium services which have different levels of service credits.

This form of MSA does not set forth the amount of service credits. Instead, the particular amounts of service credits are addressed in the individual service exhibits. Each service exhibit describes a different service offering that is available under the MSA. This form allows flexibility to associate different levels of service credits with different types of services.

> 7. Service Credits: Carrier may offer service credits related to installation intervals, Service availability, latency, and time to restore Service, which shall be set forth in the applicable Service Order Forms. These credits are the Customer's sole and exclusive remedy for Service related claims. To qualify for a service credit, Customer must not have any invoices that are past due, and must notify Carrier that a trouble ticket should be opened to document the event. In no event shall the total amount of all service credits in a month exceed the total MRC (or NRC if applicable) for the affected Service for such month. If Carrier specifies in an applicable SOF that the Service will be provided through the use of a third-party vendor, and if there is a delay in installation or interruption of such service obtained from such third party vendor, Customer shall be entitled to remedies for such delay or interruption of service only if and to the extent of the service credit to which Carrier is entitled under its agreement with such third-party vendor.

Delivery of Service:

Comment: In every service order there is a date by which the seller has agreed to make the service available for the customer's use. If the seller fails to deliver the service by that date, the customer might suffer damages. Nevertheless, a customer's recourse is typically quite limited. In all of the sample clauses below, even those in which a customer would be entitled to claim a substantial service credit, the customer would nevertheless be precluded from asserting a claim for damages for late delivery.

In the sample MSA, the customer may terminate the service entirely if the service is not delivered within 30 days following the agreed delivery date, provided that the service is on-net and provided that they delay was not caused by Customer or by a force majeure event. Note that in this clause there is no penalty or service credit for late delivery. The Customer's exclusive remedy is to terminate the order if the Service is not delivered within 30 days following the scheduled delivery date.

> 1(c) If the Service Activation Date for a Service has not occurred within thirty (30) days following the FOC date, ***except in the case of a delay caused by Customer or a Force Majeure Event***, and except in the case of an Off-Net Service, Customer may cancel the Service without penalty, ***provided that Customer notifies Seller in writing of cancellation before Seller has completed the acceptance testing***.

Comment: The following clause takes a similar approach to the preceding clause. Under neither clause would a customer be permitted to terminate for late delivery of an off-net service. The only difference is that in the clause below a customer is foreclosed from terminating after receipt of a completion notice, rather than completion of testing.

Example Clause #1:

> If Seller's installation of Service is delayed for more than thirty (30) business days beyond the Customer Commit Date for reasons other than an Excused Outage, ***Customer may terminate and discontinue the affected Service*** upon written notice to Seller and without payment of any applicable termination charge; ***provided such written notice is delivered prior to Seller delivering to Customer the Connection Notice*** for the affected Service. This Section ____ shall not apply to any Off-Net Service.

Comment: The following clause takes late delivery a step further. It would allow a customer to claim a credit for late delivery.

Example Clause #2:

> If, other than as caused by a Force Majeure Event, Seller has not delivered a Completion Notice (in good faith) respecting a Segment (or if a Segment is part of a Loop, then with respect to all Segments comprising such Loop), Lateral Segment or Riser Segment within sixty (60) days after the Customer Commit Date with respect thereto, then either party shall have the right to terminate the Customer Order with respect to such Segment and ***Seller shall, upon such termination, pay Customer a termination charge in the amount equal to the Installation Fee for that Segment***. Notwithstanding any other provisions related to late delivery under the Agreement, this Section sets forth the sole and exclusive remedies of Customer respecting a failure of Seller to complete installation of the Customer Fibers within any Segment, Lateral Segment or Riser Segment on or before the Customer Commit Date.

The following clauses provide penalties for late delivery. More often a credit for late delivery would be more highly qualified. It would typically be available only for on-net services, after a longer delay, and would be subject to force majeure events. In this example, the seller would be liable to pay service credits at an escalating level depending on how late a service delivery occurs.

Example Clause #3:

Credit Remedy: Within a given Reporting Period, should Supplier fail to deliver Services (new or rearrangements), on or before the Start of Service Date specified in the FOC by 24 hours, and in accordance with the objectives above, ***Customer will be entitled to a delay credit equal to one month's MRC*** for each Service that is delayed beyond the above objectives. Delay credits do not apply where installation is delayed by or at the request of Customer.

Comment: The exhibit that follows is a very detailed treatment of delivery intervals. Delivery intervals are defined in several different ways depending on the types of services. After the installation intervals have been defined, the seller sets forth a credit schedule. Note that the service credit for late delivery is defined as a percentage of the respective service's nonrecurring charge (NRC).

INSTALLATION INTERVALS.

Seller is developing systems and processes to maintain and deliver highly reliable and predictable installation intervals. As described in the Exhibit A Service Descriptions, Seller's classifies buildings as "On-Net," "Near-Net," or "Off-Net."

Installation Interval Objectives

Seller's Installation Interval objectives for all buildings/facilities are shown in the tables below. These intervals begin with the receipt of a complete and accurate Seller order form. The interval is complete when Seller has finished its service activation process including the Seller acceptance testing. The installation intervals are exclusive of any acceptance testing that the Customer and/or its End Users may require.

Table B5 – On-Net Service Delivery Interval

	On-Net Building Installation Interval Objectives			
	Add/Change Orders		Inside Plant Construction Required*	
	Additional Service to Existing End User[1]	Change Bandwidth of Existing End User[2]	Basic Inside Plant Work[3]	Extensive Inside Plant Riser Work[4]
Interval Components				
Site Assessment	4 Days	4 Days	4 Days	4 Days
Firm Order Commitment (FOC)	1 Day	1 Day	1 Day	1 Day
Installation	N/A	N/A	5 Days	25 Days
Seller Turn up	5 Days	5 Days	5 Days	5 Days
Total	10 Days	10 Days	15 Days	35 Days

*Where inside plant construction is required, Seller may, based on the complexity of the build, adjust service delivery intervals prior to the issuance of a FOC date. [1]Add Orders-Additional Service to a building/facility that already has one or more of Seller's Services terminating in said building/facility and the Customer seeks to order a new circuit of the same or different type in the same building/facility.

[2]Change Orders –changing bandwidth of an existing Service occurs when the Customer already has one of Seller's Services terminating in an applicable building/facility, but it seeks to change the bandwidth of the Service terminating in that building/facility without changing the type of Service. [3]Basic Inside Plant Only Work-fiber has been or can easily be run from the Point of Demarcation to the telecommunications closet serving the floor or area where the termination is located. Fiber must then be run from this telecommunications closet to the actual termination point. [4]Inside Plant Riser Work-Seller will need to perform significant work to either construct riser space or to pull fiber through existing, congested riser space in order to connect the Point of Demarcation to the telecommunications closet serving the floor or area where the termination is located. Fiber must also be run from this telecommunications closet to the actual Point of Demarcation.

Table B6 – Near-Net Service Delivery Intervals

	Near-Net Building Installation Interval Objectives		
	Conduit Lateral/No Fiber[1]	New Lateral Build Required[2]	Extensive Lateral Build Required[3]
Interval Components			
Site Assessment	4 Days	4 Days	4 Days
Firm Order Commitment (FOC)	1 Day	1 Day	1 Day
Installation	35 Days	45 Days	60 Days
Seller Turn up	5 Days	5 Days	5 Days
Total	45 Days	55 Days	70 Days

[1]Conduit Lateral to Building (No Fiber)-Seller has a conduit system adjacent to the building/facility and Seller has conduit available from the property line to the building/facility. Fiber must be pulled from the Seller Network, through the conduit lateral, to the building/facility. [2]New Lateral Build Required-Seller has a conduit system adjacent to the building/facility, but Seller does not have conduit available from the property line to the building/facility. [3]Extensive Lateral Build Required-Seller has a conduit system adjacent to the building/facility, but significant work may be required to build conduit to the building/facility Point of Demarcation.

Table B7 – Off-Net Service Delivery Intervals

	Off-Net Building Installation Interval Objectives	
	Building in Seller Jurisdiction[1]	Building Not in Seller Jurisdiction[2]
Interval Components		
Site Assessment	ICB	ICB
Firm Order Commitment (FOC)	ICB	ICB
Installation	ICB	ICB
Seller Turn up	ICB	ICB
Total	ICB	ICB

[1]Building in Seller Jurisdiction-The building/facility requiring connectivity is located within a jurisdiction where Seller offers Service, but the building/facility is not adjacent to the existing Seller Network. [2]Building Not in Seller Jurisdiction-The building/facility requiring Service is not located within a jurisdiction where Seller offers connectivity.

Installation Interval Conditions

The above tables represent Seller's Service Level Objectives for installation of its Services. Seller's ability to meet these objectives is contingent on multiple variables including the Customer's completion of prescribed tasks as well as specific assumptions regarding the installation process. These conditions include:

- Local/State/Region has no jurisdiction over private property (requires no permitting in this environment).
- As required, jurisdiction permit cycle takes no longer than 30 calendar days.
- Complete and signed Tenant Request for Building Entrance and Owner/Management Agency contact information received as a component of a complete Seller order form.
- Installation intervals are handled on an individual case basis (ICB) if Seller does not have in place, or cannot obtain upon commercially reasonable terms and pricing within thirty (30) days of execution of a Seller order form, an appropriate building access agreement allowing for Seller's installation of the Service to the requested building/facility.
- Maximum distance from an end building to a Seller Hub will be no longer than 15 route miles.
- Rack space, accessibility, and power (if necessary) are available in the building/facility to be installed at the time the order is received, or can be obtained upon commercially reasonable terms and pricing within thirty (30) days of execution of an Seller Standard Circuit Order Form .
- Optical handoff at the building/facility to be installed is assumed.
- All intervals indicated are in Business Days.
- Add/Change orders must terminate in the same location/facility as the original Standard Circuit Order Form.

The FOC represents Seller's committed date for the Installation Interval objectives. Actual Service delivery intervals committed at FOC may be different from standard intervals if any conditions for installation are not met.

Installation Interval Credit Structure

For any service installation in which Seller fails to meet the Installation Interval objectives, the following credit structure will be applied to the Non-Recurring Installation Charges (NRCs) associated with the affected circuit(s).

Table B8 – Service Delivery Credit Structure

Days Exceeding Installation Interval Objective	Installation Interval Credit Structure (% Of NRC Credited)
10 business days	25%
30 business days	50%
60 business days or more	100% or the Customer may terminate the affected circuit without penalty

Example Clause #4:

Implementation Intervals.

Seller's standard service implementation interval objective for DS_N and OC_N Service is set forth below in Table A.2. Third Party Provider service implementation intervals shall be determined on an individual case basis. Seller shall make reasonable efforts to provide Seller Services within its standard service implementation interval.

Table A.2 Implementation Intervals	
Service Type	Standard Interval POP to POP
DS-1	20 business days
DS-3	30 business days
OC-3	45 business days
OC-12	45 business days
OC-48	60 business days

With respect to any Seller Private Line Service, in the event Seller fails to turn up Customer's Service within ten (10) calendar days after the completion of the standard service implementation interval, as set forth in Table A.2, for services between Tier A cities, and within twenty (20) calendar days after the completion of the standard service implementation interval, as set forth in Table A.2, for services involving a Tier B city, Customer shall have the option to receive a credit in the amount of 1/30th of the Monthly Recurring Charge for such Seller Private Line Service for every day that such circuit is not turned up in accordance with this Section. Such credit shall not apply until Customer's Service has not been provided by the tenth (10th) day, for Tier A cities, or twentieth (20th) day, for Tier B cities, after Seller's applicable standard service implementation interval, as set forth in Table A.2. Such credit shall not exceed one (1) month's Monthly Recurring Charge for such Seller Private Line Service. If service is not turned up within thirty (30) days of the standard implementation interval for any Private Line Service, Customer shall have the right to cancel the order without liability for a cancellation charge or any other charge applicable to Seller Services, but Customer will be responsible for any Third Party cancellation or termination liability charges that are assessed to Seller as a result of such cancellation. A current list of Seller's Tier A and Tier B cities are attached hereto as Exhibit A. Such list may be modified from time to time at Seller's sole discretion.

Notwithstanding anything to the contrary in this Agreement, in no event shall Seller's failure to deliver Private Line Service within the periods set forth in subsection (b) above constitute a default under the TSA. Customer agrees that Customer's remedies set forth in subsection (b) above shall be Customer's sole remedies and Seller's sole liability in the event of any such delay. Notwithstanding the foregoing, Customer shall not have the right to receive any credits or have the right to cancel for Seller's delay in delivering Seller Private Line Service, as set forth in subsection (b) above, if the delay is caused by Customer, the Local Access Service provider or a Force Majeure event as defined in the TSA.

In the event any such failure to turn up Customer's Service is caused by a Third Party Provider, Seller shall pass through to Customer any delay credits available to Seller from the Third Party Provider as a result of such delay.

Example Clause #5:

Private Line and Wavelength Service Levels. The following service levels are applicable where Customer orders Private Line Service or Wavelength Service.

(A) Installation Service Level.

(1) Seller will exercise commercially reasonable efforts to install any Private Line Service or Wavelength Service on or before the Customer Commit Date specified for the particular Service. This Installation Service Level shall not apply to Customer Orders that contain incorrect information supplied by Customer, Customer Orders that are altered at Customer's request after submission and acceptance by Seller. In the event Seller does not meet this Installation Service Level for a particular Service for reasons

other than an Excused Outage Customer will be entitled to a service credit off of the NRC and/or MRC for the affected Service as set forth in the following tables:

For any Private Line Service:

Installation Delay Beyond Customer Commit Date	Service Level Credit
1 – 5 business days	Amount of NRC
6 – 20 business days	Amount of NRC plus charges for one (1) day of the MRC for each day of delay
21 + business days	Amount of NRC plus one (1) months' MRC

For any Wavelength Service:

Installation Delay Beyond Customer Commit Date	Service Level Credit
1 - 5 business days	5% of the MRC
6 – 20 business days	10% of the MRC
21 + business days	15% of the MRC

(2) The Installation Service Level and associated credits set forth in sub-Section (1) above shall not apply to Off-Net Local Loop Service, including, without limitation, Metropolitan Private Line (Off-Net) Service, provisioned by Seller through a third party carrier for the benefit of Customer. Seller will pass-through to Customer any installation service level and associated credit (if applicable) provided to Seller by the third party carrier for such Off-Net Local Loop Service.

Part 1, Chapter 7 - System Maintenance

A telecommunications network requires continuous monitoring and frequent maintenance. Maintenance consists of many elements and is generally classified as either scheduled, routine or planned maintenance, or unscheduled or emergency maintenance. Scheduled maintenance includes routine inspections and service of network components, including preventive inspections. Scheduled maintenance also includes regularly scheduled patrol of the network, maintenance of a "call before you dig," "CBUD" or "One Call" program, establishment and maintenance of cable markers along the entire network route, and assignment of personnel at regular intervals along the route who can monitor the network and respond quickly to threats to the network.

Unscheduled or emergency maintenance consists of all maintenance activities that do not fall within the scope of scheduled maintenance.

While the definitions of scheduled and unscheduled maintenance are often stated in detail in a dark fiber agreement, that are not set out at length in a lit services agreement. Unlike other kinds of telecommunications services agreements, the costs of all maintenance are typically borne by the seller and are included in the schedule of recurring charges. A customer typically pays a single charge for lit services, which includes the costs of all maintenance. Therefore, it is unnecessary to have an understanding of the distinct categories of maintenance. (For a more extensive discussion of scheduled and unscheduled maintenance, consult the relevant chapter under Fiber Use Agreements.)

What is more important in a lit services agreement is the schedule when the seller will conduct maintenance. Unscheduled maintenance will be performed whenever needed. This can occur on short notice. Scheduled maintenance will be conducted continuously, but scheduled maintenance that has a potential for interference with a customer's use of the service will be performed at times when it will have the least impact on the network and customers' services. Each customer will in turn schedule maintenance on its own system at times that are coordinated with the seller's maintenance windows. Here is the maintenance clause for a lit services agreement that is taken from our sample MSA.

> 9(l) Planned System Maintenance: Seller usually conducts Planned System Maintenance outside of normal working hours, on weekdays between 8:00 p.m. and 3:00 a.m. Eastern time, and on weekends after 5:00 p.m. Eastern time on Friday and before 5:00 p.m. Eastern time on Sunday. Seller will use reasonable efforts to minimize any Service interruptions that might occur as a result of Planned System Maintenance.

The clauses that follow set forth windows for scheduled maintenance that are different from the preceding clause. These clauses also require the carrier to give notice of scheduled maintenance to the customer.

Note that periods of scheduled maintenance are excluded from the definition of "Outage." That is, a Customer would not be entitled to service credits if service is interrupted for scheduled maintenance. (See Example Clause #4 and Example Clause #7 below.)

Example Clause #1:

> "Planned System Maintenance" means maintenance on a network facility that is related to service delivery, either directly (e.g. maintenance of transmission equipment, fiber, fiber cable) or indirectly (e.g. maintenance of power, environmental systems), other than maintenance that is performed in response to a threat of Service interruption.
>
> Planned System Maintenance may result in interruption of Service. The Seller NOC usually conducts Planned System Maintenance outside of normal working hours, between 12:00 AM to 6:00 AM seven (7) days a week. When Planned System Maintenance is not expected to result in an interruption of service, Seller will notify Customer of Planned System Maintenance at least three (3) days in advance. When Planned System Maintenance is expected to result in an interruption of service, Seller will notify Customer of Planned System Maintenance at least seven (7) days in advance.

Example Clause #2:

> "Maintenance Window" means 11:00 p.m. to 7:00 a.m. (local time) and certain scheduled weekends, as required;
>
> Maintenance and Repair. From and after each Acceptance Date, the relevant Service shall be provided in good working order. Seller shall use commercially reasonable efforts to perform all scheduled maintenance (which may include, without limitation, substituting, changing, converting and reconfiguring equipment and facilities with respect to a Service) during a Maintenance Window. In the event Seller determines that it is necessary to interrupt a Service for the performance of scheduled maintenance, Seller will use commercially reasonable efforts to notify Customer at least forty-eight (48) hours prior to such interruption. In no event shall interruption for maintenance constitute a failure of performance by Seller of Service in any manner.

Comment: The following clause does not include a particular maintenance window. Nevertheless, it does require seven days' notice of scheduled maintenance.

Example Clause #3:

> Maintenance support will be on a circuit level basis between Seller POPs, or, where Seller has arranged local access, between Customer Interfaces. Seller may perform scheduled or emergency maintenance (including temporary suspension of Service as necessary) to maintain or modify the Network, Network Terminating Equipment or the Services. Seller will give Customer such notice of the maintenance as is reasonably practicable in the circumstances, provided that, in the event of scheduled maintenance, Seller will give Customer at least ***seven days' notice***.

The following clause sets forth scaled notice periods. Different types of scheduled maintenance require different levels of notification. Furthermore, this clause requires the seller to inform the customer how long a maintenance activity is expected to take. Typically a customer is not entitled to service credits for interruptions of service that result from scheduled maintenance activities. However, the passage "Customer may utilize ***the remedies for Outages or Excessive***

Outages" means that if the maintenance activity exceeds the period for which the Customer was given notice, the Customer would be entitled to outage credits and invoke a termination clause.

Example Clause #4:

> Planned Network Maintenance Activity. Seller shall avoid performing network maintenance between ***0600 to 2200*** Central Time (or local time with respect to facilities comprising international Service), Monday through Friday, inclusive, that will have a disruptive impact on the continuity or performance level of Customer's Service. However, the preceding sentence does not apply to emergency maintenance including, but not limited to, restoration of continuity to a severed or partially severed fiber optic cable, restoration of dysfunctional power and ancillary support equipment, or correction of any potential jeopardy conditions. Seller will use commercially reasonable efforts to notify Customer prior to any such emergency maintenance and provide a maintenance window to Customer within which such emergency maintenance will occur. In the event that Customer's Service experiences an Outage condition during such maintenance window, Customer may utilize the remedies for Outages or Excessive Outages, as applicable, set forth in Section 3 above. Seller shall provide Customer with electronic mail, facsimile, or written notice of all non-emergency, planned network maintenance (a) not less than ***three (3) business days*** prior to performing maintenance that, in its reasonable opinion, has ***a substantial likelihood of affecting Customer's traffic for up to fifty (50) milliseconds***, and (b) not less than ***ten (10) business days*** prior to performing maintenance that, in its reasonable opinion, has a substantial likelihood of affecting Customer traffic for ***more than fifty (50) milliseconds***. Seller shall provide a maintenance window to Customer within which such planned network maintenance will occur. In the event that Customer's Service experiences an ***Outage condition outside of such maintenance window, Customer may utilize the remedies for Outages or Excessive Outages***, as applicable, set forth in Section ___ above. If Seller's planned activity is canceled or delayed, Seller shall promptly notify Customer and shall comply with the provisions of this Section to reschedule any delayed activity.

The following clause is very favorable to the customer. In this example, the seller is required to scheduled maintenance in a manner to accommodate the customer.

Example Clause #5:

> Regular System maintenance shall be performed by Provider, at Provider's sole cost and expense, and shall ordinarily not result in an Interruption of Service. All System maintenance (except in emergencies) must be scheduled according to Customer's then effective Digital Data Network Maintenance schedule ("DDN Schedule"). Provider shall submit a written request for approval of such maintenance to Customer's Network Management Center (see Exhibit E) at least thirty (30) days prior to the date appearing in the DDN Schedule on which Provider plans to perform such maintenance. Provider's written request for approval shall indicate whether the maintenance will involve any Interruption of Service. Any Customer approved and scheduled System maintenance requiring Interruption of Service shall be completed as soon as practicable, but in no

event shall such Interruption for maintenance exceed two (2) hours or extend beyond the time period approved for such maintenance. Provided that (a) Customer is notified in advance of the Interruption as set forth above, (b) the maintenance is scheduled and approved by Customer as set forth above, and (c) the Interruption begins and ends during the time period approved for such maintenance and does not exceed the two (2) hour limit, then Customer shall not be entitled to receive a Section ___ credit for such Interruption. Provider may obtain information regarding Customer's DDN Schedule by contacting Customer's Network Management Center (see Exhibit __). The DDN Schedule may be revised from time to time by Customer in its sole discretion.

The clause that follows is similar to the preceding clauses. However, it does make an important distinction that is favorable to the customer. There is a firmer commitment on the part of the Provider to conduct maintenance only during agreed maintenance windows that are convenient to the customer.

Example Clause #6:

MAINTENANCE AND REPAIR OF ON-NET SERVICES

Except as provided in Section ___ above, any maintenance and repair required on the Service Provider's system, on Service Provider or Customer End User Premises, shall be performed by Service Provider or its designated contractor(s) at no additional cost to Customer only if the failure is due to the failure of facilities or employees of Service Provider.

Service Provider shall perform all trouble maintenance and repair functions on its system and facilities from the End User Premises to the demarcation point at the Customer facilities twenty-four (24) hours per day, seven (7) days per week. Service Provider scheduled maintenance ***will be performed during specified Customer maintenance windows***, except in the case of emergencies, in which case as much notice as is practicable will be given. Customer must also be prepared to provide a maintenance window with seventy-two (72) hours' notice from the Service Provider.

Specifications

Maintenance and repair of the system will be performed so as to meet the manufacturer's specification and other specifications as set forth herein.

Any maintenance or repair function performed by Service Provider on the system which will or could affect Service provided to Customer End Users ***will be coordinated and scheduled with Customer surveillance system operations*** as practical and feasible for Service Provider. Customer shall provide and update a list of Customer contacts for maintenance and escalation purposes.

Maintenance Spares

Service Provider will provide all maintenance spares plus repair and return Service of defective parts. In general, Customer will not provide equipment storage space in Customer facilities over and above storage space available in Service Provider's equipment racks.

Scheduled Maintenance

Scheduled routine maintenance will be performed during specified Customer maintenance windows and ***will be coordinated between Service Provider and Customer***.

Maintenance which may place the system in jeopardy or require system down time will normally be performed during the "Maintenance Window" during a time mutually agreed to by Customer and Service Provider. Jeopardy and down time must be requested from the Customer surveillance system operations, up to seventy-two (72) hours prior to the requested maintenance time unless otherwise agreed to by Customer.

Service Provider maintenance personnel will notify Customer prior to beginning scheduled maintenance work and ***must receive concurrence***, which shall not be unreasonably withheld, to proceed. Service Provider personnel will make reasonable efforts to notify Customer upon completion of scheduled maintenance work and receive concurrence that all Service is fully operational.

The clause that follows sets out a more detailed description of the maintenance services that will be performed by the Carrier. However, as in the previous clause, this clause does include a particular maintenance window within which maintenance activities must be performed, and requires the Carrier to coordinate its activities with the Customer. In addition, this clause requires a very substantial fourteen days' notice of scheduled maintenance.

Example Clause #7:

MAINTENANCE AND REPAIR OF ON-NET SERVICE

APPENDIX NO. 2
MAINTENANCE AND REPAIR

1.0 Performance Monitoring and Reporting.

1.1 Carrier will be responsible for performing surveillance on its major systems. However, Customer may also perform monitoring of Customer's leased bandwidth from Carrier's demarcation point at the expense of Customer.

1.2 Carrier will sectionalize faults occurring within the system localized to the Customer system elements as follows: Carrier Transmission equipment on the End User Premises; equipment between Carrier and Carrier's demarcation point.

1.3 Customer shall have the right, prior to Circuit acceptance, to monitor any Circuit for up to two (2) hours to determine if Circuit acceptance requirements have been met.

2.0 Maintenance and Repair of On-Net Service.

2.1 Any maintenance required on the Carrier Equipment, CPA Equipment or Carrier's system, on Carrier or Customer End User premises, shall be performed by Carrier or its designated contractors at no additional cost to Customer if the failure is due to the failure of facilities or employees of Carrier.

2.2 Carrier shall perform all maintenance functions on its system and facilities from the End User Premises to the demarcation point at the Customer facilities twenty-four (24) hours per day, seven (7) days per week. This includes only trouble maintenance (Service restoration) functions. Carrier ***scheduled maintenance*** will be performed after at least ***fourteen (14) days' notice*** to Customer, except in the case of emergencies, in which case as much notice as is practicable will be given.

2.3 Specifications: Maintenance of the Carrier Network will be performed so as to meet the manufacturer's specification and Appendix __. Customer shall have the right to review Carrier's maintenance procedures and policies and edited maintenance records.

2.4 Any maintenance or service function performed by Carrier on the Carrier Network which will or could affect service provided by Customer End Users will be coordinated and scheduled through Customer surveillance system operations center whenever possible for Carrier. Customer shall provide and update a list of Customer contacts for maintenance and escalation purposes.

2.5 Response & Repair Times. In the event of a Service failure, Carrier shall have repair personnel on site within 30 minutes after receiving notification of the failure from Customer. Carrier shall restore the Service on the failed system as follows:

> (a) Electronic Restoration. In the event of an electronic failure, Carrier shall use its best efforts to restore service to the affected electronics within one and one-half (1.5) hours of arrival of maintenance personnel on site.
>
> (b) Cable Restoration. Carrier shall use its best efforts to restore the cable no later than one and one-half (1 ½) hours after failure.

2.6 Carrier shall maintain a twenty-four (24) hours a day, seven (7) days a week point-of-contact for Customer to report to Carrier system trouble reports or faults.

2.7 Equipment Spares. Carrier will provide all maintenance spares plus repair and return service of defected parts. In general, Customer will not provide equipment storage space in Customer facilities over and above storage space available in Carrier's equipment racks.

2.8 Scheduled Maintenance.

(a) Scheduled routine maintenance will be performed during specified Customer maintenance windows and will be coordinated between Carrier and Customer.

(b) Maintenance which may place the system in jeopardy or require system down time will normally be performed during the ***"Maintenance Window" of 12:00 midnight and 6:00 a.m.*** or a time mutually agreed to by Customer and Carrier. Jeopardy and down time must be requested from the Customer surveillance system operations, fourteen (14) days prior to the requested maintenance time_unless otherwise agreed to by Customer.

(c) Carrier maintenance personnel will notify Customer prior to beginning scheduled maintenance work and must receive concurrence, which shall not be unreasonably withheld, to proceed. Carrier personnel will notify Customer upon completion of scheduled maintenance work and receive concurrence that all Service is fully operational.

2.9 Access to Equipment and Facilities.

(a) Employees or agents of Carrier shall have escorted access to any Carrier equipment or facilities at a Customer End User Premises or Customer Premises, subject to End User's or Customer's access and security regulations at all times. These shall include, but not be limited to:

Proper Identification
Carrier Authorized Personnel List
Restricted Area Access Provisions
Accompaniment by End Users/Customer personnel

Carrier employees or agents, while on Customer End User Premises or Customer Premises, shall comply with applicable state, federal, End User and/or Customer plant rules and regulations.

(b) Upon request, employees or agents of Customer shall be given escorted access, for viewing only, to areas at Carrier locations containing facilities and/or equipment associated with Customer's bypass Service, subject to Carrier's access and security regulations. These shall include, but not be limited to:

Proper Identification
Customer Authorized Personnel List
Restricted Area Access Provisions
Accompaniment by Carrier personnel

Customer employees or agents, while on Carrier premises, shall comply with Carrier's plant rules and regulations. This access shall be coordinated through the Carrier sales management team assigned to Customer.

Emergency Maintenance. "Emergency Maintenance" shall mean maintenance that if not accomplished promptly by Carrier, could result in a serious degradation or loss of service to the Customer or the End User.

Planned Outage. "Planned Outage" shall mean any a disruption or degradation of Service caused by scheduled maintenance or planned enhancements or upgrades to the Carrier Network.

Service Outage. "Service Outage" shall mean a disruption or degradation of Service, ***which is not caused by scheduled maintenance*** or planned enhancements or upgrades to the Carrier Network.

Part 1, Chapter 8 - Default

Not every breach of a duty under the contract will rise to the level of a Default under the contract. "Default" has a particular, defined meaning. A Default arises after a notice by either party to the other of a material breach of the agreement and a failure by the breaching party to remedy the breach within an allowed period of time.

The sample MSA form of contract defines two categories of Defaults, monetary defaults and nonmonetary defaults. A failure to pay an amount due under the contract would always constitute a Default. However, other breaches of the contract may or may not constitute a default depending on whether the breach is material. And whether a nonmonetary breach is material would depend on the circumstances of the breach.

In addition to defining Default, the contract prescribes remedies in the event of a Default. First, either party may terminate the Agreement if the other party is guilty of a Default.

Additional remedies are available to the Seller in the event of a Default by the Customer. The Seller may suspend services, terminate the Agreement, and may recover early termination charges as a remedy. The early termination charges prescribed in the event of Default under Subsection 3(a) are calculated in the same manner as if the Seller were to terminate for one of the reasons set forth in Subsections (d)(1) or (d)(4).

A Customer may elect to terminate the Agreement or any particular Service at any time. Such an elective early termination will give rise to early termination charges under Subsection 3(c). The Default remedies are integrated with the early termination charges so that if the Seller elects to terminate the Agreement because of a Default by the Customer, this would also give rise to early termination charges.

3. Termination:

(g) Either Party may terminate a Service for Default by the other Party. In the event of a Default by Customer, Seller shall have the right to any or all of the following remedies: (1) suspend Service(s) to Customer; (2) cease accepting or processing orders for Service(s); and (3) terminate this Agreement and any applicable SOF. If Seller terminates this MSA due to a Default by Customer, Customer shall immediately pay to Seller the early termination charge as described below or in the applicable Service Order Form. If the Service provided under any Service Order hereunder has been suspended or terminated by Seller for Default and thereafter Seller agrees to restore such Service, Customer first must pay all past due charges, plus a re-connect charge equal to one (1) month's MRC, plus a deposit equal to two (2) months' MRCs.

(h) Except as otherwise provided in Subsection 1(c), if Customer terminates a Service or SOF before the expiration of its applicable Service Term for any reason other than a Default by Seller, in addition to any charges incurred for Service provided herein, Customer shall incur an early termination charge. The

Parties acknowledge and agree that the early termination charge is a reasonable estimate of the likely loss and damage suffered by Seller and is not a penalty. Notice of early termination must be delivered to Seller in writing, and will be effective thirty (30) days after receipt. The early termination charge shall be an amount equal to: (1) 100% of the MRC for each terminated Service multiplied by the number of months remaining in the first year of the Service Term, if any; plus (2) 50% of the MRC for each terminated Service multiplied by the number of months remaining in the second and succeeding years of the Service Term; plus (3) all third party charges incurred by Seller which are directly related to the installation or termination of the Service; plus (4) all supplemental charges and NRC charges (including all nonrecurring charges that were waived by Seller at the time that Customer entered into the Service Order Form), if not already paid; plus (5) in the case of collocation space, the costs incurred by Seller in returning the collocation space to a condition suitable for use by other parties. The Parties agree that if a Service is terminated, the actual anticipated loss that Seller would likely suffer would be difficult or impossible to determine and the charges set forth herein constitute a reasonable estimate of the anticipated loss from such termination and do not constitute a penalty.

(i) Either Party may terminate this Agreement, including all attachments, exhibits and SOFs, without penalty if the other Party: (1) becomes or is declared insolvent or bankrupt; (2) is the subject of any proceedings related to its liquidation, insolvency or for the appointment of a receiver or similar officer for it; (3) makes an assignment for the benefit of its creditors; (4) enters into an agreement for the composition, extension, or readjustment of all or substantially all of its obligations.

(j) Seller may suspend or terminate this MSA or a Service without penalty if: (1) Seller reasonably determines that Customer or an End User has failed to comply with any foreign, federal, state or local law or regulation related to the Service, or Seller has a reasonable belief that Customer or an End User has committed any illegal act relating to the Service, including but not limited to, use of the Services for illegal purposes; or (2) a regulatory body, governmental authority, or a court of competent jurisdiction, restricts or prohibits Seller from providing a Service on the same terms and conditions as agreed herein; or (3) Seller is unable to obtain or maintain, on acceptable terms and pricing, any access right, permit or right of way necessary to provision the Services; or (4) if such suspension or termination is necessary to protect the technical integrity of the Seller network due to actions by Customer or an End User. Any termination pursuant to subsection (1) or (4) shall constitute a Default by Customer without notice to Customer.

10(c) "Default" means: (1) in the case of a failure to pay any amount when due under this MSA or any SOF, if Customer fails to pay such amount within ten (10) days following notice specifying such failure; or (2) if Customer is in material breach of Section 4 of this Agreement; or (3) in the case of any other material breach of this MSA or a Service Order Form by either Party, a Party fails to cure such breach within thirty (30) days after notice specifying such breach;

In addition to Default, there are sometimes other events that can give rise to termination of a service. Subsection (d) would allow the Seller to terminate in the event of an unexpected cost imposed by a governmental authority that increases the cost of providing a service.

The first example clause below would apply early termination charges in the event of a default. However, this clause allows for two types of early termination. The first early termination occurs before a service is turned up, and this carries a fairly lenient early termination charge. The second early termination occurs after a service is turned up and carries a greater early termination charge. Default gives rise to one or the other termination charge depending on the timing of the customer default.

In other respects, the following clause is very similar to the standard clause, with one exception. Example Clause #1 would calculate early termination charges for collocation in a different manner.

Example Clause #1:

SECTION 4. DISCONTINUANCE OF CUSTOMER ORDERS

4.1 Discontinuance of Customer Order by Seller. Seller may terminate any Customer Order and discontinue Service without liability:

(A) if Customer fails to pay a past due balance for Service (other than amounts reasonably disputed under Section 3.6) (i) within three (3) business days after written notice from Seller respecting charges invoiced in arrears, or (ii) within seven (7) business days after written notice from Seller respecting charges invoiced in advance;

(B) if Customer violates any law, rule, regulation or policy of any government authority related to Service; if Customer makes a material misrepresentation to Seller in connection with the ordering and delivery of Service; if Customer engages in any fraudulent use of Service; or if a court or other government authority prohibits Seller from furnishing Service;

(C) if Customer fails to cure its breach (other than as addressed in sub-Sections (A), (B), (D) or (E) of this Section 4.1) of any of these Terms or any Customer Order within thirty (30) days after written notice thereof provided by Seller;

(D) if Customer files bankruptcy, for reorganization, or fails to discharge an involuntary petition therefore within sixty (60) days; or

(E) if Customer's use of Service materially exceeds Customer's credit limit, unless within one (1) day's written notice thereof by Seller, Customer provides adequate security for payment for Service.

4.2 Effect of Discontinuance. Upon Seller's discontinuance of Service to Customer, Seller may, in addition to all other remedies that may be available to Seller at law or in equity, assess and collect from Customer any applicable ***termination charge pursuant to Section 3.8***.

3.8 Termination Charges. (A) Customer may cancel a Customer Order following Seller's acceptance of the same and ***prior to the Customer Commit Date*** upon prior written notice to Seller. In the event that Customer does so, or in the event that the delivery of such Service is terminated by Seller prior to delivery of a Connection Notice due to a failure of Customer to comply with these Terms, Customer shall pay Seller a cancellation charge equal to the sum of (i) in the case of Colocation Space, the costs incurred by Seller in returning the Colocation Space to a condition suitable for use by third parties, plus (ii):

(a) Any third party cancellation/termination charges related to the installation and/or cancellation of Service;

(b) The non-recurring charges (including any nonrecurring charges that were waived by Seller at the time of the Customer Order) for the cancelled Service; and

(c) As the case may be, (i) one (1) month's monthly recurring charges for the cancelled Service if written notice of cancellation is received by Seller more than five (5) business days prior to the Customer Commit Date, or (ii) three (3) month's monthly recurring charges for the cancelled Service if written notice of cancellation is received by Seller five (5) business days or less prior to the Customer Commit Date.

Customer's right to cancel any particular Service under this Section 3.8(A) shall automatically expire and shall no longer apply upon Seller's delivery to Customer of a Connection Notice for such Service.

(B) In addition to Customer' right of cancellation under Section 3.8(A) above, Customer may terminate Service ***prior to the end of the Service Term*** upon thirty (30) days' prior written notice to Seller, subject to the following termination charges. In the event that ***after either the Customer Commit Date or Customer's receipt of the Connection Notice*** for a particular Service (whichever occurs first) and prior to the end of the Service Term, Customer terminates Service or ***in the event that the delivery of Service is terminated due to a failure of Customer to comply with these Terms***, Customer shall pay Seller a termination charge equal to the sum of (i) in the case of ***Colocation Space***, the costs incurred by Seller in returning the Colocation Space to a condition suitable for use by third parties, plus (ii):

(a) Any third party cancellation/termination charges related to the installation and/or termination of Service;

(b) The non-recurring charges (including any nonrecurring charges that were waived by Seller at the time of the Customer Order) for the cancelled Service, if not already paid; and

(c) The percentage of the monthly recurring charges for the terminated Service calculated from the effective date of termination as (1) ***100% of the remaining monthly recurring charges that would have been incurred for the Service for months 1-12 of the Service Term, plus (2) 50% of the remaining monthly recurring charges that would have been incurred for the Service for months 13 through the end of the Service Term.***

Comment: Example Clause #2 below is very similar to Example Clause #1 except that under Section 13.1 below the 30-day cure period for nonmonetary breaches would be extended under certain circumstances.

Example Clause #2:

13. Default.

13.1 A default shall occur under this Agreement if: (a) in the case of a failure to pay any amount when due under this Agreement or any Service Order Form, Customer fails to pay such amount within ten (10) days after notice specifying such breach; or (b) in the case of any other material breach of this Agreement, a Party fails to cure such breach within thirty (30) days after notice specifying such breach, ***provided that if the breach is of a nature that cannot be cured within thirty (30) days, a default shall not have occurred so long as such Party has commenced to cure within said time period and thereafter diligently pursues such cure to completion (a material breach of a Service Order Form by Seller shall not be considered a material breach of this Agreement and shall not affect any other Service Order Form).***

13.2 In the event of any default hereunder, the nondefaulting Party may avail itself of one or more of the following remedies: (a) take such actions as it determines, in its sole discretion, necessary to correct the default; (b) terminate this Agreement; and/or (c) pursue any legal remedies it may have under applicable law or principles of equity, including specific performance. Without limiting the foregoing, if the default consists of a failure of Customer to pay to Seller any Fees or expense due under any Service Order Form such default shall be ***considered a voluntary termination pursuant to Section 5.3 or Section 5.4*** and, Seller may, in addition to all other remedies, terminate or suspend any and all of its obligations in respect of such Service Order Form under this Agreement and any other Service Order Form, and apply any and all amounts previously paid by Customer hereunder toward the payment of any other amounts then or thereafter payable by Customer hereunder.

5.3 Subject to the following, Customer may terminate a Service ***prior to the applicable Service Term start date*** by providing Seller notice at least fifteen (15) days prior to the start date set forth in the applicable Service Order Form. In the event Customer so terminates a Service, Customer shall pay Seller all applicable charges incurred to that date, including any third party costs. In the event Customer so terminates a Service by notifying Seller less than fifteen (15) days but at least ten (10) days prior to the start date set forth in the applicable Service Order Form, then, in addition to paying Seller all applicable charges incurred to that date, including any third party costs, Customer shall upon such termination pay Seller a termination fee equal to any and all of the Fees that would have become due and payable during the first two (2) months of the Service Term. Further, in the event Customer terminates a Service by notifying Seller less than ten (10) days prior to the specified start date, then, in addition to paying Seller all applicable charges incurred to that date, including any third party costs, Customer shall upon such termination pay Seller a termination fee equal to any and all of the Fees that would have become due and payable during the first four (4) months of the Service Term. The Parties acknowledge and agree that if Customer so terminates a Service, the termination fee set forth in this Section 5.3 is a genuine pre-estimation of the loss and damage likely to be suffered by Seller and not a penalty.

5.4 Subject to the following, Customer may terminate a Service ***prior to the end of the applicable Service Term***. In the event Customer terminates a Service prior to the end of its Service Term, then, in addition to paying Seller all applicable charges incurred to that date, Customer shall upon such termination pay Seller a ***termination fee equal to one-hundred percent (100%) of any and all of the remaining Fees and other amounts payable with respect to the first year of the Service Term and fifty percent (50%) of the remaining Fees and other amounts payable for all years (if any) after the first year with respect to the Service Term for such Service.*** The Parties acknowledge and agree that if Customer so terminates a Service, the termination fee set forth in this Section 5.4 is a genuine pre-estimation of the loss and damage likely to be suffered by Seller and not a penalty.

Example Clause #3: The remedies in the following two clauses are simpler and more straightforward than previous clauses. The early termination charges are not calculated as a discounted value of the future revenue stream. A default by customer accelerates all amounts payable under the agreement.

TERMINATION AND TERMINATION CHARGES.

8.2. Seller may terminate this Agreement, or any Service, or both, immediately on notice if Customer (a) fails to make any payment due under this Agreement, (b) fails to provide security or additional security within the timeframe required under Section 3, or (c) fails to meet any monthly minimum charge or threshold as set out in this Agreement or the applicable Service Appendix/Exhibit.

8.3 Either Party may terminate this Agreement, any Service, or both, immediately on notice, if the other (a) commits a material breach of this Agreement (other than a failure to pay or provide security which is covered under Section 8.2), which is capable of remedy, and fails to remedy the breach within thirty days written notice, (b) commits a material breach of this Agreement which cannot be remedied, or (c) is repeatedly in breach of this Agreement.

8.4 Seller may at its sole option, upon the occurrence of any of the events detailed in Section 8.2 or Customer's breach as outlined in Section 8.3 giving Seller a right to terminate this Agreement or any Service(s): (a) cease accepting or processing orders for Service(s) and suspend Service(s) without prejudice to its right to terminate this Agreement or the Service(s); (b) cease all electronically and manually generated information and reports (including any CDR not paid for by Customer); (c) draw on any letter of credit, security deposit or other assurance of payment and enforce any security interest provided by Customer; (d) collect from Customer an amount equal to any Minimum Charge agreed between the Parties for the remaining portion of the unexpired term of this Agreement, minus any amounts collected under this Section 8.4(c); and/or (e) pursue such other legal or equitable remedy or relief as may be appropriate.

8.5 If this Agreement or any Service is terminated prior to the expiration of the initial or any renewal term of any Service (except if properly terminated by Customer for Seller's breach under Section 8.3), then Customer shall pay to Seller upon demand an ***early termination fee in an amount equal to the aggregate sum of each existing Service's MRC times the number of months remaining of the applicable term and all third party termination liability for Customer's termination***. Customer agrees that such a termination fee is based on an agreed revenue expectation based on actual Customer Service data and is not a penalty.

<u>Example Clause #4</u>:

14. Termination by Seller: (a) Seller may terminate this Agreement or any Service Order hereunder or suspend Services, with prior written notice, upon: (i) Customer's failure to pay any amounts as provided herein; (ii) Customer's breach of any provision of this Agreement or any law, rule or regulation governing the Services; (iii) any insolvency, bankruptcy assignment for the benefit of creditors, appointment of trustee or receiver or similar event with respect to Customer; or (iv) any governmental prohibition or required alteration of the Services.

15. Termination Liability: If, prior to the end of the term, Seller terminates this Agreement or any Service Order(s) hereunder under section 14, above, or if Customer terminates this Agreement or any Service Order(s) hereunder for any reason other than Seller's material breach of this Agreement that remains uncured after written notice and a reasonable cure period, ***Customer must pay immediately to Seller all monthly recurring charges associated with the terminated Service(s) for the balance of the term*** in such Service Order(s).

<u>Example Clause #5</u>: Under all of the previous clauses, a failure to pay an amount owed would always constitute a Default. Any other breach of the agreement would constitute a default only if the breach is "material." The clause below expands on this principle to define certain nonmonetary breaches as always constituting Defaults. A violation of the law by Customer, Customer's failure to abide by the Seller's Acceptable Use Policy would always be a deemed material breach.

> Default and Remedies. A "Default" shall occur if (a) Customer fails to make payment as required under this Agreement and such failure remains uncorrected for ten (10) calendar days after written notice from Seller; or (b) either party fails to perform or observe any material term or obligation contained in this Agreement, and any such failure remains uncorrected for thirty (30) calendar days after written notice from the non-defaulting party. If Customer uses the Services for any unlawful purpose or in any unlawful manner, Seller shall have the right immediately to suspend and/or terminate any or all Services hereunder without notice to Customer. If Customer violates Seller's Acceptable Use Policy ("AUP") posted on Seller's AUP website at www.____________, which is incorporated herein by reference, and if such activity is affecting Seller's network, or other parties on Seller's network, Seller shall have the right immediately to suspend and/or terminate any or all Services hereunder without notice to Customer. Customer affirms that it has reviewed and assented to the AUP. For any violations of Seller's AUP that are not affecting either Seller's network or third parties on Seller's network, Customer shall, upon three (3) business days' notice, have the opportunity to cure such violation prior to suspension or termination. In the event of a Customer Default for any reason, Seller may in addition to its right available to it at law or in equity: (i) suspend Services to Customer; (ii) cease accepting or processing orders for Services; (iii) withhold delivery of Call Detail Records (if applicable); and/or (iv) except in the event of an AUP violation, terminate this Agreement. If this Agreement is ***terminated because of a Customer Default, such termination shall not affect or reduce Customer's minimum monthly commitments required under this Agreement, if applicable; and, all Early Termination Charges shall apply***.

> Early Termination Charge. Customer may disconnect any Seller Service after installation by providing written notification to Seller thirty (30) calendar days in advance of the effective date of the disconnection and paying to Seller an "Early Termination Charge" in an amount equal to (i) ***100% of the monthly charge for each such disconnected On-Net circuit, multiplied by the remaining payments through month 12; (ii) 50% of the monthly charge for each such disconnected On-Net circuit, multiplied by the remaining payments for months 13-60 if applicable; plus (iii) any termination liability associated with local access or any other Third Party Service***.

Part 2 - Service Descriptions and SLA

A service description will explain with specificity the characteristics of a service, including the performance specifications and the contract terms and conditions that pertain to the service. The contract terms and conditions that apply to all services will be found in the Master Services Agreement. The contract terms and conditions that do not pertain to all services will be found in the service descriptions to which they apply.

A service level agreement (or SLA) is an element of a service description that describes a level of performance that is expected from a service, including service credits to which a customer may be entitled if there is an interruption of service or a failure to maintain the expected level of performance.

Part 2, Chapter 1 - Optical Wavelength Service

While the Master Services Agreement sets forth general contract terms and conditions that are applicable to all services provided under the MSA, an individual service description is prepared for each service that is offered under the MSA. A service description is agreed to by the parties and is integrated with the MSA by means of a service order form which makes reference to and incorporates both the MSA and the service description. Each service description and service order form should identify the MSA to which it is associated.

Service Description:

Each service description is unique. The technical performance specifications, the commercial terms such as recurring and nonrecurring charges, and credits for interruptions of service, may be different for each service.

The example service description below describes a typical optical wavelength service. This particular service offering is available between two identified end points only, called the "Seller Termination Points." Other service descriptions might not have this limitation. The service may be available between any two points that may be found on Seller's network. Even though a service may be offered and available between many points, every service order must identify the two termination points to which that particular service order applies.

Protected and Unprotected Service:

Another characteristic of this service example is that the service s "unprotected." That means that the Seller's network used to provide this service consists of a single network path between the Seller Termination Points. Other service offerings found in the examples below may be provided by means of a "dual path" or "protected" service. When a service is offered by means of a dual path, it is sometimes also said that the Seller's network or the service is "diverse."

Network diagrams are sometime use in describing the Seller's network and the manner in which a service will be provided. Example Clause #3 includes a series of diagrams that depict each service in graphic form.

Example Clause #3 also describes what are referred to as "scalable" services. Subject to certain limitations, a customer may purchase services within a range of bandwidth, and adjust or scale the amount of bandwidth from time to time according to its needs. This flexibility allows a customer to purchase bandwidth only as needed, and better control its costs.

These Exhibits A and B are entered into as of the ___ day of ____, 20__ and are incorporated into and are subject to all of the terms and conditions of that certain Service Order dated _______________ by and between _____________ ("**Seller**") and Customer.

Exhibit A
To
Seller Telecommunications Service Order Form
Service Description – Optical Wavelength Service

1. SERVICE DESCRIPTION, OPTICAL WAVELENGTH SERVICE

The Optical Wavelength Service (the "Service") provides point-to-point optical connectivity between two locations.

Optical Wavelength Service: The "Seller Network" consists of the telecommunications facilities that are owned or operated by Seller to provide the Service between the following termination points (the "Seller Termination Points"):

(a) _______________.
(b) _______________.

The Service is available at ____ (__) Gbps line rate, and is offered as a single optical circuit path, unprotected service between Seller Demarcation Points. The Service is delivered with a two-fiber handoff at the Seller Demarcation Point. Dual entrance facilities are not available for the Service.

TECHNICAL SPECIFICATIONS.

	Optical Wavelength Service
Service Characteristics	Single Path Unprotected
Line Rate	___ Gbps
Handoff	__ fibers
Availability*	____%

* See Service Level Agreement for applicability and exclusions.

PRICING: Pricing for the Service is set forth below. Pricing is subject to change at Seller's sole discretion on fourteen (14) days' notice. Changes in pricing do not apply to any effective SOFs, are not retroactive or cumulative, and apply only to new SOFs entered into after the effective date of the notice of price change.

Service Term (Months	MRC	NRC

)		
___ ________	$_____ $_____	$______ $______

Example Clause #1:

Interconnect Specifications within Seller PoP:

Optical Interface Parameters		
SONET/SDH Equivalent Interface	OC48/STM16	OC192/STM64
Synchronous Throughput	2488.32 Gb/s (Better known as2.5 Gb/s)	9.953 Gb/s (Better known as10 Gb/s)
Fiber Type	Single Mode	Single Mode
Optical Interface	FC or SC	FC or SC
Wavelength range	1290-1560 lambda	1290-1560 lambda
Transmitter Requirements	ITU-T G.957 S-16.1	ITU-T G.691 S-64.1
Receiver Requirements	ITU-T G.957 S-16.1	ITU-T G.691 S-64.1
Line Code	Binary NRZ (scrambled according to recommendation G707)	Binary NRZ (scrambled according to recommendation G707)
Transmitter Level	+0 dBm to -5 dBm	+5dBm to 0 dBm
Receiver Overload At 1E-10 BER	> 0dBm	> -1 dBm
Receiver Sensitivity At 1E-12 BER	< -18dBm	< -11dBm
Extinction Ratio	> 9.0 dB	> 7 dB
Minimum Side mode suppression	30dB	30dB

2. Performance Objectives:

2.5G and 10G circuit performance will be measured using two parameters: Availability and BER. The following assumptions apply to the derived data:

- 2.5Gbps and 10Gbps circuits originate and terminate (not in all segments of SELLER network) on wavelength translators interfacing to Seller optical DWDM backbone
- MTTR for Terrestrial DWDM equipment: 4 hours
- MTTR for Terrestrial fiber optic cable: 12 hours

3. Transparency:

The current product is a 2.5G and 10G "Semi-Transparent" wavelength service. This means that the customer's signal must meet the SONET/SDH frame (as per G.707, GR-253-CORE and GR-1377-CORE) and rate. The signal will have a Section, Line and Path overhead associated with it for SONET framed signals and Regenerator, Multiplexer, and Path overhead for SDH framed signals. Seller will have complete ownership of the Section overhead in SONET and Regenerator overhead in SDH. Seller will reserve the right to write, modify or terminate any or all of the overhead byte in the section or Regenerator overhead. Within the Line and Path for SONET and Multiplexer and Path for SDH portion of the overhead Seller will reserve the right to write, modify or terminate the H1-H3, B2 andS1 bytes of the overhead in certain parts of the Seller network (mostly applicable to 2.5G circuits) (Contact PM for further details).

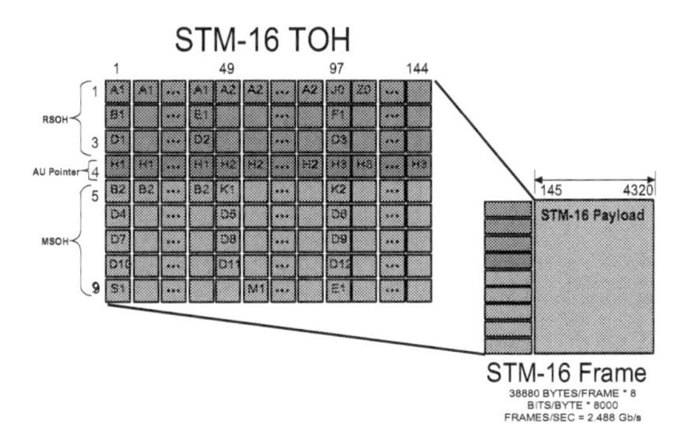

Example Clause #2:

CARRIER PRIVATE LINE SERVICES
(Not Applicable to Broadband Wireless Access)

SELLER Carrier Private Line Services provide dedicated (i.e., non-switched) connectivity for voice, data and video applications. These dedicated services typically consist of non-switched communications circuits and any required equipment, connecting one or more locations. Network configurations vary greatly and can include point-to-point connection, or various multi-point connections between locations. SELLER's Carrier Private Line Services as set forth herein include On-Net and Off-Net long haul private lines (i.e., InterLATA and interstate services). The Services will be offered in each area to the Customer by an entity ("Authorized Entity") which is either an affiliate or subsidiary of SELLER and/or which SELLER manages or with whom it is otherwise contractually affiliated. SELLER will notify Customer from time to time of the identity of each Authorized Entity. The terms and conditions of this Agreement are, and shall be, applicable to the Services provided to the Customer by each Authorized Entity.

1. SERVICE DESCRIPTIONS FOR ON-NET SERVICES

1.1 Dedicated

Each Carrier Private Line is dedicated to Customer and is billed on a fixed monthly basis. The entire usable bandwidth for each Carrier Private Line is available to Customer for its exclusive use, twenty-four (24) hours a day, seven (7) days a week.

1.2 Point-To-Point

Carrier Private Line Services are available between Customer designated locations on a point-to-point basis. Service may be ordered between the Customer's POP and an End User location, between two (2) Customer POP's, or between two (2) End User locations. There are two (2) basic configurations for Dedicated Transport Service: Hubbed Service and Two Point Service.

Hubbed Service allows Customer to aggregate multiple lower capacity private lines terminating at multiple locations onto one higher capacity service terminating at one other Customer location.

Two Point Service allows for two (2) Customer designated locations to be connected by one carrier private line. The service terminated at both locations must be the same speed/capacity.

1.3 Long-Haul Service

Long-Haul Service refers to services not provided entirely within a local calling area, i.e., InterLATA and interstate services.

1.4 Local Loop Service

A Local Loop refers to the connection from the End User location to the Central Office. For the purposes of this Exhibit, a Local Loop is the connection from the Interexchange Carrier ("IXC") POP to the End User location.

1.5 DS-1 Service

DS-1 Service is a dedicated, high capacity, full duplex channel with a line speed of 1.544 Mbps isochronous serial data having a line signal format of either Alternate Mark Inversion ("AMI") or Binary 8 Zero Substitution ("B8ZS") and either Superframe ("D4") or Extended Superframe formats. DS-1 Service has the equivalent capacity of twenty-four (24) Voice Grade ("VG") services or twenty-four (24) DS-0 services. AMI can support twenty-four (24) individual 56 Kbps channels and B8ZS can support twenty-four (24) individual 64 Kbps channels.

1.6 DS-3 Service

DS-3 Service is a dedicated, high capacity, full duplex channel with a line speed of 44.736 Mbps isochronous serial data having a line code of binary with three zero substitution ("B3ZS"). DS-3 Service has the equivalent capacity of twenty-eight (28) DS-1 Services at 1.544 Mbps or 672 VG services or 672 DS-0 Services at 56/64 Kbps.

1.7 OCN Services

SONET ("Synchronous Optical Network") is a standard for optical telecommunications transport. It was formulated by the American National Standards Institute ("ANSI"), which sets industry standards in the U.S. for telecommunications and other industries.

The first step in the SONET multiplexing process is the generation the lowest base signal. In SONET, this base signal is referred to as Synchronous Transport Signal level-1, or simply STS-1, which operates at 51.84 Mb/s. Higher-level signals are integer multiples of STS-1 ("DS-3"), creating the family of STS-N signals in Table 1. This table also includes the optical counterpart for each STS-N signal, designated OC-N (Optical Carrier level-N).

OC-3 Optical Carrier level 3. A SONET channel equal to three (3) STS1's, equal to 155.52 Mbps (STS-3 equivalent).

OC-12 Optical Carrier level 12. A SONET channel of 622.08 Mbps (STS-12 equivalent).

OC-48 Optical Carrier level 48. A SONET channel of 2.488 Gpbs (STS-48 equivalent).

2. INTERFACE REQUIREMENTS

2.1 DS-3 Interface Requirements

The interface at Network Interface and Customer Interface will be at DSX-3 cross connect located in the Customer DSX-3 environment and be of the type equaling the current Customer standard.

2.2 DS-1 Interface Requirements

The interface at Network Interface and Customer Interface will be at DSX-1 cross connect. The signal format shall be B8ZS. AMI shall be provided as an option. The frame format shall be that of the Extended Superframe ("ESF"). Super Frame ("SF") frame format shall be provided as an option.

2.3 OCN Interface Requirements

The interface at Network Interface and Customer Interface will be at the fiber distribution panel located in the Customer Optical environment and be of the type equaling the current Customer standard.

Example Clause #3: Often the parties will enter into several service orders and perhaps more than one MSA. The introductory paragraph below is intended to avoid confusion by making specific reference to the particular MSA to which this service description is associated.

This Exhibit A, which provides expanded Service Level Descriptions, is hereby incorporated into the Master Service Agreement ("MSA") dated the ____ day of ______________________, 20__, between ____ ("Customer") and __________ ("Seller").

SELLER SONET SERVICE.

1. DESCRIPTION.

Protected SONET Services
These Protected SONET Services deliver traffic along a primary and secondary path within the Seller network, and include automatic protection switching. Customer can select either a two fiber connection (1:0 handoff) or a four fiber connection (1+1 handoff) to connect its equipment to the Seller network. All DS3 services are provided with an electrical handoff.

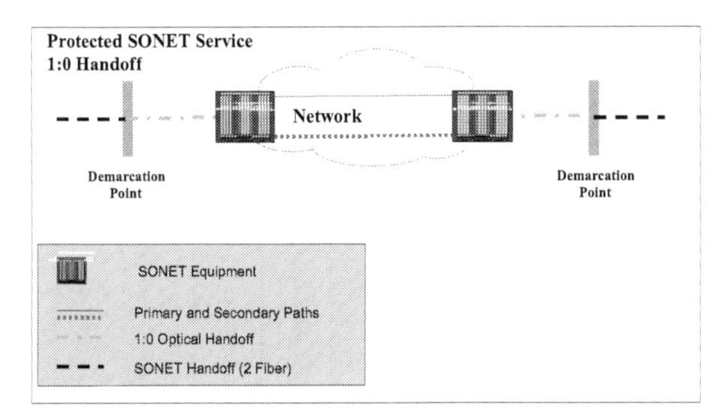

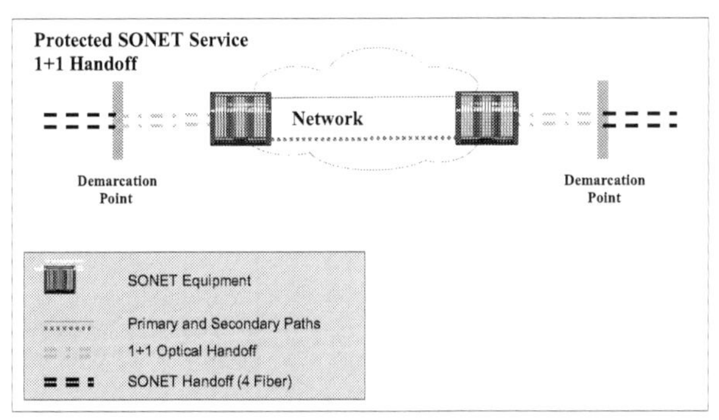

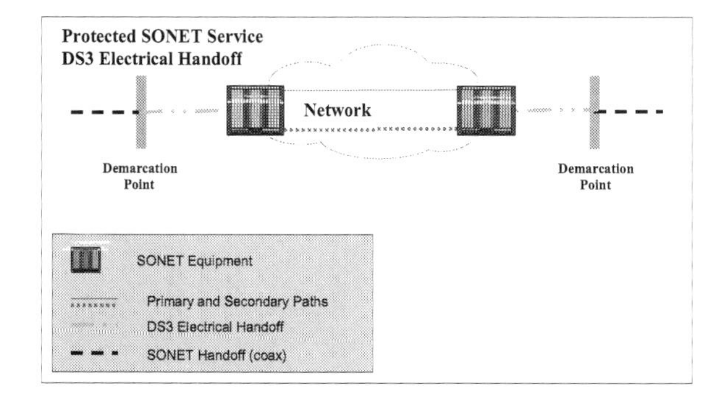

2. TECHNICAL SPECIFICATIONS.

Service Characteristics	Protected SONET Services						
Line Rate	44.7 Mbps	155.5 Mbps		622.1 Mbps		2.488 Gbps	
Optical Signal or Digital Signal	DS3	OC-3/OC-3c		OC-12/OC-12c		OC-48/OC-48c	
SDH Equivalent	N/A	STM-1		STM-4/STM-4c		STM-16/STM-16c	
Wavelength	N/A	1310 nm		1310 nm		1310 nm	
Handoff	Coax	1:0	1+1	1:0	1+1	1:0	1+1(ICB)
Availability*	99.999%	99.9%	99.999%	99.9%	99.999%	99.9%	99.999%

* See MSA for applicability and exclusions.

SELLER OPTICAL WAVELENGTH SERVICE.

1. DESCRIPTION.

Dual Path Optical Wavelength Services
These Dual Path Optical Wavelength Services provide two optical circuit paths between the Seller demarcation points. These optical circuit paths will be diverse from each other. These Services are delivered with a four-fiber handoff at the Seller demarcation point.

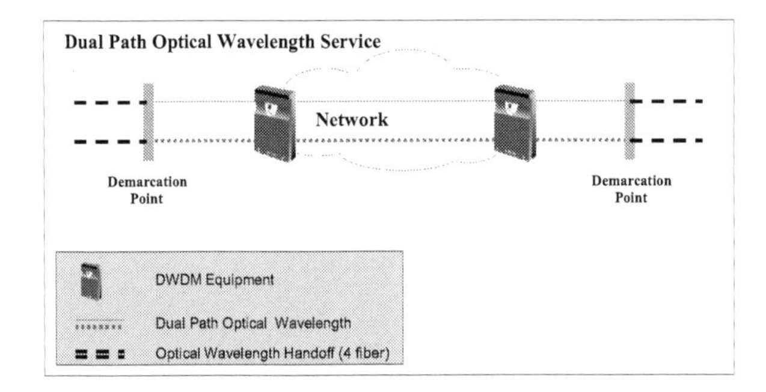

Core Protected Optical Wavelength Services

These Core Protected Optical Wavelength Services provide two optical circuit paths within the Seller network. These optical circuit paths will be diverse from each other. In the event of a failure along the primary path, the traffic will be automatically re-routed to the secondary path. These Services are delivered with a two-fiber handoff at the Seller demarcation point.

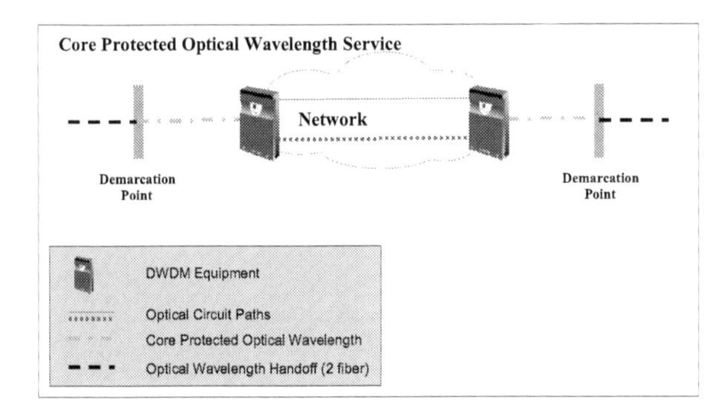

<u>*Single Path Optical Wavelength Services*</u>
These Single Path Optical Wavelength Services provide a single optical circuit path between the Seller demarcation points. These Services are delivered with a two-fiber handoff at the Seller demarcation point.

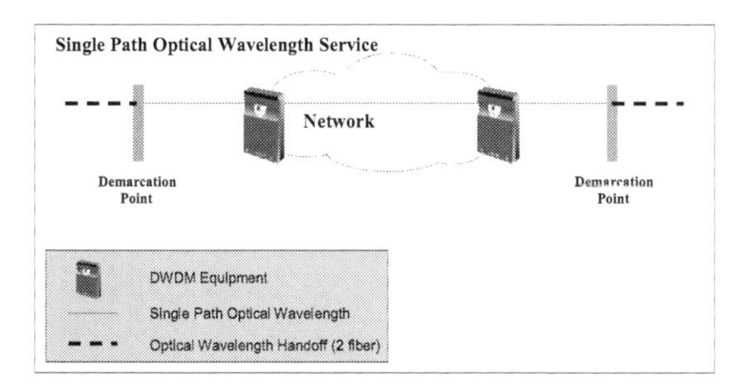

2. <u>TECHNICAL SPECIFICATIONS.</u>

Service Characteristics	Dual Path			Core Protected			Single Path		
Line Rate	1.25 Gbps	2.5 Gbps	10 Gbps	1.25 Gbps	2.5 Gbps	10 Gbps	1.25 Gbps	2.5 Gbps	10 Gbps
SONET Equivalent	N/A	OC-48	OC-192	N/A	OC-48	OC-192	N/A	OC-48	OC-192
Comparable SDH	N/A	STM-16	STM-64	N/A	STM-16	STM-64	N/A	STM-16	STM-64
Wavelength	1310 nm	1310 nm	1310 nm	1310 nm	1310 nm	1310 nm	1310 nm	1310 nm	1310 nm
Handoff	4 fiber	4 fiber	4 fiber	2 fiber	2 fiber	2 fiber	2 fiber	2 fiber	2 fiber
Availability*	99.999 %	99.999 %	99.999 %	99.9%	99.9%	99.9%	99%	99%	99%

* See MSA for applicability and exclusions.

SELLER ETHERNET SERVICE.

1. DESCRIPTION.

A. Services.

Scalable Ethernet Service
This Scalable Ethernet Service delivers IP (Internet Protocol) data packets between the demarcation points using Ethernet frames. Each Scalable Ethernet Service will have a Bandwidth Profile, which indicates its maximum data throughput. Customer may change a Bandwidth Profile provided it does not exceed the physical line rate of the port on which the circuit is provisioned (e.g. Customer could not have a 150 Mbps Bandwidth Profile on a circuit delivered via 10/100BaseT connection).

Fixed Ethernet Service
This Fixed Ethernet Service delivers IP (Internet Protocol) data packets between two demarcation points using Ethernet frames. Each Fixed Ethernet Circuit is offered at the 1 Gbps line rate and may be provided in an unprotected or a protected version. The Bandwidth Profile of this service may not be changed.

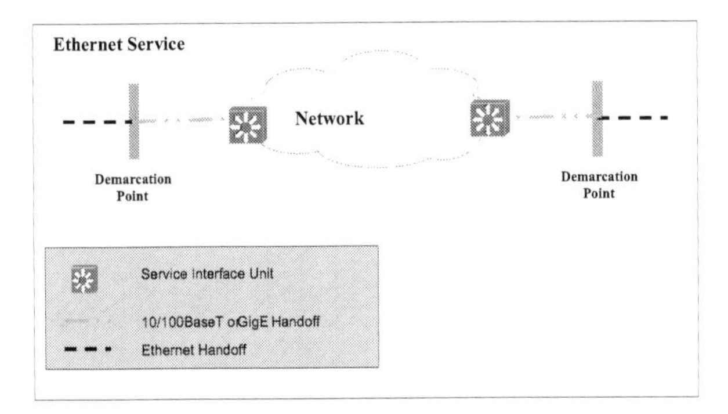

B. Interface Options.

7

Service Interface Unit
To provide Ethernet service, Seller must install, at or near the demarcation point, a Service Interface Unit ("SIU"), which it will own and maintain. If Customer is unable to provide a suitable environment for the SIU, Seller may terminate the Service Order without penalty and/or decline to award service credits due to a degradation in Service resulting from Customer's acts or omissions.

STANDARD SIU
The standard interface for Ethernet service is an untagged interface. This interface provisions one Ethernet Service per physical port on the network. Multiple customers may be provided service from one Standard SIU.

DEDICATED SIU
A Dedicated SIU provides Service to only one customer. A Dedicated SIU is required if Customer orders a Service with a Bandwidth Profile of greater than 400 Mbps. If Customer requests a Dedicated SIU for a Service with a Bandwidth Profile of less than 400 Mbps, an additional charge will apply and will be included in the price listed on the SOF. The maximum Bandwidth Profile for each Dedicated SIU is 1 Gbps.

DEDICATED SIU -- Tagged Interfaces
For an additional charge, which will be included on the price stated on the SOF, Customer can receive multiple Services over one Gigabit Ethernet handoff by selecting a Dedicated SIU with a tagged interface. This will permit Customer to assign its own VLAN tags to each Ethernet Service. The sum of Bandwidth Profiles of the Ethernet Services provisioned to a Dedicated SIU with a tagged interface shall not exceed 1 Gbps.

2. TECHNICAL SPECIFICATIONS.

	Scalable Ethernet	Fixed Ethernet	
		Protected GigE	Unprotected GigE
Service Characteristics			
Bandwidth Profile	10,15,20,30,40,50,60,70,80,90,100, 150,200,300,400,500,600,800,1000 Mbps	1 Gbps	1 Gbps
Handoffs	10/100BaseT, GigE	GigE	GigE
Scalable	Yes	No	No
Availability*	99.9%	99.9%	99%
Service Quality Objective (packet loss)	.1%	.1%	.1%

* See MSA for applicability and exclusions.

3. SERVICE PROVISIONING.

Bandwidth Profile Changes.

Customer may request a change in the Bandwidth Profile of a scalable Ethernet service at any time during the Service Term. Seller reserves the right to refuse a Bandwidth Profile change request at Seller's sole discretion.

4. ETHERNET PRICING.

Ethernet Service will be billed one month in advance based on the Bandwidth Profile in place at the end of the previous month. The monthly charge is calculated by aggregating daily charges, which are measured at the peak Bandwidth Profile in effect during a midnight-to-midnight day. Following the first month's bill, each subsequent bill will contain the amount for the next month as well as a debit or credit based on the changes in the Bandwidth Profile during the previous month. Disconnect charges will be based on the peak Bandwidth Profile during the Service Term.

If Customer reduces the Bandwidth Profile, there will be a $500 fee for each reduction. If Customer reduces the Bandwidth Profile below the original Bandwidth Profile listed on the Service Order Form, in addition to the $500 fee, the MRC will be changed to equal the new Bandwidth Profile plus 75% of the difference between the rates for the new and the original Bandwidth Profile.

Part 2, Chapter 2 - Service Interruptions

Service Level Agreement:

Every lit fiber service is subject to a standard of reliability, a set of specifications or service level, and a definition of "outage" or "unavailable" that is associated with the service. That standard is referred to as a "service level agreement" or "SLA."

Sometimes "outage" or "unavailable" is defined very simply.

Definition #1:

Committed Availability – means the percentage of time in a Measurement Period that the private line circuit is ***available*** for use by Customer.

Definition #2:

Outage Credits. Customer acknowledges the possibility of an unscheduled period of time during which Private Line Service is ***unavailable*** ("Outage"). An Outage shall begin upon the earlier of Seller actual knowledge of the Outage or Seller receipt of notice from the Customer of the Outage.

However, more often "outage" or "unavailable" is defined by reference to technical performance specifications. A service is unavailable any time that the technical specifications have not been met.

Definition #3:

For purposes of this Section ____, "unavailable" or "unavailability" means the duration of a break in transmission measured from the first of ten (10) consecutive severely erred seconds ("SESs") on the affected Service until the first of ten (10) consecutive non-SESs. An SES is a second with a bit error ratio of greater than or equal to 1 in 1000.

Definition #4:

"Service Outage" means a period during which there is a break in transmission. For these purposes a break in transmission is signalled by the first of ten (10) consecutively severely errored seconds ("SESs"), and the end is signalled by the first of ten (10) consecutive non-SESs. A SES is defined in ITU Standard (G.821).

Severely Errored Seconds ("SES"). An SES is any second in which the Bit Error Rate ("BER") is worse than 1 x 10-6.

Definition #5:

Service Interruption or Outage. A Service is considered interrupted or unavailable when there has been a loss of signal, 100% packet loss, or when tests confirm the observation of any ten (10) consecutive SES. These ten (10) consecutive seconds are considered to be part of the unavailable time. The period of unavailable time ends when the BER in each second is better that 10-6 for a period of ten (10) consecutive seconds. These ten (10) consecutive seconds are considered to be available time.

Definition #6:

OUTAGE AND OUTAGE CREDITS
For SONET Service, "Outage Time" is the period between the first of ten consecutive Severely Errored Seconds (SES > ~10-6 BER) and the first of ten consecutive non-SES. For 2.5 Gbps and 10Gbps Optical Wavelength Service, "Outage Time" is the number of SES's in the month. For 1.25 Gbps Optical Wavelength Services and Ethernet Service, "Outage Time" is any period when there is 100% packet loss. A lost packet is one with a valid Frame Check Sequence and a valid destination MAC address that is not delivered to its destination. For Dual Path Optical Wavelength Service Outage Time includes only the amount of time that both paths experience concurrent Outage Time. For all Services, Outage Time resulting from Planned System Maintenance, Force Majeure Events, acts or omissions of Customer, End User or a third party vendor or caused by their respective network equipment, or inability to access Premises are excluded from Service credit calculations. Subject to the foregoing, in the event that Outage Time, less applicable exclusions, exceeds a continuous period of thirty (30) minutes or more (hereafter an "Outage"), Provider shall provide a credit (the "Outage Credit") for Service in the amount of 1/440 of the monthly recurring charge for the Service for each half hour or fraction thereof in excess of the first thirty (30) consecutive minutes that the affected Service fails to conform to the Technical Specifications.

Comment: In addition to the definition of "outage" or "unavailable," an SLA will set forth a standard of reliability. That is, it is assumed that a service will not be available 100% of the time. Nevertheless, it is reasonable to expect that a service will be available almost all of the time.

For example, a service may be described as a "five nines service." That means that the service is expected to perform in accordance with its specifications 99.999% of the time. Since there are 24 hours in a day, a month consisting of 30 days would have 720 hours. A "five nines" service would be expected to operate in accordance with its specifications during at least 719.99 hours of a month. A "four nines" service would be expected to operate in accordance with its specifications during at least 719.93 hours of a month.

A service level agreement usually includes outage credits or service credits. If a "five nines" service experiences an interruption of service that continues for more than .01 hours in a month, the customer may be entitled to a service credit. If a "four nines" service experiences an interruption of service that continues for more than .07 hours in a month, the customer may be entitled to a service credit. In many agreements a failure to meet the applicable service level does not immediately give rise to a service credit. Some service level agreements are written so that there is a short grace period before service credits would apply.

Furthermore, there are usually exceptions to the allowance of outage credits. In the first example below, there are two major exceptions. First, outage credits are granted only if a service becomes "Unavailable" for a reason other than an Excused Outage. An Excused Outage is defined in the Master Services Agreement as "any outage, unavailability, delay or other degradation of Service related to, associated with or caused by Planned System Maintenance, Customer actions or inactions, Customer provided power or equipment, any third party other than a third party directly involved in the operation and maintenance of the Seller network, including, without limitation, Customer's end users, third party network providers, traffic exchange points controlled by third parties, or any power, equipment or services provided by third parties, or a Force Majeure Event."

Second, outage credits are frequently not available for Off-Net Services. Off-Net Service is defined in the Master Services Agreement. "Off-Net" means a service that originates from or terminates to a location that is not on the Seller network."

In the first example below, an interruption of service is referred to as an "Unavailability." In other agreements an interruption of service may be referred to as an "Outage."

A service is a "protected service" if the service has redundant or diverse paths to protect against an outage. An "Outage" is not deemed to have occurred unless both paths experience an interruption at the same time.

> "For Dual Path Optical Wavelength Service Outage Time includes only the amount of time that both paths experience concurrent Outage Time."

Limitations on Credits:

Note that an interruption of service does not constitute a breach or default under a service agreement. Although a seller will make strenuous efforts to avoid interruptions of service, it is understood that outages will occur from time to time, and service credits are available to mitigate some of the effects of a loss of service. A seller will take measures to avoid or minimize any interruption of service because an interruption may entitle the customer to a service credit. Nevertheless, it does not follow that the occurrence of an interruption of service or an allowance of service credits means that the seller has breached the agreement. A service level agreement will often expressly limit a customer's remedy for interruptions of service to outage credits. The following sentences appear in the standard Master Services Agreement. Similar passages appear in several of the examples below.

Example Clause #1:

> "Seller may offer service credits related to installation intervals, Service availability, latency, and time to restore Service, which shall be set forth in the applicable Service

> Order Forms. These credits are the Customer's sole and exclusive remedy for Service related claims."

Comment: Service credits are often available not only for interruptions of service, but also for late delivery and other obligations relating to provision of a service. When a seller offers multiple types of credits, a clause such as the following is often used to limit the total potential liability of the seller. In this example, the seller's liability is limited to the monthly recurring charge for the month in which a credit is earned.

> "The maximum credit allowance for all Service Outages during any thirty (30) day period shall not exceed the Service Charges due Seller from Customer for the Service experiencing the Service Outage during such thirty (30) day period."

In several of the forms below, service credits are available according to an escalating scale. In these examples the amount of the credit is not simply a pro rata part of the monthly recurring charge during which an outage is suffered. Although the amount of the service credit is calculated as a percentage of the monthly recurring charge, an interruption of service for a short period of time will give rise to a small service credit. An interruption for a longer period of time will give rise to a much larger credit.

Chronic Outage Termination:

In addition to "outage credits" or "service credits," many agreements give the customer an ability to terminate the service order or the agreement entirely if an interruption of service continues for an extended period of time. Just as the right to claim service credits is usually qualified and limited, so is a right to terminate for chronic outage. Chronic outage termination may not be available at all for an unprotected service or for an off-net service, or, if available, may be subject to stricter standards and a lesser service level agreement. (See Example Clause #2.)

Note that Example Clause #9 would allow SLA credits and a right to terminate for chronic outage for on-net services only, and neither remedy would be available for off-net services.

Also, Example #9 sets forth a schedule of service credits for protected service. It does not explicitly say that service credits are not available for unprotected services, but the MSA that accompanies this exhibit does say that the stated service credits are Customer's sole and exclusive remedies for interruptions of service. Therefore, this clause would apparently allow no service credit for an outage on an unprotected service.

Note that in Example Clause #2 and Example Clause #9 one of the exceptions to the allowance of service credits is unscheduled or emergency maintenance. The use of that phrase in this context is questionable. Any interruption of service that is not caused by planned maintenance can be expected to give rise to unplanned or unscheduled maintenance in order to restore service. That is the essence of a service outage for which service credits should be allowed. In contrast, Example Clause #2 and Example Clause #15 state the principle correctly. The Customer should be entitled to service credits whenever there is an "unplanned outage" (unless caused by customer)(Example Clause #2) or "any material degradation of Service or any unscheduled, continuous and/or interrupted period of time when a Service(s) fails to be available or comply with the Technical Specifications." (Example Clause #15.)

Force Majeure:

One of the exceptions for service outages or Unavailability is a force majeure event. The same exception may apply to chronic outage. That is, a force majeure event is an exception to the definition of chronic outage, and therefore the customer will not be entitled to terminate the applicable service if the outage is caused by a force majeure event. This can cause inconvenience to a customer if the outage continues for a long period of time. The solution to this would be either to exclude force majeure from the definition of chronic outage, or to limit the period for which force majeure may apply. Here is an example clause that would allow the customer to terminate a service if a chronic outage continues for thirty days.

> *Force Majeure.* Neither party will be liable for any failure or delay in its performance under this Agreement due to any cause beyond its reasonable control, including acts of war, acts of God, earthquake, flood, embargo, riot, sabotage, labor shortage or dispute, governmental act or failure of the Internet (not resulting from the actions or inactions of Carrier), provided that the delayed party: (a) gives the other party prompt notice of such cause, and (b) uses its reasonable commercial efforts to promptly correct such failure or delay in performance. The foregoing notwithstanding, **if Carrier is unable to provide Service(s) for a period of thirty (30) consecutive days as a result of a continuing force majeure event, Customer may cancel the Service(s).**

Time to Restore or MTTR:

Sometimes outage credits are combined with other service related standards and credits, such as service delivery intervals and standards for prompt restoration of service. One of these criteria is variously referred to as "time to restore," "time to repair," "mean time to restore" or "mean time to repair." That is, the service agreement will require that, in the event of a service, interruption, the seller must take action to remedy the outage within a limited period of time. Mean time to restore or repair is the average response time and might be measured over a period of a month, a calendar quarter, or a year.

A similar performance commitment is sometimes given for "time to respond." That is, the seller will not guarantee that service will be restored within a certain interval, but will agree that in the event of a service interruption they will normally have appropriate repair personnel at the site of a service disruption within a specified period of time.

Sometimes a service level agreement will state that if the seller fails to maintain the required standard of response times, the customer will be entitled to enhanced service credits, in addition to the normal outage credits. In other agreements, MTTR is stated merely as a repair time or response time "objective," and there are no service credits specifically associated with a failure to meet the agreed MTTR.

Exhibit B
To
Seller Telecommunications Service Order Form

Service Level Agreement (SLA)

Availability Service Level for Unprotected Fast 10G Optical Wavelength Service.

If a __G Optical Wavelength Service becomes Unavailable (as defined above) ***for reasons other than an Excused Outage***, Customer will be entitled to a service credit. The amount of the credit shall be based on the cumulative Unavailability for the affected Service in a given calendar month, as set forth in the following table:

Cumulative Monthly Unavailability	Service Level Credit for Unprotected Fast 10G Optical Wavelength Service (% of MRC)
Less than __ hours	No Credit
__ hours or longer but less than __ hours	__% of the MRC
__ hours or longer but less than __ hours	__% of the MRC
__ hours or longer but less than __ hours	__% of the MRC
__ hours or longer but less than __ hours	__% of the MRC
__ hours or longer but less than __ hours	__% of the MRC
__ hours or longer but less than __ hours	__% of the MRC
__ hours or longer	100% of the MRC

Service Unavailability

"Unavailable" or "Unavailability" for wavelength Circuits means the duration of an interruption in transmission measured from the first of ten (10) consecutive severely erred seconds ("**SES**") on the affected Service until the first of ten (10) consecutive non-SES. "**SES**" means the point when the Bit Error Ratio (BER) exceeds 10^{-6} for a period of ten consecutive seconds and ends when the BER drops below 10^{-6} for a period of one second.

For purposes of calculating a credit, Unavailability is to be measured from the time that Seller receives a verified notification from Customer that a Circuit is Unavailable until the time that Seller determines that the Service has been restored.

The Availability Service Levels and associated credits set forth in this Section ***shall not apply to Off-Net Service*** provisioned by Seller through a third party carrier for the benefit of Customer. Seller will pass through to Customer any availability service level and associated credit (if applicable) provided to Seller by the third party carrier for such Off-Net Service.

Example Clause #2: In the first example clause below the amount of a service credit varies depending on several criteria. One of the criteria is whether the service is protected (i.e., dual path), core protected, or unprotected (single path). By "core protected" is meant that the seller backbone network is diverse, but the network segments that branch off from the seller backbone network to the end points are single path only.

TECHNICAL SPECIFICATIONS

	Dual Path			*Core Protected*			*Single Path*		
Service Characteristics									
Line Rate	1.25 Gbps	2.5 Gbps	10 Gbps	1.25 Gbps	2.5 Gbps	10 Gbps	1.25 Gbps	2.5 Gbps	10 Gbps
SONET Equivalent	N/A	OC-48	OC-192	N/A	OC-48	OC-192	N/A	OC-48	OC-192
Comparable SDH	N/A	STM-16	STM-64	N/A	STM-16	STM-64	N/A	STM-16	STM-64
Wavelength	1310 nm	1310 nm	1310 nm	1310 nm	1310 nm	1310 nm	1310 nm	1310 nm	1310 nm
Handoff	4 fiber	4 fiber	4 fiber	2 fiber	2 fiber	2 fiber	2 fiber	2 fiber	2 fiber
Availability*	99.999%	99.999%	99.999%	99.99%	99.99%	99.99%	99%	99%	99%
Service Credits (Percent of MRC)									
Cumulative Unavailability (Monthly)									
0 min to less than 4 Min	10%	10%	10%	0%	0%	0%	0%	0%	0%
4 Min to less than 30 Min	25%	25%	25%	10%	10%	10%	0%	0%	0%
30 Min to less than 1 Hour	50%	50%	50%	25%	25%	25%	0%	0%	0%
1 Hour to less than 4 Hours	60%	60%	60%	50%	50%	50%	0%	0%	0%
4 Hours to less than 7 Hours	70%	70%	70%	60%	60%	60%	10%	10%	10%

7 Hours to less than 24 Hours	100%	100%	100%	75%	75%	75%	25%	25%	25%
24 Hours or more	100%	100%	100%	100%	100%	100%	50%	50%	50%

Example Clause #3: In the following example chronic outage termination is available for on-net services if service is interrupted for an extended period of time. Chronic outage termination is not available for off-net services.

Furthermore, service credits are calculated differently depending on whether a service is protected or unprotected. An unprotected service accrues service credits at a rate that is half that of a protected service.

ARTICLE ___ - CREDIT ALLOWANCES FOR SERVICE OUTAGES FOR ON-NET SERVICES

In the event that Carrier is unable to restore a portion of the Service as required hereunder, or in the event of a Service Outage, Customer shall be entitled to a credit for the affected Circuits for all unplanned outages. Credit allowances shall be deducted from the charges payable by Customer hereunder and shall be expressly indicated on a subsequent bill to Customer. A Service Outage begins when Carrier is notified or becomes aware of the failure, whichever first occurs. A Service Outage ends when the affected line and/or associated station equipment is fully operative. If the Customer reports Services or a facility or Circuit to be inoperative but declines to release it for testing and repair, it is considered to be impaired, but shall not be deemed a Service Outage. ***If a Service Outage continues for a period of twelve (12) continuous hours, or twenty-four (24) cumulative hours, in any six (6) month period, Customer may terminate the affected Service provided under this Agreement upon written notice to Carrier.***

(a) Credit Allowances do not apply to Service Outages (1) caused by the negligence or acts of Customer and/or End User or its agents, (2) due to failure of power; (3) due to failure or malfunction of non-Carrier equipment or systems; (4) due to circumstances or causes beyond the control of Carrier or its agents; (5) during any period in which Carrier is not given access to the Service Premises; or (6) during a *Planned Outage*, unscheduled Emergency Maintenance, scheduled maintenance, alteration or implementation as described in Appendix No. __ herein.

(b) Customer must request a credit allowance for a Service Outage within one hundred and fifty (150) days after delivery of an invoice respecting the affected Service or any claim for an allowance is waived.

(c) ***On-Net*** Dedicated Transport Services will be credited as follows for cumulative Service Outage(s) within any continuous 30 day period:

Service Outage(s) Length	Credit Per Affected Circuit
3 minutes or less	None
More than 3 minutes but less or equal to 1 hour	5% of monthly recurring revenue of the Circuit.

Each hour above 1 hour	an additional 10% of the monthly recurring revenue of the Circuit, capped at 100% in any month

1. Whenever a Customer reports to the Carrier (or vice versa) that a Circuit has experienced Recurring Service Outages, the Carrier shall immediately perform a detailed investigation and report the findings to the Customer.

2. Notwithstanding Carrier's obligation to provide credit allowances in the event of Service Outages, the Customer is granted the option to disconnect a Circuit without termination liability if that specific Circuit experiences Recurring Service Outages.

(1) Availability Service Level for ***Protected*** Seller Private Line Service. (1) the Availability Service Level for Protected Seller Private Line Service delivered over Seller's network is 99.99% for Protected Terrestrial Seller Private Line Service and 99.9% for Protected Submarine Seller Private Line Service. In the event that any Protected Seller Private Line Service becomes unavailable (as defined below) for reasons other than an Excused Outage, customer will be entitled to a service credit off of the MRC for the affected Service based on the cumulative unavailability of the affected Service in a given calendar month as set forth in the following table.

Cumulative Unavailability	Service Level Credit
0 – 5 minutes	No Credit
5:01 minutes – 45 minutes	5% of the MRC
45:01 minutes – 4 hours	10% of the MRC
4:01 – 8 hours	20% of the MRC
8:01 – 12 hours	30% of the MRC
12:01 – 16 hours	40% of the MRC
16:01 – 24 hours	50% of the MRC
24:01 + hours	100% of the MRC

(2) The Availability Service Levels and associated credits set forth in this Section ____ shall not apply to Off-Net Local Loop Service, including, without limitation, Seller Metropolitan Private Line (Off-Net) Service, provisioned by Seller through a third party carrier for the benefit of Customer. Seller will pass-through to Customer any availability service level and associated credit (if applicable) provided to Seller by the third party carrier for such Off-Net Local Loop Service.

(C) Availability Service Level for Unprotected Seller Private Line Service and Seller Wavelength Service.

(1) In the event that any ***unprotected*** Seller Private Line Service or Seller Wavelength Service becomes unavailable (as defined in Section ____ above) for reasons other than an Excused Outage, Customer will be entitled to a service credit off of the MRC for the affected Service based on the cumulative unavailability for the affected Service in a given calendar month as set forth in the following table:

Cumulative Unavailability	Service Level Credit
0 – 24 hours	No Credit
24:01 – 30	2.5% of the MRC
30:01 – 36 hours	5% of the MRC
36:01 – 42 hours	7.5% of the MRC
42:01 - + hours	10% of the MRC

(2) The Availability Service Levels and associated credits set forth in this Section _____ ***shall not apply to Off-Net*** Local Loop Service, including, without limitation, Seller SM Metropolitan Private Line (Off-Net) Service, provisioned by Seller through a third party carrier for the benefit of Customer. Seller will pass through to Customer any availability service level and associated credit (if applicable) provided to Seller by the third party carrier for such Off-Net Local Loop Service.

Definitions:

For purposes of this Section ____, "unavailable" or "unavailability" means the duration of a break in transmission measured from the first of ten (10) consecutive severely erred seconds ("SESs") on the affected Service until the first of ten (10) consecutive non-SESs. An SES is a second with a bit error ratio of greater than or equal to 1 in 1000.

On-Net: "On-Net" shall refer to Premises which are served by the Carrier Network and is also referred to as "Type I." Service provided to an On-Net location utilizes solely the Carrier Network.

Service Outage: "Service Outage" shall mean a disruption or degradation of Service, which is not caused by scheduled maintenance or planned enhancements or upgrades to the Carrier Network.

Planned Outage: "Planned Outage" shall mean any a disruption or degradation of Service caused by scheduled maintenance or planned enhancements or upgrades to the Carrier Network.

Excused Outage: Any outage, unavailability, delay or other degradation of Service related to, associated with or caused by scheduled maintenance events, Customer actions or inactions, Customer provided power or equipment, any third party, excluding any third party directly involved in the operation and maintenance of the Seller network but including, without limitation, Customer's end users, third party network providers, traffic exchange points controlled by third parties, or any power, equipment or services provided by third parties, or an event of force majeure as defined in Section ____.

"Recurring Service Outage" means: (i) a particular Service Outage, for which trouble tickets have been opened, occurring twice or more within a thirty (30) day period; (ii) Service Outage(s) resulting in cumulative service credits equaling $____ or more during any continuous twelve (12) month period; or (iii) a single Service Outage of eight (8) continuous hours or more.

Force Majeure: Neither party shall be liable, nor shall any credit allowance or other remedy be extended, for any failure of performance or equipment due to causes beyond such party's reasonable control. In the event Seller is unable to deliver Service as a result of force majeure,

Customer shall not be obligated to pay Seller for the affected Service for so long as Seller is unable to deliver.

Example Clause #4: The following clause would allow no outage credit at all for an interruption of protected service lasting less than two hours, or for an unprotected service lasting less than eight hours. Service descriptions sometimes describe service level "objectives" that do not correspond exactly to the schedules of outage credits. The Performance Standards below, which are stated as two-nines or five-nines service, are performance "objectives," and do not correspond to the schedules of outage credits.

The list of exceptions to outage credits that appears in the third paragraph is very expansive. And the maximum credit for any month may not exceed the total monthly recurring charge for the applicable service during that same month.

> Outage Credits: Unless otherwise stated in any applicable Service Order or for other reasons stated herein, in the case of any Service Outage, Customer's sole remedy shall be to seek in writing within thirty (30) days of the end of the month in which the Service Outage occurred, and ***Seller's sole liability shall be to provide, a credit allowance*** for the Service Outage in accordance with the following:
>
> ***No credit*** shall be allowed for any Service Outage of ***two (2) hours*** or less. For Service Outages for ***Protected Services*** that are greater than two (2) hours, Customer shall be credited at the rate of ten percent (10%) of the Service Charges (as specified in the Service Order) applicable to the Service which is subject to the Service Outage for each two (2) hour period or major fraction thereof that a Service Outage continues. For Service Outages for ***Unprotected Services*** that are greater than ***eight (8) hours***, Customer shall be credited at the rate of ten percent (10%) of the Service Charges (as specified in the Service Order) applicable to the Service which is subject to the Service Outage for each two (2) hour period or major fraction thereof that a Service Outage continues. However, ***the maximum credit allowance for all Service Outages during any thirty (30) day period shall not exceed the Service Charges due Seller from Customer for the Service experiencing the Service Outage during such thirty (30) day period.*** Each Service Outage is to be measured from the time Customer notifies Seller that a Service Outage has occurred to the time of restoration of Service, as determined by Seller.
>
> No credit allowances shall be made for Service Outages arising from or relating to: (i) any acts or omissions of an entity other than Seller, including, but not limited to, Customer, Customer's agents, employees, end users or other service providers connected to Seller's Services, system, network, equipment or facilities; (ii) Customer's noncompliance with this Agreement; (iii) any emergency or routine maintenance; (iv) any failure of any Service provided by others (including, without limitation, any non-Seller equipment or facilities used in connection with the affected Service); (v) any period in which Seller is not given full access to its equipment or facilities for the purpose of investigating and correcting a Service Outage; (vi) any period in which Customer continues to use Service on an impaired basis or releases Service to Seller for maintenance or installation purposes; or (vii) any Force Majeure Event(s).
>
> Performance Standards:

General. Service standards apply on a one-way basis between Seller's POPs and exclude nonperformance due to any Force Majeure Event, planned interruptions for maintenance purposes, or other reasons set forth below. Seller's Private Line Services will be provided according to the following performance standards and objectives:

Type of Service	Unprotected Availability	Protected Availability
DS-1	99%	99.999%
DS-3	99%	99.999%
OC-n	99%	99.999%
Wavelengths	99%	99.999%
EPL	99%	99.999%

Availability. Availability is a measurement in seconds of the percent of total time that a Service is operative when measured over a calendar month period.

Severely Errored Seconds ("SES"). An SES is any second in which the Bit Error Rate ("BER") is worse than 1 x 10-6.

Service Interruption or Outage. A Service is considered interrupted or unavailable when there has been a loss of signal, 100% packet loss, or when tests confirm the observation of any ten (10) consecutive SES. These ten (10) consecutive seconds are considered to be part of the unavailable time. The period of unavailable time ends when the BER in each second is better that 10-6 for a period of ten (10) consecutive seconds. These ten (10) consecutive seconds are considered to be available time.

Repair Objective. In the event of a Service Outage, Seller will use commercially reasonable efforts to meet a performance objective for repair of four (4) hours.

Reporting. Customer will promptly notify Seller of any Availability or BER performance problems with a Service. During problem diagnosis, Seller will accept Customer's measurements of the performance of a Service.

Example Clause #5: The following example defines outage credits according to two standards of availability. For one type of service, outage credits are awarded one-half hour increments for outages that last longer than four hours. For the second type of service, outage credits are awarded on a pro rata basis, beginning immediately on the occurrence of an outage.

CREDIT FOR INTERRUPTION OF SERVICE

A credit shall be applied when Service is interrupted or does not meet performance standards for any period lasting ***four (4) or more consecutive hours***. No credit will be applied if the interruption is caused by (i) the negligence of Customer; (ii) the failure of facilities or equipment provided by Customer or other third party; (iii) Company's inability to gain access to Customer's equipment and facilities; or, (iv) Customer's failure to release the Service, when requested by Company, to perform testing and maintenance.

The amount of the credit shall be determined by dividing Customer's total monthly charge by 1440, and then multiplying the result by the number of ***1/2-hour increments***, or major fraction thereof, of interruption, after the initial four (4) hours.

In the event that Service is provisioned with SMART Access, then the amount of the credit shall be 100% of one month's charges for the interrupted Service. Partial credit will be given in the event that a portion of the Service is interrupted. In the case of DS-1 Service, the amount of partial credit shall be determined by dividing the monthly rate for Service by the number of equivalent DS-0 circuits and then multiplying the result by the number of interrupted DS-0 circuits. In the case of DS-3 Service, the amount of partial credit shall be determined by dividing the monthly rate for Service by the number of equivalent DS-1 circuits and then multiplying the result by the number of interrupted DS-1 circuits. Customer shall notify SELLER by telephone at ________________ of any interruption in Service.

No credit shall be provided unless the Customer provides documentation of the outage as reported to Seller's Network Operations Center ("NOC") (through, trouble ticket numbers).

Example Clause #6: The following clause represents the simplest approach to outage credits. A customer is allowed a credit for any outage that exceeds five minutes. And outage credits are calculated simply according to the duration of the outage multiplied by the monthly recurring charge.

Credit Allowances for Service Outages:

Wavelength Service: Supplier will deliver Service Availability greater than 99.99% for each ***Unprotected*** Wavelength Circuit, each month, and 99.999% for each ***Protected*** Wavelength Circuit, each month. Availability will be calculated by subtracting the total outage time from the total time in a given month and dividing the result by the total time. Outage time is computed from the time the trouble is reported to the Supplier Network Operations Center (NOC) and the time the trouble is reported as cleared by the NOC. Availability will be computed with the aggregate outage for each Circuit in the calendar month.

Supplier will issue a credit for any Wavelength Service Outage as defined in the previous paragraph according to the following schedule:

Unprotected Circuit Availability:	Credit Issued Will Be:
100% - 99.99%	0% credit of MRC
≤ 99.98%	100% credit of MRC

Protected Circuit Availability:	Credit Issued Will Be:
100% - 99.999%	0% credit of MRC
≤ 99.998%	100% credit of MRC

Each credit will be calculated on a per-circuit, per-month basis. Circuit Availability will be based on a seven hundred and twenty (720) hour month. Each Outage credit will be calculated from the time Supplier receives notice from Seller of actual circuit unavailability (established by a "Trouble Ticket") until Supplier restores circuit availability.

Bit Error Rate (BER)

In addition to the above remedy, a Wavelength service will be deemed unavailable (that is, experiencing a "Service Outage") for the relevant period, if the circuit displays a BER greater than 1×10^{-12}. If the BER exceeds a measurement of 1×10^{-12}, Seller will be entitled to a credit equal to twenty percent (20%) of the MRC for the affected circuit.

All Other Services: Supplier will deliver Service Availability greater than 99.99% for each Circuit, each month. Availability will be calculated by subtracting the total outage time from the total time in a given month and dividing the result by the total time. Outage time is computed from the time the trouble is reported to the Supplier Network Operations Center (NOC) and the time the trouble is reported as cleared by the NOC. Availability will be computed with the aggregate outage for each Circuit in the calendar month.

Each credit will be calculated on a per-circuit, per-month basis. Circuit Availability will be based on a seven hundred and twenty (720) hour month. Each Outage credit will be calculated from the time Supplier receives notice from Seller of actual circuit unavailability (established by a "Trouble Ticket") until Supplier restores circuit availability.

Supplier will provide an outage credit in the amount of one month's recurring charges (MRC) for any outages, which are ***five (5) minutes or more in duration***.

General terms applying to SLAs

SLA credits are not applied to usage charges or any third party charges passed through to the Customer, including charges for any local access circuits provided to Customer by Seller.

SLA credits are calculated after deduction of all discounts and other special pricing arrangements, and are not applied to governmental fees, taxes, surcharges and similar additional charges.

If an incident affects the performance of the Service and results in a period of Service Unavailability entitling Customer to one or more credits under different SLA parameters, only the single highest credit applying in respect of that incident will be applied.

In no event will SLA credits in any calendar month exceed either 100% of the total MRC(s) payable by Customer for the Service in that month, or, in the case of Waves provided on a Pre-Paid Lease basis, the Implied MSC for a given month.

As a condition of entitlement to SLA credits, Customer shall cooperate with Seller in addressing any reported Service problems.

SLA credits are applied only upon Customer's written request, which must be submitted within (15) business days of the end of the month in which entitlement to an SLA credit arose.

For Waves provided on an Annual Lease basis, all agreed SLA credits claimed by Customer for a given month will be totaled and applied to Customer's next following invoice for the Service, or as promptly thereafter as is practical in the event of a dispute. For Waves provided on a Pre-Paid Lease basis, all agreed SLA credits claimed by Customer (i) shall be issued after receipt of

Customer's written request for credit, (ii) may be applied by Customer only against charges for new Waves ordered by Customer from Seller, or against charges for Waves whose initial Service Term is extended by Customer and (iii) shall accrue on a monthly basis and must be used within twenty-four (24) months of issuance.

The SLAs provided for in these terms apply only in respect of Waves that are provisioned on Seller's Network and, in the case of End-to-End Service, to local access circuits provided by Seller (via third party providers).

SLA credits provided for in these terms and conditions are Customer's exclusive remedy with respect to items covered in these terms and conditions.

Exclusions:

No SLA credit shall apply to the failure of a Wave to comply with these SLA terms, or to any period of Service Unavailability, caused, in whole or part, by any of the following:

(a) a failure of Customer's premises equipment or equipment of a Customer's vendor;
(b) a failure in local access facilities connecting the Customer to Seller's Network which are not provided by Seller;
(c) force majeure events as defined in the Master Agreement;
(d) any act or omission of Customer or any third party (including but not limited to, Customer's agents, contractors or vendors), including, but not limited to (i) failing to provide Seller adequate access to facilities for testing, (ii) failing to provide access to Customer premises as reasonably required by Seller (or its agents) to enable Seller to comply with its obligations regarding the Service, (iii) failing to take any remedial action in relation to a Service as recommended by Seller, or otherwise preventing Seller from doing so, or (iv) any act or omission which causes Seller to be unable to meet any of these SLA terms;
(e) customer's negligence or willful misconduct, which may include Customer's failure to follow agreed-upon procedures;
(f) any scheduled maintenance periods when Customer has been informed of such maintenance, and any emergency maintenance;
(g) disconnection or suspension of the Service by Seller pursuant to a right to do so under the Master Agreement or these terms and conditions; or
(h) Outages attributable to (i) long-haul domestic local access circuits in the US between a Seller POP and a Customer's premises which is in a different Local Access Transport Area (LATA) and/or (ii) long-haul international local access circuits between a Seller POP and a Customer premises in a different country.

Example Clause #7: Outage credits are usually not available during periods of scheduled maintenance. Therefore, the parties will sometimes agree that scheduled maintenance will require a minimal prior notice. (See Section 4 below.) If a seller must perform emergency maintenance in order to restore service, the customer will nevertheless be entitled to an outage credit. The seller would not be excused from extending outage credits merely because they are conducting maintenance during the outage period. A maintenance period will not excuse the seller from paying outage credits if the minimal notice was not given before maintenance is performed.

This example clause also includes credits for late delivery. There is a grace period of ten days after which the customer would be entitled either to claim a service credit for late delivery or to terminate the service without penalty or other charges.

OUTAGES

GENERAL

As described below, Provider offers Service Credits related to installation intervals and Service Availability. ***These credits are the Customer's sole and exclusive remedy for Service-related claims.*** To qualify for a service credit, Customer must not have any invoices that are past due, and must notify Provider that a trouble ticket should be opened to document the event. ***In no event shall the total amount of service credits per month exceed the total MRC (or NRC if applicable) for the affected Service for such month.*** If at Customer's request Provider provides through a third-party vendor a service that is not a standard Provider Service, and if there is a delay in installation or interruption of such service, Customer shall be entitled to a Service credit only if and to the extent of the service credit to which Provider is entitled under its agreement with such third-party vendor.

2. OUTAGE AND OUTAGE CREDITS

For SONET Service, "Outage Time" is the period between the first of ten consecutive Severely Errored Seconds (SES > ~10-6 BER) and the first of ten consecutive non-SES. For 2.5 Gbps and 10Gbps Optical Wavelength Service, "Outage Time" is the number of SES's in the month. For 1.25 Gbps Optical Wavelength Services and Ethernet Service, "Outage Time" is any period when there is 100% packet loss. A lost packet is one with a valid Frame Check Sequence and a valid destination MAC address that is not delivered to its destination. ***For Dual Path Optical Wavelength Service Outage Time includes only the amount of time that both paths experience concurrent Outage Time.*** For all Services, Outage Time resulting from ***Planned System Maintenance***, Force Majeure Events, acts or omissions of Customer, End User or a third party vendor or caused by their respective network equipment, or inability to access Premises are excluded from Service credit calculations. Subject to the foregoing, in the event that Outage Time, less applicable exclusions, exceeds a continuous period of thirty (30) minutes or more (hereafter an "Outage"), Provider shall provide a credit (the "Outage Credit") for Service in the amount of 1/440 of the monthly recurring charge for the Service for each half hour or fraction thereof in excess of the first thirty (30) consecutive minutes that the affected Service fails to conform to the Technical Specifications.

3. OUTAGE START/END TIME FOR CREDIT CALCULATION PURPOSES.

An Outage will begin upon Provider's receipt of notice from Customer of the Outage. An Outage is concluded upon Provider's confirmation that the Service has been restored.

4. EXCESSIVE OUTAGE

If Customer experiences a ***Chronic Outage*** with respect to this Service, Customer shall be entitled to terminate the affected Service without further obligation by providing Provider with written notice within (5) five business days following such Chronic Outage (a "Chronic Outage Termination"). For purposes of this Section, a Service suffers an "Outage" if there is Outage Time (as described above) of two (2) or more hours. A "Chronic Outage" is if such

Service, measured over any thirty (30) consecutive day period, experiences more than five (5) unrelated Outages.

SCHEDULED MAINTENANCE

Provider usually conducts ***Planned System Maintenance*** outside of normal working hours, i.e. 12:00 AM to 6:00 AM local time. Provider will use reasonable efforts to minimize any Service interruptions that might occur as a result of Planned System Maintenance. ***Provider agrees to provide Customer with electronic mail, telephone, facsimile or written notice of all non-emergency Planned System Maintenance (a) no later than three (3) days prior to performing maintenance*** that, in its reasonable opinion, has a substantial likelihood of affecting Customer traffic for up to 50 milliseconds, and (b) no later than ten (10) days prior to performing maintenance that, in its reasonable opinion, has a substantial likelihood of affecting Customer traffic for more than 50 milliseconds. If Provider's planned activity is canceled or delayed, Provider shall promptly notify Customer and shall comply with the provisions of the previous sentence to reschedule any delayed activity. Should such maintenance exceed the time scheduled materially affecting the Service, Customer will be entitled to Outage Credits.

CIRCUIT ANNIVERSARY

On the annual anniversary date of a Service, If a circuit was less reliable than 99.95% in any given month, Provider will provide Customer a full month's credit for the monthly recurring charge on that circuit.

CREDITING

All Outage Credits shall be credited on Customer's next monthly invoice for the affected Service.

RECURRING SERVICE ERRORS.

If any aspect of the Services contain recurring errors or deficiencies, then Provider promptly shall perform a root cause analysis, at no additional charge to Customer, to determine the cause of such recurring errors. Within five (5) days of receiving notice from Customer of a Provider recurring error or Provider's discovery of such recurring errors, Provider shall provide Customer with a written copy of its analysis, which shall include an action plan containing a reasonably detailed description of corrective action to be taken by Provider and the date by which such corrective action shall be completed, all subject to the reasonable approval of Customer. Provider shall correct such recurring errors, at no additional charge to Customer and to the reasonable satisfaction of Customer in accordance with its action plan. Provider's obligation to remedy recurring errors is in addition to its obligation to correct other errors and deficiencies in accordance with this Agreement, and nothing contained in this Article shall relieve Provider of such obligations.

INSTALLATION INTERVALS AND DELAY CREDITS

Regarding installation intervals, if Provider completes its acceptance testing on a newly installed Circuit ***after the FOC date, excluding delays caused by Customer or Force***

Majeure Events, Provider will provide payment credits against the NRC for the Service as follows: ***11 to 30 business days after the FOC date Customer receives twenty five percent (25%) NRC credit; 31 to 60 business days after the FOC date Customer receives fifty percent (50%) NRC credit; more than 60 business days after the FOC Customer receives either one hundred percent (100%) NRC credit or option to cancel the Service without penalty***, provided that Customer notifies Provider in writing of cancellation before Provider's acceptance testing is completed.

Example Clause #8: The following clause would also allow service credits for late delivery, but would distinguish between different types of services. Credits for late delivery for one type of service would be greater than for another type of service.

INSTALLATION INTERVALS AND DELAY CREDITS

INSTALLATION INTERVALS
Provider will meet the following Installation Intervals:

DS1	15 days
DS3	30 days
OC3	30 days
OC12	45 days
OC48	60 days

FAILURE TO MEET INSTALLATION INTERVALS
Failure of Provider to deliver a Service within the Implementation Intervals will be cause for Provider to compensate Customer for damages ("Delivery Credit") as follows:

Delivery Credit for Provider's failure to meet the Implementation Interval for Services with a 1 year term	Delivery Credit for Provider's failure to meet the Implementation Interval for Services with a term of 3 years or more
Three (3) days credit for each day the Services are late	One (1) week credit for each day the Services are late.

Total Delivery Credits shall not exceed sixty (60) days for each affected Service. Delivery Credits will not be provided for Provider's failure to meet the Implementation Interval if the delay was a result of a force majeure event, actions of Customer. Unless otherwise agreed in writing, where Provider has exceeded the Implementation Interval by five (5) days or more, Customer may cancel the Service Order without charge or liability.

Example Clause #9: The example that follows includes service description and outage credits for both domestic and international services.

Availability:

The availability objective for all wavelength circuits between Seller's Network Interface points specified above is to provide performance levels over any consecutive 12 month period is as follows:

Linear Path, ***Unprotected*** Wavelengths

2.5G and 10G
99.0%

Linear Path, **Unprotected** Wavelengths: Wavelengths will be sold in pairs, one transmit and one receive, so the Customer will receive a minimum of 2 fibers one carrying the transmit and the other the receive wavelengths between point A and point Z. In this case the two fibers will be carried in the same cable and there is no protection in case of a fiber cut. In this scenario Seller provides no protection on the optical layer or electrical layer.

Availability information for 2.5G and 10G wavelength services assumes individual wavelengths from Seller's POP to Seller's POP. This excludes any Customer provided access links to the Seller Transport network.

Outages attributable to incidental damage to or severance of outside fiber optic cable plant, or scheduled maintenance and or Force Majeure is excluded from the performance objective stated above.

Error-Free Seconds (EFS) and Error Seconds (ES) are the primary measure of error performance. An Error-Free Second is defined as any second in which no bit errors are received. Conversely, an Error Second is any second in which one or more bit errors are received.

2.5G and 10G Wavelengths: The general performance objectives for 2.5G and 10G wavelengths are as follows:

BER: 1×10^{-12}
BBER: 2×10^{-6}

5. Acceptance Criteria:

The acceptance criteria for 2.5G and 10G circuits between Seller's Network Interface points is to provide the performance levels shown below during a 24 hour test period. If no errors are observed during the 24 hours of the test, the facility may be considered acceptable.

6. Definitions:

Availability: Availability is a measure of the relative amount of time during which the circuit is available for use, according to ITU definitions.

DWDM: Dense wavelength division multiplexing (DWDM) is the higher capacity version of the WDM, which is a means of increasing the capacity of fiber-optic data transmission systems through the multiplexing of multiple wavelengths of light.

OC-48/STM-16: Optical Carrier level 48 / Synchronous Transfer Mode Level 16 signal transmitting at 2.488 Gb/s. This is only used as Reference for comparison of rate (waves do not offer the same protection as SONET/SDH).

OC-192/STM-64: Optical Carrier level 192 / Synchronous Transfer Mode Level 64 signal transmitting at 9.953 Gb/s. This is only used as Reference for comparison of rate (waves do not offer the same protection as SONET/SDH).

Point of Presence (POP): A physical location where a long distance carrier terminates lines before connecting to the Metro Carrier, local exchange carrier, another carrier, or directly to a customer.

Exhibit A – Private Line Outage Policy

Service Level Agreement for Domestic and International Private Line Services

1. Service Level Objectives

Supplier will make commercially reasonable efforts to meet a Service Availability of 99.99% for domestic Protected (1+1) Circuits and 99.9% for domestic Unprotected (1+0) Circuits pursuant to the Service Levels set forth in its performance specifications. This attachment sets forth the credit(s) that Customer will receive if the Service Levels are not met, in addition to those rights and remedies available under the Agreement.

2. Allowance for Service Outage Periods

(a) A Circuit shall be deemed to be in an outage condition if, while Customer is using or attempting to use such Circuit, such Circuit loses continuity, becomes unavailable or fails to comply with the applicable specifications for such Circuit ("Outage"). Subject to the restrictions herein, Customer is entitled to an "Outage Credit" in the event that the Service Levels described in the Service Level Agreement are not met. An "Outage Period" begins when a report is made to Supplier's Network Control Center from Customer by telephone (or via Supplier's on-line trouble ticketing interface, if applicable) that Service has been impaired, lost or interrupted. Customer must agree that such Circuit is released for repair by Supplier or its agent. An Outage Period ends when the Circuit is restored. Supplier will notify customer by telephone and Customer will confirm that Service has been restored. Any additional time necessary for Customer's confirmation shall not operate to extend the calculation of the Outage Period. Events that cause an Outage but involve simultaneous multiple failures, shall be treated as one single Outage for purposes of calculation of Outage Credits. In the event of any dispute between the parties in respect of a Service being available or subject to an Outage Credit, Supplier shall retain the sole right to determine the period of such availability for the purpose of calculating any Outage Credits due under the terms of this Agreement.

All Outage Credits shall be subject to the following restrictions:

(i) No credit shall be allowed with respect to any period during which Customer fails to afford access to any facilities provided by Supplier for the purpose of investigating and correcting an interruption to Service.

(ii) The Monthly Lease Rates used to determine any credit hereunder shall be the then current Monthly Lease Rates being assessed.

(iii)In no event shall any credit be allowed hereunder (1) in excess of the then current Monthly Lease Rate for the applicable Circuit or (2) with respect to any Circuit for which Customer (i) fails to make or (ii) is excused from making any payment because of operation of law or any other reason.

(b) The duration of the Outage Period and Outage Credits will be determined at the sole discretion of Supplier, based upon Supplier's internal records. Customer shall have the right to request credit(s) for a period of one hundred and eighty (180) days after the occurrence of an outage or alleged outage. Customer shall have the right to contest any calculations of credit(s) for a period of thirty (30) days after Customer's receipt of invoice on which said credit(s) appear.

(c) No Outage Credits are allowed for Outage Periods:

(1) Caused directly or indirectly by the acts or omissions of Customer;
(2) Caused by the failure of equipment or systems provided by Customer or any third party (not under the direction or control of Supplier), including any provider of local access service to Supplier contracted for, by, or on behalf of Customer (in such case, Supplier will coordinate with such local access service provider to cure such failure as quickly as practicable);
(3) Caused by a Force Majeure event;
(4) Occurring with respect to a Circuit released by Customer to Supplier (i) to perform maintenance, (ii) to make rearrangements at the direction of Customer, or (iii) to implement an order from Customer for a change in the Circuit; or
(5) Occurring with respect to a Circuit that Customer elects not to release for testing or repair and continues to use on an impaired basis.
(6) Interruption of Service on a Circuit for maintenance. Supplier shall use its best efforts to give Customer ten (10) days prior notice thereof by telephone, facsimile or E-mail. Supplier will use its best efforts to schedule such Service interruptions between midnight and 6:00 a.m. for domestic circuits or during local off-peak hours for international circuits. Credits will not be allowed with respect to such Service interruptions if Supplier has used its best efforts to so notify Customer in accordance with this paragraph.

(d) The credits and/or cancellation of a Circuit in the case of chronic outage problem provided for hereunder shall be Supplier's sole liability and Customer's sole remedy in the event of any outage period or interruption of Service.

3. Service Level Outage Credits

Domestic Service Level Outage Credits will be calculated and granted based upon the following Service Availability Objective:

Domestic Outage Credit Schedule – DS-X and OC-X Protected (1+1) Circuits		
e Levels	e Time Period	e Credits
0 Outage	tes to less than 4 minutes	No credit
1 Outage	tes to less than 30 minutes	1 day credit
2 Outage	utes to less than 60 minutes	3 days credit
3Outage	utes to less than 4 hours	4 days credit
4 Outage	s to less than 8 hours	7 days credit
5 Outage	s to less than 24 hours	7 days credit
6 Outage	rs +	Full month credit
Two events of Level 1 or greater outage in one month		Double credit
Three or more events of Level 1 or greater outage in one month		Triple credit

Domestic Outage Credit Schedule – OC-X Unprotected (1+0) Circuits			
Outage Levels	Outage Time Period	Outage Credits	
Level 0 Outage	0 minutes to less than 30 minutes	=	No credit
Level 1 Outage	30 minutes to less than 60 minutes	=	4 hours credit
Level 2 Outage	60 minutes to less than 4 hours	=	8 hours credit
Level 3 Outage	4 hours to less than 8 hours	=	24 hours credit
Level 4 Outage	8 hours to less than 24 hours	=	5 days credit
Level 5 Outage	24 hours +	=	7 days credit
Two events of Level 1 or greater outage in one month		=	Double credit
Three or more events of Level 1 or greater outage in one month		=	Triple credit

(b) International Service Level Outage Credits will be calculated and granted based upon the following Service Availability Objective:

International Full Circuit Outage Credit Schedule			
Outage Levels	Outage Time Period	Outage Credits	
Level 0 Outage	0 minutes to less than 20	=	No credit

	minutes		
Level 1 Outage	20 minutes to less than 60 minutes	=	4 hours credit
Level 2 Outage	60 minutes to less than 4 hours	=	8 hours credit
Level 3 Outage	4 hours to less than 8 hours	=	24 hours credit
Level 4 Outage	8 hours to less than 24 hours	=	2 days credit
Level 5 Outage	24 hours +	=	Actual plus 2 days credit
Two events of Level 1 or greater outage in one month		=	Double credit
Three or more events of Level 1 or greater outage in one month		=	Triple credit

International Half Circuit Outage Credit Schedule			
Outage Levels	Outage Time Period	Outage Credits	
Level 0 Outage	0 minutes to less than 60 minutes	=	No credit
Level 1 Outage	60 minutes to less than 4 hours	=	8 hours credit
Level 2 Outage	4 hours to less than 8 hours	=	24 hours credit
Level 3Outage	8 hours to less than 24 hours	=	2 days credit
Level 4 Outage	24 hours +	=	Actual plus 2 days credit
Two events of Level 1 or greater outage in one month		=	Double credit
Three or more events of Level 1 or greater outage in one month		=	Triple credit

4. Chronic Outage. In the event that a single ***On-Net Protected*** (1+1) Circuit experiences three (3) or more Outages of a Level 1 event during any period of ninety (90) consecutive days, or a single ***On-Net Unprotected*** (1+0) Circuit experiences three (3) or more Outages of a Level 2 event during any period of thirty (30) consecutive days, (and such Outages are the type for which credit would have been given to Customer pursuant to the terms herein), Customer may declare that the Circuit has a "***chronic outage***" problem. Customer will notify Supplier, and Supplier will have seventy-two (72) hours to identify and correct the chronic outage problem. If the problem is not corrected, ***Customer may cancel the affected*** Circuit without any additional or further liability to Supplier. Cancellation shall be in addition to any credits due to Customer for outages.

Example Clause #10: In the following example the service level agreement for protected service is greater than for unprotected service. Furthermore, service credits are not available at all for off-net services.

(1) Availability Service Level for ***Protected*** Private Line Service. (1) The Availability Service Level for Protected Private Line Service delivered over Seller's network is 99.99% for Protected Terrestrial Private Line Service and 99.9% for Protected Submarine Private Line Service. In the event that any Protected Private Line Service becomes unavailable (as

defined below) for reasons other than an Excused Outage, customer will be entitled to a service credit off of the MRC for the affected Service based on the cumulative unavailability of the affected Service in a given calendar month as set forth in the following table.

Cumulative Unavailability	Service Level Credit
0 – 5 minutes	No Credit
5:01 minutes – 45 minutes	5% of the MRC
45:01 minutes – 4 hours	10% of the MRC
4:01 – 8 hours	20% of the MRC
8:01 – 12 hours	30% of the MRC
12:01 – 16 hours	40% of the MRC
16:01 – 24 hours	50% of the MRC
24:01 + hours	100% of the MRC

For purposes of this Section ____, "unavailable" or "unavailability" means the duration of a break in transmission measured from the first of ten (10) consecutive severely erred seconds ("SESs") on the affected Service until the first of ten (10) consecutive non-SESs. An SES is a second with a bit error ratio of greater than or equal to 1 in 1000.

(2) The Availability Service Levels and associated credits set forth in this Section ____ shall not apply to Off-Net Local Loop Service, including, without limitation, Metropolitan Private Line (Off-Net) Service, provisioned by Seller through a third party carrier for the benefit of Customer. Seller will pass-through to Customer any availability service level and associated credit (if applicable) provided to Seller by the third party carrier for such Off-Net Local Loop Service.

Availability Service Level for Unprotected Private Line Service and Wavelength Service.

(1) In the event that any ***unprotected*** Private Line Service or Wavelength Service becomes unavailable (as defined in Section _____ above) for reasons other than an Excused Outage, Customer will be entitled to a service credit off of the MRC for the affected Service based on the cumulative unavailability for the affected Service in a given calendar month as set forth in the following table:

Cumulative Unavailability	Service Level Credit
0 – 24 hours	No Credit
24:01 – 30	2.5% of the MRC
30:01 – 36 hours	5% of the MRC
36:01 – 42 hours	7.5% of the MRC
42:01 - + hours	10% of the MRC

(2) The Availability Service Levels and associated credits set forth in this Section ____ ***shall not apply to Off-Net*** Local Loop Service, including, without limitation, SM Metropolitan Private Line (Off-Net) Service, provisioned by Seller through a third party

carrier for the benefit of Customer. Seller will pass-through to Customer any availability service level and associated credit (if applicable) provided to Seller by the third party carrier for such Off-Net Local Loop Service.

Example Clause #11:

The following is the first of several examples in which the seller has combined credits for delays in restoring service with standard outage credits.

MTTR:

TIME TO RESTORE

Time to Restore (TTR) is the time required to restore the failure and resume availability of a Service. The time is measured from the moment an outage is reported until the Service is available.

TTR Objectives:

Seller's TTR objective is to restore a fiber cut that occurs within twelve (12) hours, and to resolve Unavailability from any other cause within one (1) hour.

Credits:

For each full hour by which Seller's response to an Unavailability of a Service exceeds the applicable TTR objective, Seller will credit ten percent (10%) of the MRC for the applicable Circuit. Credits for failure to meet this SLA may be applied only to the MRC for the applicable Circuit.

Exceptions:

The Seller TTR SLA is not available for:

Delays caused by Customer, including, but not limited to: (1) acts or omissions by the Customer, its agents or vendors; (2) inaccurate, incomplete or changes to previously accepted orders; (3) inaccessible customer premises, faulty customer equipment or facilities necessary to install the Services; or (4) extension of access to a Circuit demarcation point; or
Force Majeure Events; or

Delays resulting from Customer's failure to comply with its obligations under this Agreement, including failure to pay valid amounts past due; order suspensions due to Customer's credit worthiness; or

Failure of any network elements other than Seller's network; or

Time attributed to Customer's delay in responding to Seller's requests for assistance to repair an outage; or

Trouble tickets held open at the End-User's request once a fix has been implemented and service has been restored; or

Unavailability occurring during a scheduled maintenance window.

Example Clause #12: Sometimes response and repair times are stated as service "objectives" only, and are not actually associated with distinct service credits. Note that the objectives for response times are stated, but there are no service credits if the Carrier should not achieve these objectives. This does not mean that the Customer would not be entitled to its normal credits for interruption of service; it means only that there are no service credits specifically associated with the response times.

Response & Repair Times. In the event of a Service failure, Carrier shall have repair personnel on site within 30 minutes after receiving notification of the failure from Customer. Carrier shall restore the Service on the failed system as follows:

(a) Electronic Restoration. In the event of an electronic failure, Carrier shall use its best efforts to restore service to the affected electronics within one and one-half (1.5) hours of arrival of maintenance personnel on site.

(b) Cable Restoration. Carrier shall use its best efforts to restore the cable no later than one and one-half (1 ½) hours after failure.

Carrier shall maintain a twenty-four (24) hours a day, seven (7) days a week point-of-contact for Customer to report to Carrier system trouble reports or faults.

Example Clause #13: The following example adopts a modified approach to service credits. Times to respond and repair are stated as service objectives only, with no associated service credit, while late delivery would give rise to a service credit.

Response & Repair Times. In the event of a Service Outage, as that term is defined in Appendix 1, Service Provider is to have repair personnel on site within two (2) hours after receiving notification of the Service Outage from Customer or from its internal alarms. Service Provider will restore the Service on its failed system as follows:

(a) Electronic Restoration. In the event of an electronic failure, Service Provider will use its best efforts to restore service to the affected electronics within one and one-half (1) hours of arrival of maintenance personnel on site.

(b) Cable Restoration. In the event of a cable failure, Service Provider shall ***begin cable restoral within two (2) hours*** after the faulty cable is identified. The Mean Time to Restore ("MTTR") for ***cable cuts is eight (6) hours after failure***. Measurement of MTTR begins when the failure occurs and ends when service is accepted by Customer.

(c) Emergency Reconfiguration. If the Service Provider's system has the capability to provide route, reconfiguration to maintain service between the Customer's facility and the

Customer's customer, Service Provider will provide reconfiguration if other means of restoral will not restore Service within the time frames stated in subparagraph (i) and (ii) above. Reconfiguration will begin within one (1) hour after the need to reconfigure is determined and will be completed within two (2) hours thereafter.

Installation Delay Credits. If a DS-1 DS-3, OCN, or Wavelength is ordered with an agreed upon Start of Service Date and the due date is missed due to the sole fault of Service Provider by more than twenty-four (24) hours, Customer will be entitled to a delay credit of an amount equal to five percent (5%) of the MRC for that Circuit for each business day of delay. The delay credits for any Circuit will be capped at one hundred percent (100%) of the MRC for the Circuit. If for any reason a Service is not provided within thirty (30) days of its Start of Service Date, Customer may cancel the Service Order Form without cost or liability. To the extent that all or any part of the Services is delayed upon installation, Customer's sole remedy will be that which is set forth hereunder. Delay credits do not apply in those instances where installation is delayed by or at the request of Customer.

Example Clause #14: The following example combines service credits for late installation, interruptions of service, and failure to respond within agreed time constraints. MTTR standards vary according to geographic location and are typically calculated on a monthly basis.

SERVICE DELIVERY COMMITMENTS

Seller realizes that business depends on effective communications, and that's why Seller provides SLAs for service installation, circuit availability, and mean-time-to-repair ("MTTR") with every contract on key data products. If for any reason Seller misses these metrics, an Eligible Customer will receive credits to compensate for the unavailability of service.

MEAN-TIME-TO-REPAIR ("MTTR")

MTTR Commitment.

If the average monthly MTTR for all Priority 1 and Priority 2 trouble tickets is greater than the applicable amount set forth in Table 3, then Seller will provide a Service Credit to the Eligible Customer. The MTTR measurement is based on the location of Customer's site and the associated country class. Country class designations are provided in Table 3. Internationally, MTTR commitment applies only to locations that are within fifty (50) km or thirty (30) miles from a network node site.

MTTR Credit.

If an Eligible Customer believes Seller has failed to meet its MTTR Commitment, Eligible Customer must contact its Seller representative in writing within 15 business days of such failure. Upon Seller's verification that the actual MTTR over the Measurement Period was greater than the Committed MTTR, Seller will issue a Service Credit to the Eligible Customer. The Service Credit will equal six (6) days of the monthly recurring charges for the affected circuit in the applicable month, not to exceed the limits in Section 5. Approved Service Credit(s) will be applied to an Eligible Customer's invoice during the next billing cycle. Any decision made by Seller concerning this SLA or associated credits will be final, binding and conclusive, and is within Seller's sole discretion.

MANAGED NETWORK MEAN TIME TO REPAIR ("MTTR")

Managed MTTR Commitment.

Seller will meet the below MTTR timeframes or Seller will provide *Eligible Customer* with a service credit. To be eligible for this SLA, customer's network must meet the following criteria: (1) minimum of 11 sites, (2) no more than 50% of sites can be international; (3) each site must be configured with out of band access; and (4) each site must be under a same-day maintenance contract. This SLA guarantees the MTTR of all customer premise devices in customer's network, averaged over a month.

Continental United States Sites and Alaska, Hawaii, and Puerto Rico: 4 hour MTTR

International Sites (excluding Alaska, Hawaii, and Puerto Rico): 10 hour MTTR

Managed MTTR Credit.

If Seller fails to meet the MTTR commitment in a ***month***, Seller will automatically credit the following month's invoice 10% of the total monthly recurring management charges.

EXCLUSIONS FOR SELLER PRIVATE LINE SERVICE LEVEL AGREEMENTS.

Exclusions. The Seller Frame Relay Service SLAs are not valid for:

Customer delays including, but not limited to: (i) acts or omissions by the customer, his agents or vendors; (ii) inaccurate, incomplete or changes to previously accepted orders; (iii) unavailability/faulty customer premises, customer premise equipment (CPE) and/or facilities necessary to install the services; or, (iv) extension of access circuit demarcation point; or,

Force majeure events, as defined in the applicable Seller Private Line services agreement between the parties; or,

Customer's failure to materially comply with its obligations as defined in customer's Agreement for Seller Private Line Services, including failure to pay valid past-due amounts; order suspensions due to customer's credit worthiness; or,

Failure of any network elements other than Seller's or a foreign carrier partner's private line network; or

Troubles resolved as "No Trouble Found"; or,

Time attributed to customer's delay in responding to Seller's requests for assistance to repair an outage; or,

Trouble tickets held open at the End-User's request once a fix has been implemented and service has been restored; or

Outages and calculations due to local in-country practices, any national laws, customs, or regulations (e.g. time intervals in which the local loop problem occurs outside of the local loop provider's standard maintenance window; or

Managed IP Telephony customers; or

Small Medium Business (SMB) bundle customers.

5. MAXIMUM SERVICE CREDITS.

Monthly Service Credit.

Service Credits issued in any month under any Service Delivery Performance Commitment SLA will not exceed Eligible Customer's total monthly recurring charges for the affected Private Line circuit, or 150% of the total non-recurring charges for the affected Private Line circuit.

Service Credits issued in any month under any Managed Performance Service Commitment SLA will not exceed 20% of customer's total monthly Managed Network Services management fee.

Yearly Service Credit. The combined cumulative total of Service Credits issued during a Contract Year under these SLAs will not exceed 20% of an Eligible Customer's total monthly recurring charges for all Private Line circuits and MNS charges invoiced during the Contract Year.

6. DEFINITIONS.

The following definitions are used in this SLA:

Committed Availability – means the percentage of time in a Measurement Period that the private line circuit is available for use by customer.

Contract Year – means the 12-month billing period commencing on the first day of the month after customer's Seller services agreement is effective and each successive 12 month billing period.

Continental United States – means any location within the 48 contiguous states of the United States and the District of Columbia.

Customer Commit Date – means the scheduled service installation date as determined by Seller and communicated to customer.

Eligible Customer – means any customer who has purchased Private Line Services from Seller on or after January 1, 2004 with a minimum 1 year term commitment (or an existing Seller Private Line services customer that renews its existing agreement for an additional term of 1 year or longer), is in full compliance with the terms of its Seller Private Line services agreement, and meets any specific eligibility criteria set forth in the particular service commitment.

End-to-End – means the network call path between originating customer premise and terminating customer premise, including access but excluding Customer Premises Equipment.

International – means any location except those within the 48 contiguous states of the United States and the District of Columbia.

Measurement Period – means a calendar month

Mean Time to Repair - monthly average of the time taken between opening trouble ticket and restoring service for all Trouble Tickets designated as Priority 1 and Priority 2 for a particular circuit (total repair time divided by number of applicable tickets during the calendar month)

Mean Time to Repair Calculation Methodology

Sum of minutes between opening & restoring/clearing
Monthly
of all P1 & P2 trouble tickets during the calendar month =
MTTR
Total number of trouble tickets during the calendar month

No Trouble Found – means a Seller Private Line customer reports a problem that cannot be duplicated by Seller. For example, Customer reports an out-of-service condition, but Seller sees its service up and active with no evidence of a recent outage.

Service Credit – means 1/30 of the applicable monthly recurring charge for the affected private line.

Example Clause #15: In the following is a simple example of an MTTR stated as a service objective only. There are no service credits associated with these service objectives.

SLO

Mean Time to Restore. Mean Time to Restore (MTTR) ***objective*** shall be the average time required to restore service and resume availability in a one month (720 hour) period and is stated in terms of equipment failure and cable outages. The time is measured from the moment the outage is reported until the latter of (i) restoration of the first fiber on a cable cut or (ii) equipment is repaired and service is available. With respect to this Service, Seller's ***objective*** will be to repair network equipment within an average of two (2) hours and the first fiber on a cable cut within an average of four (4) hours. Seller's ***objective*** is to undertake repair efforts on equipment or fiber upon the earlier of Seller's actual knowledge of the problem, or receipt of notice from Customer of the problem, provided that Customer has released all or part of the Service for testing. The maintenance standards in this Section only apply for equipment or fiber on Seller's owned and operated network and from the Demarcation Points.

Example Clause #16:

PERFORMANCE OBJECTIVES:

DS1, DS3, OC-3, OC-12, OC-48, OC-3c, OC-12c, and OC-48c circuit performance will be measured using two parameters: Availability and Error-Free Seconds.

The following assumptions apply to the derived data:

(a) The circuits originate and terminate on the SONET OC-48 backbone
(b) High speed protection switching: 1 for N, where N=2
(c) MTTR for SONET equipment: two (2) hours
(d) MTTR for fiber optic cable: twelve (12) hours (Bellcore Standard)
(e) Cable cut rate: 4.39 /year/1,000 sheath miles (Bellcore Standard)

Availability is a measure of the relative amount of time during which the circuit is available for use. According to CCITT and ANSI definitions, unavailability begins when the Bit Error Ratio (BER) in each second is worse than 1.0 E-3 for a period of ten (10) consecutive seconds.

The availability objective for all circuits between Provider's Network Interface points specified above is to provide performance levels over a twelve (12) month period as follows:

V&H Miles	DS1, DS3, OC-3, OC-12, OC-48, OC-3c, OC-12c, and OC-48c
0-2500	99.99%
2501-4000	99.99%

This excludes any Customer provided access links to the Provider's digital network. Outages attributable to incidental damage to or severance of outside fiber optic cable plant, or scheduled maintenance is excluded from the performance objective stated above.

Error-Free Seconds (EFS) and Error Seconds (ES) are the primary measure of error performance. An EFS is defined as any second in which no bit errors are received. Conversely, an ES is any second in which one or more bit errors are received.

1. OUTAGES

OUTAGE AND OUTAGE CREDITS

In the event of any material degradation of Service or any unscheduled, continuous and/or interrupted period of time when a Service(s) fails to be available or comply with the Technical Specifications attached hereto as Exhibit C for a continuous period of thirty (30) minutes or more (hereafter an "Outage"), Provider shall provide a credit (the "Outage Credit") for Service in the amount of 1/440 of the monthly recurring charge for the Service for each half hour or fraction thereof in excess of the first thirty (30) consecutive minutes that the affected Service fails to conform to the Technical Specifications.

OUTAGES OVER SIX (6) HOURS

If the Outage is greater than six (6) hours, then Customer will have the option to either (i) terminate the affected Service without charge or liability; or (ii) continue to receive an Outage Credit in the amount of 1/1440 of the monthly recurring charge.

At Customer's option, Provider will roll the affected Service to either (i) another On-Net circuit or (ii) obtain a short-term substitute service from another provider for the remaining duration of the Outage

2. OUTAGE START/END TIME FOR CREDIT CALCULATION PURPOSES.

An Outage will begin upon the earlier of Provider's actual knowledge of the Outage or Provider's receipt of notice from Customer of the Outage. An Outage is concluded upon Customer's confirmation that the Service has been restored.

3. EXCESSIVE OUTAGE

In the event that Customer experiences either two (2) or more Outages, or eight (8) aggregate hours of Outages in any one (1) month (the "Excessive Outage"), Customer shall be entitled, in addition to the applicable Outage Credit, if any, to terminate such Services as are affected by the Excessive Outage without liability.

Example Clause #17: In the following clause the seller has set forth a detailed explanation of what is meant by "Mean Time to Repair." Response times are to be measured on a quarterly basis. Notwithstanding explicit MTTR standards, the seller specifically disclaims any liability for a failure to meet the MTTR specifications. The MTTR standards are objective, and no service credits are associated with these standards.

Mean Time to Repair. "Mean Time to Repair" or "MTTR" shall be the average time required to repair a Service and restore its availability and is stated in terms of equipment and cable outages. The time is measured from the time that the Service Outage is reported by Buyer to Seller until the service is available. With respect to Services provided on Seller's Network Facilities, Seller will use all commercially reasonable efforts to (a) repair network equipment within an average of two (2) hours of when Seller's technical representative arrives on the applicable site where the equipment is located, not to exceed a total of four (4) hours from the time the Service Outage is reported and (b) have the first fiber on a cable cut restored within an average of six (6) hours of when Seller's technical representative arrives on the applicable site where the cable cut is located, not to exceed a total of eight (8) hours from the time the Service Outage is reported. ***Such averages will be calculated over a calendar quarter basis***. Seller will undertake repair efforts on equipment or fiber when Seller first becomes aware of the problem, or when notified by Buyer and Buyer has released all or part of the Service for testing, at which point a Trouble Ticket will be established. ***Notwithstanding the above, the failure of Seller to meet such standards shall not constitute a default under this Agreement and Seller shall not be liable to pay Buyer any penalties or damages or credit any portion of the Recurring Charges or Non-Recurring Charges under this Agreement as a result of such failure, except for Outage Credits as set forth in Section ___ above.***

Example Clause #18: Here again, the seller has specifically disclaimed any liability for failure to meet the MTTR objectives.

Mean Time to Restore (MTTR) Objectives

MTTR is defined as the amount of time in hours it takes for Seller to restore the service. MTTR is an objective, not a guaranteed parameter. Seller's objective is to restore the service after an outage as quickly as feasible. The objectives only pertain to Seller provided services.

Outage	MTTR
Mesh	10 seconds
Non-Protected	24 hours

Example Clause #19: As in several of the previous examples, the MTTR standards are stated as an "objective," not as a guarantee of performance. And even though the response standards are stated explicitly, no service credit is associated with a failure of seller to meet these standards.

Response and Repair Times

Except as provided in Article ___ of the Agreement, in the event of a Service Outage, Service Provider will respond within thirty (30) minutes after receiving notification of the Service Outage from Customer and a Trouble Ticket is opened. After receiving notification of the Service Outage from Customer, Service Provider shall begin work to restore the Service on its failed system.

Mean Time to Repair ("MTTR") is the time it takes to fix a problem and close out the associated Trouble Ticket. Trouble Ticket(s) kept open at the Customer's request shall not be included in this calculation. Service Provider's MTTR objective is a yearly average of two (2) hours per occurrence, with no single occurrence lasting greater than four (4) hours.

Service Provider shall maintain a twenty-four (24) hours a day, seven (7) days a week point-of-contact for the Customer to report to Service Provider system Troubles and open a Trouble Ticket.

Example Clause #20: MTTR is measured monthly in the following clause.

Technical Specifications-Service Level Objectives. Although Outage Credits are provided as set forth above, Seller's objective is to provide Private Line Service that meets the following service level objectives ("SLO") with respect to these technical specifications. ***Seller shall have no liability for its failure to achieve these objectives***. The Outage Credits provided above are Customer's sole and exclusive remedy for any failure, interruption or degradation of the Service.

Mean Time to Restore. Mean Time to Restore (MTTR) objective shall be the average time required to restore service and resume availability in a ***one month (720 hour) period*** and is stated in terms of equipment failure and cable outages. The time is measured from the moment the outage is reported until the latter of (a) restoration of the first fiber on a cable cut or (b) equipment is repaired and service is available. With respect to Seller's Private Line Service, Seller's objective will be to repair network equipment within an average of two (2) hours and the first fiber on a cable cut within an average of four (4) hours. Seller's objective is to undertake repair efforts on equipment or fiber upon the earlier of Seller's actual knowledge of the problem, or receipt of notice from Customer of the problem, provided that Customer has released all or part of the Service for testing. The maintenance standards in this Section only apply for equipment or fiber on Seller's owned and operated network and from Seller POP to Seller POP.

Example Clause #21: The following clause requires the seller to furnish to the customer a monthly report to explain the number and duration of service interruptions. The MTTR standards are integrated with outage credits to afford to the customer with enhanced credits in the event of a failure to meet the MTTR standards. Furthermore, the amount of service credits for failure to meet the MTTR standards escalate according to Supplier's lack of responsiveness.

Upon completion of a Reporting Period, Supplier shall provide to Customer an Objective Status Report by the third (3rd) business day of the following Reporting Period. A Reporting Period is represented by a normal calendar month. Should Supplier fail to meet any of the parameters set forth in this SLA within a given Reporting Period, this will be remedied accordingly. Any Credit Remedies due Customer shall automatically be applied to Customer's invoice immediately following such Reporting Period. Supplier will also be evaluated based on its overall yearly results during the term of this SLA. Dependent on the results of such yearly evaluation, Customer would be entitled to additional discounts off its total annual billing.

MAINTENANCE AND REPAIR

1. Mean Time to Repair (MTTR)

Description: The measure of timeliness in repairing circuits to its normal operational level.

Objectives:

YR 1: 85% of all outages with a MTRR $\leq$ 2 hours
YR 2: 87% of all outages with a MTRR $\leq$ 2 hours
YR 3: 89% of all outages with a MTRR $\leq$ 2 hours

Calculation: (Total Number of outages in a Reporting Period with a MTTR of 2 hours or less ÷ the total number of outages in the same reporting period) X 100

2. Failure Frequency (FF)

Description: Percent of total circuit failures.

Objectives:
YR 1: 1%
YR 2: 0.99%
YR 3: 0.98%

Calculation: (Total number of circuit failures in the Reporting Period ÷ the total number of embedded circuits for the same Reporting Period) X 100)

Credit Remedy: Within a given Reporting Period, should the circuit Failure Frequency exceed the above percentages in its respective year, the total Monthly Recurring Charge (MRC) for the circuits that exceed such objectives will be credited by ten percent (10%).

Repeat Trouble (RT) NEW

Description: The number of Repeat Troubles a given circuit consistently experiences during a Reporting Period, thus causing outages.

Objectives:
YR1: Two (2)
YR2: One (1)
YR3: Zero (0)

Calculation: After the first outage occurrence count the number of outages for each circuit during a Reporting Period.

Credit Remedy: Within a given Reporting Period, should a circuit experience the same type of trouble, after the first occurrence, in excess of the above objectives, in its respective year, the total Monthly Recurring Charge (MRC) for the circuits that exceed such objectives, will be credited by ten percent (10%).

<u>Example Clause #22</u>: In the following example MTTR reports are furnishes to the Customer but for information purposes only. The parties agree to periodic reports or meetings to review MTTR performance. But there are no consequences associated with MTTR.

Performance Reviews. If requested by Customer, Vendor and Customer will meet no less frequently than semi-annually to review Vendor's Network performance. Vendor will provide year-to-date Mean Time to Repair ("MTTR") reports for all affected Services provided to Customer.

<u>Example Clause #23</u>: In the following example, the concept of MTTR has been incorporated into, and has become a part of, the definition of outage or service interruption.

Performance Standards.

Circuit Availability is a percentage of the actual amount of time a given circuit is meeting the prescribed criteria to the total amount of time in one month and is usually an expression that takes into account two factors: Mean Time Between Failures (MTBF) and Mean Time to Restore (MTTR). Both of these performance metrics are commonly measured in units of Hours. Circuit Availability can be calculated by the following expression:

Circuit Availability (%) = (MTBF – MTTR) divided by MTBF.
MTTR includes the following four activities:

Network Failure notification by Customer to Seller

Confirmation of failure by Customer upon notification by Seller

Repair of equipment or fiber by Seller

Notification by Seller to Customer that Service has been restored
Acknowledgement by Customer that Service has been restored

Some network components and elements that qualify as valid performance variables are used in the Circuit Availability equation above and some do not. Below are the elements that are included and those that are not included in this calculation.

The calculation includes the following:

(a) Hardware components such as optical line cards, media converter devices, network routers and switches
(b) Fiber cuts and fiber damage that result in a loss of service
(c) Provisioning changes that result in a loss of service
(d) Any other condition that result in a loss of Service

The calculation does not include:

(a) Time required for application software upgrades and fixes (up to one per month)
(b) Scheduled Preventive Maintenance (up to one per month)
(c) Some Network Elements must be shut down when new cards are installed (scheduled ICB)
(d) Complete Network Element shutdown to install new software (scheduled ICB)
(e) Extended outages due to dangerous situations caused by fire or electrical wires

Circuit Availability (SONET). Notwithstanding any of the provisions detailed in Section 6.1, the "Circuit Availability Rate" is defined as the percentage of the amount of time any given circuit is available for transmitting Customer's data over a consecutive 30-day period, measured on an individual circuit basis during the term of the Service of that circuit. Seller Service(s) will be provided to Customer with a Circuit Availability Rate equal to or greater than the requirements outlined in SLA Credit Outage Chart in Section 8.2.

Packet Loss (Ethernet Connectivity). Ethernet service will be provided with the Service not exceeding 99.85% packet loss averaged over any consecutive 24-hour period, and will be maintained at a minimum of 99.990% per year. The Ethernet Service will provide switched Ethernet networking between sites identified on the specific Service Orders.

Circuit Outages. A Circuit Outage is the period that is known as Mean Time to Restore (MTTR). Notwithstanding the Circuit Availability criteria specified in Sections __, Seller Service will be deemed to have suffered a "Circuit Outage" if any circuit is not available for transmitting Customer's data. The Circuit Outage begins at the time Customer notifies Seller's Tech Hotline (1-8__-___-____) or Seller recognizes such an outage or network fault, and ends when Service is deemed restored pursuant to Section 8.3, provided that no Circuit Outage shall be deemed to occur if the affected Service is restored prior to such notification.

<u>Example Clause #24</u>: In the following two example clauses, MTTR is used to describe the seller's expected response to any service outage in the context of outage credits.

Restoration.

(a) When restoring a cut Cable in the Leased Fibers, the parties agree to work together to restore all traffic as quickly as possible. Service Provider, promptly upon arriving on the site of the cut, shall determine the course of action to be taken to restore the fibers and shall begin restoration efforts.

(b) The goal of emergency restoration splicing shall be to restore service as quickly as possible. This may require the use of some type of mechanical splice to complete the temporary restoration.

(c) In the event of a notification by the Service recipient of any failure, interruption or impairment in the operation of the Leased Fibers, the Service Provider agrees make all reasonable efforts, subject to Force Majeure Events, to repair traffic-affecting discontinuity within four (4) hours after the Service Provider maintenance employee's arrival at the location of the discontinuity or damage. In addition, Service Provider agrees to follow the following "Mean Time to Restore" objectives. If, subject to Force Majeure Events, Service Recipient experiences a period of continuous interruption in the use of the Leased Fibers, Service Recipient shall receive a credit against the MRC according to the following schedule:

Period of Continuous Interruption	Level Credit
Less than 8 hours	No Credit
≥ 8 and < 24 hours	50% of the MRC
> 24 hours	100% of the MRC

In no event may the total amount of credits per month exceed the total MRC for such month.

Example Clause #25: This seller uses the term "GNMC," which means Global Network Management Center. Other sellers might instead use the term Network Operations Center or "NOC." In this clause response times vary according to the level of threat to the service.

Service Failure Response/Repair Objectives for On-Net Service.

Seller shall diligently respond to (i.e., identify problem and initiate corrective plan of action) and repair any On-Net Service failure, based upon the grade of the failure assigned by Seller's GNMC, and shall use its commercially reasonable efforts to respond to On-Net Service failures in accordance with the table set forth below.

Grade	Severity Level	Mean Time to Respond
1	CRITICAL FAILURE (E.G., PRIMARY *AND* SECONDARY CIRCUIT OUTAGE)	15 MINUTES
2	Major failure (e.g., primary *or* secondary circuit outage)	1 hour
3	MINOR FAILURE (E.G., SIGNAL PERFORMANCE DEGRADATION)	NEXT BUSINESS DAY
4	Important event (e.g., potential condition being monitored)	Second Business Day
5	Informational event (e.g., requests for documentation)	Second Business Day

Example Clause #26: The two clauses that follow have several features that are less common. First, this carrier allows outage credits only for protected services. That is, service credits are not allowed at all for single path, unprotected services. That is fairly unusual. More often, service credits will be associated with an unprotected service but will be calculated according to a lesser standard than a protected service. (See Example Clause #1 above.) But to deny any service credit at all for an unprotected service is not common.

Second, interruptions of service are calculated on a daily basis. There are two limitations on the amount of credits to which a customer may be entitled. The amount of credit is limited to all charges applicable to the services for that day, as well as the total monthly charges for that month. The significance of the daily limitation is that "interruption" of service is sometimes defined to include more than just an outage. "Interruption" can be defined to include both a service outage and a latency failure. There could be a service credit for an outage and another for a latency failure. Therefore, according to this form, if in one day the service fails the latency specification and also suffers an interruption of service, the total service credit for both would be limited to the charge for the service for that day.

Third, in order to claim a credit, the Customer must report the interruption of service to the Carrier.

CREDIT ALLOWANCES FOR CARRIER SERVICES

(a) Carrier shall use commercially reasonable efforts to maintain the Services in accordance with applicable performance standards. In the event of an i***nterruption of both primary and redundant services at the same time*** ("**Interruption**"), Carrier shall provide credit allowances in the amounts and on the terms and conditions set forth in this Section ____.

Interruption Period	Credit	Interruption Period	Credit
Less than 30 min	None	30 min - 2 hr 59 min	1/10 day
3 hr - 5 hr 59 min	1/5 day	6 hr - 8 hr 59 min	2/5 day
9 hr - 11 hr 59 min	3/5 day	12 hr - 14 hr 59 min	4/5 day
15 hr - 24 hr	One day		

(b) An Interruption period shall be deemed to begin ***when Customer reports to Carrier*** at 1-800 -___-____ a verifiable Interruption in Services. Credit allowances shall not apply to Interruptions (1) caused by Customer, any End User or any Affiliated Person, (2) due to failure of services, equipment or facilities provided by Customer or others, (3) during any period in which Carrier is not given access to Customer Facilities or (4) due to scheduled maintenance and repair. **(c)** Carrier shall be responsible for no more than ***one full day's credit*** for any period of 24 hours. In no event shall the credit allowance in any month exceed the actual monthly charges for the service interrupted. **(d)** All claims are subject to review and verification by Carrier. Interruption periods shall be based solely on Carrier measurements and ***shall only apply to redundant services, where all such redundant services are not available at the same time***. This Section __ ***shall not apply to Customers that have not purchased redundancy services***. **(e)** The credit allowances set forth in this Section ___ are Customer's sole and exclusive remedy with regard to any service outage or any Carrier failure to meet the Service obligations.

Example Clause #27: According to this clause, service credits are available only for services that are on-net and protected.

SERVICE OUTAGES FOR ***ON-NET SERVICES***

In the event that Service Provider is unable to restore a portion of the Service as required hereunder, or in the event of a Service Outage, Customer shall be entitled to a credit for the prorated monthly recurring charges for the affected Circuits for all unplanned outages in excess of thirty (30) minutes. A Service Outage occurs when the Customer experiences a complete loss of connectivity. Credit allowances, if any, shall be deducted from the charges payable by Customer hereunder and shall be expressly indicated on a subsequent bill to Customer. A Service Outage begins when Service Provider is notified or becomes aware of the failure, whichever occurs first. A Service Outage ends when the affected line and/or associated station equipment is fully operative. The total outage time is calculated by taking the difference between the start time and end time less any delay time associated with Service Provider's inability to access the Customer or End User Premise. If the Customer reports Services or a facility or Circuit to be inoperative but declines to release it for testing and repair, it is considered to be impaired, but shall not be deemed a Service Outage.

(a) Credit Allowances do not apply to Service Outages: (i) caused by the negligence or acts of Customer and/or End User or its agents; (ii) due to failure of power; (iii) the failure or malfunction of non-Service Provider equipment or systems; (iv) circumstances or causes beyond the control of Service Provider or its agents; (v) during any period in which Service Provider is not given access to the Service Premises; or (vi) a Planned Service Outage, unscheduled Emergency Maintenance, scheduled maintenance, alteration or implementation as described herein.

(b) To be eligible for an Outage Credit, Customer must submit a request for credit in writing within sixty (60) days following the occurrence that includes the Trouble Ticket Number and Circuit ID or any claim for an allowance is waived. Unless otherwise specifically stated, Service Outages are not aggregated for purposes of determining the credit allowance.

(c) Service Outages for On-Net, ***protected*** SONET Transport Services:

Service Outage Length	Credit Per Circuit
30 minutes or less	None
Between 30 minutes and 1 hour	5% of MRC of the Circuit.

Each hour above 1 hour an additional 5% of the MRC of the Circuit, capped at 50% of the MRC for any single Service Outage. Credits are capped at 100% of the MRC for all Service Outages to that same Circuit in any month.

A Circuit is considered to have Chronic Trouble if it experiences five (5) or more related Service Outages occurring over any thirty (30) consecutive day period or if it has more than forty-eight (48) aggregate hours of outages over any thirty (30) consecutive day period. The Customer shall have the option to either (a) obtain Service Outage credits as set forth above or (b) terminate the affected Circuit provided under this Agreement without liability upon written notice to Service Provider. In addition, in the event that a Circuit continues to experience Chronic Trouble within a thirty (30) day period after clearing the most recent Chronic Trouble for the same Circuit, the Customer may disconnect the specific Circuit without incurring termination liability. Whenever a Customer reports to the Service Provider (or vice versa) that a Circuit has a Chronic Trouble, the Service Provider shall immediately perform a detailed investigation and report the findings to the Customer. Customer's credit and termination rights as set forth herein shall not apply, however, in the event that any Trouble is caused or contributed to, directly or indirectly, by any act or omission of Customer and/or End User, affiliates, agents or representatives.

The provisions of this Section ___ shall apply only to Service Outages of On-Net Services.

Part 2, Chapter 3 - Chronic Outage

If an interruption of service continues for an extended period of time, a customer may have a right to terminate the service or the entire MSA. This right may apply to all services provided under a Master Services Agreement or it may apply to some services but not to others. This topic is discussed also in the chapter entitled "Service Interruptions."

The concept of "Chronic Outage Termination" arises from the interrelation of at least three definitions. First, there is a definition of an "Outage." Second, there are exceptions to the definition of "Outage." Finally, there is a definition of "Chronic Outage."

Except for service credits for interruption of service, a right to terminate for chronic outage is typically a customer's only remedy in the event of a chronic outage.

Example Clause #1: In the first example "Outage Time" is defined as a period during which there are ten consecutive "severely errored seconds." However, if this condition occurs during a planned maintenance, a force majeure event, or results from any of several other causes, this would not be considered as "Outage Time." If "Outage Time" continues for a period of two hours or longer, that would constitute an "Outage" and would entitle the customer to an outage credit. A "Chronic Outage" would occur if there are more than five "Outages" within any consecutive 30-day period.

A Customer that experiences chronic outage and wishes to terminate a service without penalty must notify the Seller within five business days following the occurrence of a chronic outage. If a customer fails to notify the seller within that time, the right to terminate would lapse. The reason for the five day limitation is so that if a chronic outage occurs, the customer's right to terminate does not continue indefinitely.

> All Services will be provided to Customer in accordance with, the terms of this Agreement and the mutually agreed Service Level Agreements ('SLAs') for each Service. Seller will remedy any delays, interruptions and degradation in the Service and restore the Service in accordance with the SLA. Seller and Customer shall report any delays, interruptions, degradation in the Service to the other's designated program management and/ or maintenance control center and if Seller is unable to restore the Service as required in the SLA, Customer shall receive a billing credit at the rate set forth in the SLA applicable to the affected Service. ***If Customer experiences a Chronic Outage with respect to a Service, Customer shall be entitled to terminate the affected Service without penalty or any further obligation or liability whatsoever by providing Seller with written notice within five (5) business days (excluding holidays) of the Chronic Outage (a "Chronic Outage Termination")***. For purposes of this Section, ***a Service suffers an Outage if the Service experiences continuous Outage Time (as described in the applicable SLA) of two (2) or more hours. A "Chronic Outage" is if a Service measured over any thirty (30) consecutive day period, experiences more than five (5) unrelated Outages***. Customer's credit and termination rights shall not apply, however, in the event the delays, interruptions, degradation in the Service are caused or contributed to, directly by any act or omission of the Customer. Within 30 days of a notice of intent to terminate due to failure to meet SLAs, Customer will

set the final Service termination date that is not sooner than 30 days after the date of such notice and no later than 90 days from the date of such notice. This period shall be deemed the Service Transfer Period. During the Service Transfer Period, Seller will continue to provide the Services, provided Customer continues to pay for the Services, pursuant to the terms of this Agreement and provide its commercially reasonable efforts assistance to Customer in its transfer efforts to Customer or an alternate provider and until the end of the Service Transfer Period. As described in the SLA, Seller offers Service credits related to installation intervals and Service provisioning. These credits are the Customer's sole and exclusive remedy for Service-related claims. To qualify for a service credit, Customer must not have any invoices that are past due, and must notify Seller that a trouble ticket should be opened to document the event. In no event shall the total amount of service credits per month exceed the total MRC or NRC (as applicable) for the affected Service for such month.

Each Service's Availability objective is listed on the applicable SOF. Availability is calculated monthly, and if the Service fails to attain the applicable Availability objective, Customer shall receive a credit equal to twenty five percent (25%) of the MRC of that particular Service, which shall be applied against the following month's invoice. "Availability" means the measure of Service availability calculated as (Total Time minus Outage Time) divided by Total Time. For purposes of this definition, "Total Time" is the number of seconds in a calendar month. For SONET Service, ***"Outage Time" begins from the first of ten consecutive Severely Errored Seconds (SES > ~10-6 BER)*** and ends upon the first of ten consecutive non-SES. **Outage Time does not include Service interruptions that result from Planned System Maintenance, Force Majeure Events, scheduled or unscheduled Outages caused by Customer, End User or a third party, or their respective network equipment, or lack of Premises access**. "Planned System Maintenance" means maintenance on a network facility that is related to service delivery, either directly (maintenance of transmission equipment, fiber cable, etc.) or indirectly (maintenance of power, environmental systems, etc.); Planned System Maintenance: Seller usually conducts Planned System Maintenance outside of normal working hours, i.e. 12:00 AM to 6:00 AM local time.

Example Clause #2: The following example draws several distinctions. First, chronic outage termination is available only for one identified service, not for all services under the master services agreements. Second, chronic outage termination is not available at all for off-net services. Finally, chronic outage is defined differently for protected on-net services and unprotected on-net services.

Chronic Outage. Customer may elect to terminate an affected Seller Private Line Service prior to the end of the Service Term without termination liability if, ***for reasons other than an Excused Outage,***

(1) For Protected On-Net Seller Private Line Service, such ***Protected On-Net*** Seller Private Line Service is unavailable (as defined in Section ____ above) for four (4) or more separate occasions of more than two (2) hours each OR for more than twenty four (24) hours in the aggregate in any calendar month; ***or***

(2) For Unprotected On-Net Seller Private Line Service, such ***Unprotected On-Net*** Seller Private Line Service is unavailable (as defined in Section ____ above) for three (3) or more separate occasions of more than twelve (12) hours each OR for more than forty two (42) hours in the aggregate in any calendar month.

Customer may only terminate such On-Net Seller Private Line Service that is unavailable as described above, and must exercise its right to terminate the affected On-Net Seller Private Line Service under this Section, in writing, within thirty (30) days after the event giving rise to a right of termination hereunder, which termination will be effective as set forth by Customer in such notice of termination. Except for any credits that have accrued pursuant to Section __, this Section __ sets forth the sole remedy of Customer for chronic outages or interruptions of any Seller Private Line Service.

The "Service Level" commitments applicable to the Services are found in Seller's Service Schedules for each Service. Service levels ***do not apply to off-net services*** (unless otherwise stated on a Service Schedule) or during periods of ***force majeure or Service maintenance*** ("***Excused Outages***").

Example Clause #3: In the following example there are no distinctions among services offered under the master services agreement. A chronic outage of any service occurs if there is a continuous interruption of service for twelve hours, or separate interruptions totaling twenty-four hours in any six month period.

In the event that Carrier is unable to restore a portion of the Service as required hereunder, or in the event of a Service Outage, Customer shall be entitled to a credit for the affected Circuits for all unplanned outages. Credit allowances shall be deducted from the charges payable by Customer hereunder and shall be expressly indicated on a subsequent bill to Customer. A Service Outage begins when Carrier is notified or becomes aware of the failure, whichever first occurs. A Service Outage ends when the affected line and/or associated station equipment is fully operative. If the Customer reports Services or a facility or Circuit to be inoperative but declines to release it for testing and repair, it is considered to be impaired, but shall not be deemed a Service Outage. ***If a Service Outage continues for a period of twelve (12) continuous hours, or twenty-four (24) cumulative hours in any six (6) month period, Customer may terminate the affected Service provided under this Agreement upon written notice to Carrier***.

Example Clause #4: In the following clause "Chronic Problems" is defined as more than fifteen periods of service interruption, called "Downtime," in a single month or a single continuous interruption of more than eight hours. "Downtime" is defined to exclude periods of scheduled maintenance. However, there is a specific force majeure clause that excuses Seller's performance. That is, if "Downtime" is caused by a force majeure event, the Seller's performance would be excused and the customer would not be able to cancel the Service under the "Chronic Problems" clause. However, within the force majeure exception there is a provision that if a service interruption continues for 30 consecutive days, the customer would be permitted to cancel the Service despite the force majeure event without incurring a termination charge.

Termination Option for Chronic Problems. Customer may terminate this Agreement for cause and without penalty by notifying Seller within ten (10) business days following the end of a calendar month in the event either of the following occurs: (i) Customer experiences ***more than fifteen (15) Downtime periods resulting from three (3) or more nonconsecutive Downtime events during the calendar month***; ***or*** (ii) ***Customer experiences more than eight (8) consecutive hours of Downtime due to any single event.*** Such termination will be effective thirty (30) days after receipt of such notice by Seller.

The service level warranty set forth in this section ____ shall only apply to the bandwidth and facilities Service(s) provided by Seller and, does not apply to (i) any professional services; (ii) any supplemental services; and (iii) any Service(s) that expressly exclude this service level warranty (as stated in the specification sheets for such Services). This Section ___ states customer's sole and exclusive remedy for any failure by Seller to provide Service(s).

"Downtime" shall mean sustained packet loss in excess of fifty percent (50%) within Seller' U.S. network for fifteen (15) consecutive minutes due to the failure of Seller to provide Service(s) for such period. Downtime ***shall not include*** any packet loss or network unavailability during Seller's ***scheduled maintenance*** of the Internet Data Centers, network and Service(s), as described in the Rules and Regulations.

Termination For Cause. Either party may terminate this Agreement if: (i) the other party breaches any material term or condition of this Agreement and fails to cure such breach within thirty (30) days after receipt of written notice of the same, except in the case of failure to pay fees, which must be cured within five (5) days after receipt of written notice from Seller; (ii) the other party becomes the subject of a voluntary petition in bankruptcy or any voluntary proceeding relating to insolvency, receivership, liquidation, or composition for the benefit of creditors; or (iii) the other party becomes the subject of an involuntary petition in bankruptcy or any involuntary proceeding relating to insolvency, receivership, liquidation, or composition for the benefit of creditors, if such petition or proceeding is not dismissed within sixty (60) days of filing. Customer may also terminate this Agreement without penalty in accordance with the terms set forth in paragraphs _____)("***Termination Option For Chronic Problems***") or ___ ("Force Majeure") of this Agreement.

Force Majeure. Neither party will be liable for any failure or delay in its performance under this Agreement due to any cause beyond its reasonable control, including acts of war, acts of God, earthquake, flood, embargo, riot, sabotage, labor shortage or dispute, governmental act or failure of the Internet (not resulting from the actions or inactions of Seller), provided that the delayed party: (a) gives the other party prompt notice of such cause, and (b) uses its reasonable commercial efforts to promptly correct such failure or delay in performance. ***If Seller is unable to provide Service(s) for a period of thirty (30) consecutive days as a result of a continuing force majeure event, Customer may cancel the Service(s).***

Example Clause #5: Here again, the definition of "Chronic Service Outage" is dependent on the definition of "Service Outage." In this clause the customer has a longer period in which to terminate service in the event of a chronic outage. The customer has until the end of the month following the month in which the chronic outage occurs.

Availability. If a Service Outage (as defined below) occurs with respect to Services provided entirely on Seller's Network Facilities and Seller is unable to provide the Services at the Availability Factors (as defined below), then Seller will credit Buyer's invoice for the applicable period with an amount equal to the Service Outage Credit (as defined below) in the month following the request by Buyer and determination of the applicable Service Outage Credit pursuant to the provisions set forth below; provided that Buyer must request such Service Outage Credit and such request must be made within thirty (30) days of the applicable Service Outage. If a Chronic Service Outage (as defined below) occurs for any circuit, then Buyer shall have the right to terminate the applicable circuit consistent with paragraph (e) of this Section ___.

(a) ***Service Outage***. A circuit shall be deemed unavailable during the relevant period if (i) the circuit experiences a complete loss of service or (ii) the circuit experiences a Transmission Problem. A Transmission Problem shall mean the following:

(1) For DS-1 and DS-3 Services a Transmission Problem shall mean when a DS-1 or DS-3 Service is experiencing a BBER of 10^{-3} or worse for ten (10) consecutive seconds. Such Transmission Problem shall terminate when the applicable circuit experiences ten (10) seconds of a BBER of 10^{-6} or better.

(2) For OCn Services a Transmission Problem shall mean when an OCn Service is experiencing a Severely Errored Second for ten (10) consecutive seconds. Such Transmission Problem shall terminate when the applicable circuit experiences ten (10) seconds without a Severely Errored Second.

(b) A Service Outage shall begin upon the earlier of Seller's actual knowledge of the Service Outage or Seller's receipt of notice from Buyer of the Service Outage and shall end upon the correction of the loss of service or the correction of the Transmission Problems as set forth above. In order to determine whether a Transmission Problem exists, Buyer shall be required to request and authorize an intrusive test to be taken of the applicable circuit; provided an intrusive test shall not be required to the extent that Seller has installed equipment that is capable of monitoring such Transmission Problems on an ongoing basis without the need for an intrusive test. The Parties will cooperate on the timing and manner in which any such intrusive test is conducted, taking into consideration the need to minimize the interruption of Buyer's or other customers' services. In the case of a Transmission Problem, the Service Outage shall be deemed to have commenced upon Buyer's request for an intrusive test to the extent such test determines that a Transmission Problem has occurred. Notwithstanding the above, ***a Service Outage shall not be deemed to have occurred and no Service Outage Credits will apply***:

(1) During periods (A) of less than ten (10) minutes, (B) in which Seller is not given access to its facilities or equipment that are required to provide the Services or to remedy

any Service Outage, (C) in which planned or scheduled maintenance and repair activities are occurring, (D) in which Buyer continues to use the Services on an impaired basis, or (E) that are not reported to Seller within thirty (30) days of the date the Service was affected; or

(2) For interruptions that are caused by or due to (A) acts or omissions of Buyer or a third party, including without limitation an interruption on the Third Party Facilities that may provide a portion of the Services, (B) the failure or malfunction of facilities or equipment not owned or operated by Seller, including without limitation the failure of the power supply, or (C) a ***Force Majeure Event*** or (D) disconnections by Seller for non-payment or other contract default or breaches by Buyer.

(c) Availability Factor. The following Availability Factors shall apply to the following Services that are provided entirely on Seller's Network Facilities:

(i) DS-1 Service – 99.90%

(ii) DS-3 Service – 99.95%

(iii)OCn Services – 99.99%

(iv)Ethernet Private Line Services

Unprotected – 99.0%
Protected (Network protection; no client protection) – 99.9%
Load Shared – 99.99% one of the two diverse paths will be available.

(v) Managed Wavelength Services

a. Unprotected – 99.0%
b. Protected (Network protection; no client protection) – 99.9%
c. Load Shared – 99.99% one of the two diverse paths will be available

The Availability Factors set forth above shall be measured during each calendar quarter and is a measurement of the percent of total time that Service is operative and deemed available to Buyer in accordance with the above specifications when measured over such period.

(d) Service Outage Credit. With respect to any Service Outages in excess of the Availability Factors, the Service Outage Credit shall be equal to an amount equal to (a) the Recurring Charge applicable to the affected circuit during the calendar quarter in which the Service Outage occurred multiplied by (b) the number of hours or fractions thereof that the Service Outage occurred during the applicable period divided by 2160 hours. Service Outage Credits are calculated after deduction of all discounts and other special pricing arrangements, and are not applied to governmental fees, taxes, surcharges and similar additional charges, nor are credits available for any usage based Services.

Buyer's right to receive such Service Outage Credit shall be the Buyer's sole and exclusive remedy and the Seller's sole and exclusive obligation in the event of a Service Outage.

(e) Chronic Service Outage. An affected circuit shall be deemed to have experienced a ***Chronic Service Outage*** to the extent that in any calendar month (i) ***three or more Service Outages*** have occurred with each such Service Outage having a duration of more than thirty (30) minutes ***or*** (ii) one Service Outage has occurred for a duration of more than ***forty-eight (48) hours***, in each case where the applicable Service Outage has been reported by Buyer to Seller with respect to any circuit within a calendar month. If a Chronic Service Outage occurs, then Buyer shall have the right to terminate the affected circuit upon providing written notice to Seller, without the incurrence of any Termination Charge; provided that Buyer terminates the applicable Service ***on or before the expiration of the calendar month following which the Chronic Service Outage occurred***. Buyer's right to terminate the affected circuit shall be the Buyer's sole and exclusive remedy and the Seller's sole and exclusive obligation in the event of a Chronic Service Outage.

Example Clause #6: The following example sets forth a simpler standard for chronic outage termination. There is no time limit within which a customer must exercise its right to terminate service in the event of a chronic outage. And an outage is defined merely as a failure of the service to meet the "Service Level Requirements for that Service."

Chronic Outage

Notwithstanding Service Provider's obligation to provide credits in the event of Service Level Failure, in the event a Circuit or Service experiences ***Chronic Trouble***, as set forth in this Agreement or in the applicable Exhibit(s), ***Customer may also disconnect such Circuit or Service***, which is experiencing Chronic Trouble and disconnect any affected Service(s) without applicable termination liability.

Chronic Trouble. "Chronic Trouble" shall mean a situation in which a particular Circuit has experienced the same type of Service Level Failure twice within a thirty (30) day period over two consecutive months, and the Service Level Failure is found not to be the fault of Customer or End User.

Service Level Failure. "Service Level Failure" shall mean the failure of a Service to meet the Service Level Requirements for that Service, and any disruption or degradation of the Service.

Example Clause #7: The following is a simplified approach to chronic outage termination. "Chronic Outage" is defined as five Outages within a 30-day period.

EXCESSIVE OUTAGE

If Customer experiences a ***Chronic Outage*** with respect to this Service, Customer shall be entitled to terminate the affected Service without further obligation by providing Provider with written notice within (5) five business days following such Chronic Outage (a "Chronic Outage Termination"). For purposes of this Section, a Service suffers an "***Outage***" if there is Outage Time (as described above) of ***two (2) or more hours***. ***A "Chronic Outage" is if such Service, measured over any thirty (30) consecutive day period, experiences more than five (5) unrelated Outages.***

For SONET Service, "Outage Time" is the period between the first of ten consecutive Severely Errored Seconds (SES > ~10-6 BER) and the first of ten consecutive non-SES. For 2.5 Gbps and 10Gbps Optical Wavelength Service, "Outage Time" is the number of SES's in the month. For 1.25 Gbps Optical Wavelength Services and Ethernet Service, "Outage Time" is any period when there is 100% packet loss. A lost packet is one with a valid Frame Check Sequence and a valid destination MAC address that is not delivered to its destination. For Dual Path Optical Wavelength Service Outage Time includes only the amount of time that both paths experience concurrent Outage Time. For all Services, Outage Time resulting from Planned System Maintenance, Force Majeure Events, acts or omissions of Customer, End User or a third party vendor or caused by their respective network equipment, or inability to access Premises are excluded from Service credit calculations. Subject to the foregoing, in the event that Outage Time, less applicable exclusions, exceeds a continuous period of ***thirty (30) minutes or more*** (hereafter an "**Outage**"), Provider shall provide a credit (the "Outage Credit") for Service in the amount of 1/440 of the monthly recurring charge for the Service for each half hour or fraction thereof in excess of the first thirty (30) consecutive minutes that the affected Service fails to conform to the Technical Specifications.

Example Clause #8: The definition of "Chronic Outage" in the following example is stricter in that for on-net services a Chronic Outage occurs if there are three Outages within a 90-day period. However, unlike the previous examples, this clause does require the customer to give notice of an Outage, and provides for a 72-hour period following notice from Customer of a Chronic Outage within which time the Seller may remedy the Outage before the Customer would have a right to terminate.

Chronic Outage. In the event that a single ***On-Net Protected*** (1+1) Circuit experiences three (3) or more Outages of a Level 1 event during any period of ninety (90) consecutive days, or a single ***On-Net Unprotected*** (1+0) Circuit experiences three (3) or more Outages of a Level 2 event during any period of thirty (30) consecutive days, (and such Outages are the type for which credit would have been given to Customer pursuant to the terms herein), Customer may declare that the Circuit has a "***chronic outage***" problem. Customer will notify Supplier, and Supplier will have seventy-two (72) hours to identify and correct the chronic outage problem. If the problem is not corrected, ***Customer may cancel the affected*** Circuit without any additional or further liability to Supplier. Cancellation shall be in addition to any credits due to Customer for outages.

Part 2, Chapter 4 - Low Latency Service

Latency is the measure of the time that is required to transmit a bandwidth between particular end points of a service. Here is a definition taken from Example Clause #4 below.

> **"Latency"** means the average round trip time period (rounded up to the next full millisecond) for the transmission of an IP packet between points of presence on the Provider's network, as reported by such Provider, during a Period. Latency measurements shall not include latency attributable to outages or disruptions caused by Buyer or the facilities on Buyer's side of the Demarcation Point connecting Buyer-designated termination points to the applicable Demarcation Point.

For most services the greatest importance is placed on reliability or availability of the service, and latency is not considered of greatest importance. Furthermore, service credits or outage credits are based on interruptions in service availability only, and not on latency:

> "A Circuit shall be deemed to be in an outage condition if, while Customer is using or attempting to use such Circuit, such Circuit loses continuity, becomes unavailable or fails to comply with the applicable specifications for such Circuit ("Outage"). Subject to the restrictions herein, Customer is entitled to an "Outage Credit" in the event that the Service Levels described in the Service Level Agreement are not met."

One of the most important means to achieve reliability of a service and avoid interruptions is through diversity. A protected service is provided over two parallel paths, a primary path over which the service is intended to be carried, and a secondary path that is available immediately if there is a failure of the primary path. A diverse or "protected" route is more reliable than an unprotected route. To maximize the protection of a diverse network, a minimal separation may be prescribed between the two parallel paths until they approach the end points. In this way if a catastrophic event occurs, it is less likely to affect both paths at the same time. For a more detailed discussion of diversity, please consult that chapter entitled "Diversity."

For some services a greater importance is placed on latency than on reliability. While most service level agreements are based on reliability, some service level agreements are based on latency or on both reliability and latency. Latency is largely a function of geographic distance between the relevant end points. The most direct route between the two end points will have the lowest latency. Therefore, the secondary path of a protected service will necessarily not be the most direct route between the end points and therefore will have a greater latency. In those instances in which low latency is of utmost importance, reliability will likely be compromised for the advantage of lower latency over the most direct path.

Latency is most often measured round trip. That is, a latency specification is often defined as the round trip measure of transmission time between point A to point Z plus the latency of the transmission back from point Z to point A. Because latency is primarily a function of distance, a latency specification for any particular service is set forth in the service order for that service and not in the MSA or even in the service description.

Example Clause #1: In the first example, a credit for interruption of service and a credit for latency deficiency are integrated into a single service credit. Interruption of service is defined as a packet loss. A failure in providing the service in accordance with latency specifications is referred to as "packet delay." If the service suffers either a packet loss or packet delay, the customer will be entitled to a service credit. Also, there are two categories of services, and each has its own distinct latency specification.

Latency guarantees for Standard and Expedited service:

The following latency numbers assume Fast Ethernet (100 Mbps) Service Interfaces. Latency is defined as the time interval from the arrival of the first bit at the source Service Interface until the departure of the last bit at the destination Service Interface. ***For Standard Data Service the average latency is 1.5 ms and the 95th percentile is 5 ms. For Expedited Data Service the average latency is .5 ms and the 95th percentile is 2.5 ms.*** These objectives only apply to packets that conform to the bandwidth profile on a Service Interface.

Packet-loss delivery guarantees for Standard and Expedited service:

A packet received on a Service Interface with a valid Frame Check Sequence and a valid destination MAC address that is not delivered to the proper destination Service Interface is deemed to be a lost packet. The packet loss ratio is defined as the ratio (as a percentage) of the number of lost packets divided by the number of transmitted packets. For Expedited Data Service, the packet loss ratio is .0004%. For Standard Data Service, the packet loss ratio is .004%. These objectives only apply to packets that conform to the Bandwidth Profile.

Monetary penalties for excess latency/packet-loss/down-time:

Service on a Logical Wire is considered down when the objectives for ***packet delay*** (both average and 95th percentile) ***and loss*** are not all met for a continuous period of two minutes and remains down until all objectives are met for a period of two minutes. If a Logical Wire is down for more than 30 minutes in a day and it is not during a scheduled maintenance window, the full charges for that day will be credited to your account. You should notify Seller of any such occurrences so that a case can be opened to document the event. (Note: Penalties do not apply to customer caused outages or disruptions).

Scheduled downtime:

Seller may rarely require that sections of the network be brought down in order to perform maintenance, upgrades or other functions. In the rare instances where this must occur, all affected customers will be notified in advance of the timing and expected duration. In all cases Seller will make a best effort to arrange for these scheduled maintenance windows to occur during the least intrusive time possible. The standard policy for maintenance widow scheduling is to send an email to the contact names specified on the customer's contact list at least 48 hours in advance of the scheduled maintenance window. The customer is responsible for keeping their contact information current and accurate, and must provide a minimum of two contacts for notification.

Example Clause #2: In the following example an interruption of service is defined as "Downtime"; a latency problem is referred to as "Excess Latency." The two terms are merged into a single service credit and "Performance Problem" is defined as the occurrence of either "Downtime" or "Excess Latency." The Customer will be entitled to a Service Credit if there is an interruption of Service (Downtime) or if the latency specification is not maintained (Excess Latency). The allowance of service credits, the exceptions to credit allowances, and the maximum amount of a credit are typical of service credit allowances without latency specifications. The only exception to similar treatment for both interruptions of service and latency deficiencies in this clause is that Customer's right to terminate for chronic problems is limited to "Downtime" and would not be permitted for periods of "Excessive Latency."

Service Level Warranty. In the event that Customer experiences any of the service performance issues defined in this Section ___ as a result of Seller's failure to provide bandwidth or facility services, Seller will, upon Customer's request in accordance with paragraph ___(d) below, credit Customer's account as described below (the "Service Level Warranty"). The Service Level Warranty shall not apply to any services other than bandwidth and facility services, and, shall not apply to performance issues (i) caused by factors outside of Seller's reasonable control; (ii) that resulted from any actions or inactions of Customer or any third parties; or (iii) that resulted from Customer's equipment and/or third party equipment (not within the sole control of Seller).

(a) Service Warranty Definitions. For purposes of this Agreement, the following definitions shall apply only to the Services (not including Professional Services).

(i) ***"Downtime" shall mean sustained packet loss*** in excess of fifty percent (50%) within Seller's U.S. network for fifteen (15) consecutive minutes due to the failure of Seller to provide Service(s) for such period. Downtime shall not include any packet loss or network unavailability during Seller's scheduled maintenance of the Internet Data Centers, network and Service(s), as described in the Rules and Regulations.

(ii) "Excess Latency" shall mean transmission latency in excess of one hundred twenty (120) milliseconds round trip time between any two points within Seller's U.S. network.

(iii) "Excess Packet Loss" shall mean packet loss in excess of one percent (1%) between any two points within Seller's U.S. network.

(iv) "Performance Problem" shall mean Excess Packet Loss and/or Excess Latency.

(v) "Service Credit" shall mean an amount equal to the pro-rata monthly recurring connectivity charges (i.e., all monthly recurring bandwidth-related charges) for one (1) day of Service.

(b) Downtime Periods. In the event Customer experiences Downtime, Customer shall be eligible to receive from Seller a Service Credit for each Downtime period. Examples: If Customer experiences one Downtime period, it shall be eligible to receive one Service Credit. If Customer experiences two Downtime periods, either from a single event or multiple events, it shall be eligible to receive two Service Credits.

(c) Performance Problem; Packet Loss and Latency. In the event that Seller discovers or is notified by Customer that Customer is experiencing a Performance Problem, Seller will take all actions necessary to determine the source of the Performance Problem.

(i) Time to Discover Source of Performance Problem; Notification of Customer. Within two (2) hours of discovering or receiving notice of the Performance Problem, Seller will determine whether the source of the Performance Problem is limited to the Customer Equipment and the Seller equipment connecting the Customer Equipment to the Seller LAN. If Seller determines that the Customer Equipment and Seller connection are not the source of the Performance Problem, Seller will determine the source of the Performance Problem within an additional two (2) hour period. In any event, Seller will notify Customer of the source of the Performance Problem within sixty (60) minutes of identifying the source.

(ii) Remedy of Packet Loss and Latency. If the source of the Performance Problem is within the sole control of Seller, Seller will remedy the Performance Problem within two (2) hours of determining the source of the Performance Problem. If the source of and remedy to the Performance Problem reside outside of the Seller LAN or WAN, Seller will use commercially reasonable efforts to notify the party(ies) responsible for the source of the Performance Problem and cooperate with it (them) to resolve such problem as soon as possible.

(iii) Failure to Determine Source and/or Remedy. In the event that Seller (A) is unable to determine the source of the Performance Problem within the time periods described in subsection (i) above and/or; (B) is the sole source of the Performance Problem and is unable to remedy such Performance Problem within the time period described in subsection (ii) above, Seller will deliver a Service Credit to Customer for each two (2) hour period in excess of the time periods for identification and resolution described above.

(d) Customer Must Request Service Credit. In order to receive any of the Service Credits described in this Section ___, Customer must notify Seller within ten (10) business days from the time Customer becomes eligible to receive a Service Credit. Failure to comply with this requirement will forfeit Customer's right to receive a Service Credit.

(e) Remedies Shall Not Be Cumulative; Maximum Service Credit. The aggregate maximum number of Service Credits to be issued by Seller to Customer for any and all Downtime periods and Performance Problems that occur in a single calendar month shall not exceed seven (7) Service Credits. A Service Credit shall be issued in the Seller invoice in the month following the Downtime or Performance Problem, unless the Service Credit is due in Customer's final month of Service. In such case, a refund for the dollar value of the Service Credit will be mailed to Customer. Customer shall also be eligible to receive a pro-rata refund for (i) Downtime periods and Performance Problems for which Customer does not receive a Service Credit and (ii) any Services Seller does not deliver to Customer for which Customer has paid.

(f) Termination Option for Chronic Problems. ***Customer may terminate*** this Agreement for cause and without penalty by notifying Seller within ten (10) business days following the end of a calendar month in the event either of the following occurs: (i) Customer experiences more than fifteen (15) ***Downtime*** periods resulting from three (3) or more nonconsecutive Downtime events during the calendar month; or (ii) Customer experiences more than eight (8) consecutive hours of ***Downtime*** due to any single event. Such termination will be effective thirty (30) days after receipt of such notice by Seller.

(g) THE SERVICE LEVEL WARRANTY SET FORTH IN THIS SECTION ___ SHALL ONLY APPLY TO THE BANDWIDTH AND FACILITIES SERVICE(S) PROVIDED BY SELLER AND, DOES NOT APPLY TO (I) ANY PROFESSIONAL SERVICES; (II) ANY SUPPLEMENTAL SERVICES; AND (III) ANY SERVICE(S) THAT EXPRESSLY EXCLUDE THIS SERVICE LEVEL WARRANTY (AS STATED IN THE SPECIFICATION SHEETS FOR SUCH SERVICES). THIS SECTION ___ STATES CUSTOMER'S SOLE AND EXCLUSIVE REMEDY FOR ANY FAILURE BY SELLER TO PROVIDE SERVICE(S).

Example Clause #3: In the example that follows, the service level agreement for latency is stated separately from interruptions of service. Outages and latency failures are not combined into a single service level agreement. And latency is measured on a monthly basis. A credit is due to the customer if the average monthly latency exceeds the latency specification.

SLA Definition For: Latency:

Service Level Agreement

Seller is committed to maintain an ***average monthly latency*** of <10MS ("Average Latency") for Seller MAN services within a region.

Calculation:
Latency is defined as the amount of time it takes for a packet to traverse from Seller's managed customer premise equipment at one location to Seller's Point of Presence, and from there to its destination at Seller's customer premise equipment at another location. Measurements will be taken every 5 minutes. The highest 5% of all measurements will be excluded, and an average latency calculated for each day, as well as an Average Latency for the calendar month based on the remaining 95% of measurements.

Components Included:

- All components of Seller's managed network.

Exceptions:

The following shall be excluded from the determination of latency under this SLA.

- Network downtime attributable to any actions or omissions of Customer, or Customer's equipment;

- Seller' network downtime during Normal Maintenance;

- The failure of any components that cannot be corrected due to the inaccessibility of Customer Premises to Seller's personnel at the Customer location or causes beyond the reasonable control of Seller; or

- Failure or malfunction of equipment or applications not owned, managed or controlled by Seller.

- Latency across a DS1 Type 2 access service.

Reporting Methods:

Average Latency performance reports will be available via a password-protected customer web site. This report will be provided once a month and will be contained in the Monthly SLA Report.

Compensation:

In the event that Seller fails to meet the Average Monthly Latency performance level, Seller will credit Customer's account in the amount equal to 10% of one month's recurring charges.

Example Clause #4: As with outage credits, latency specifications are defined on a service by service basis. Some services offered under an MSA may be assigned a latency specification, while other services may not have a latency specification. In this example, a service called "IP Transit" is assigned a latency specification.

Latency is measured on a monthly basis and only between the Demarcation Points. Any time period experienced beyond the Demarcation Points is not taken into account in measuring whether the Service has complied with the latency specification. That is, the latency specification does not apply to elements of the service that are on the customer side of the demarcation points.

Where a latency specification is pertinent, an "SLA Failure" is defined as either an interruption of service or latency that exceeds the agreed specification. However, credits are calculated differently for interruption of service and latency failures. An outage credit is calculated pro rata based on the contract value over the applicable month, while a credit for latency failure is calculated using an agreed monetary sum, or liquidated damage.

The Customer will have a right to terminate the agreement if there is an extended interruption of service; however, the Customer will not have a right to terminate the agreement in the event of an extended latency failure.

REMEDIES FOR SLA FAILURES

Unless excused by Force Majeure or Buyer's failure to perform, in the event of an SLA Failure during any Period, Seller shall apply, as liquidated damages, the applicable Credit to the payment due from Buyer on the next following Payment Date. Buyer shall notify Seller of the occurrence of such SLA Failure no later than twenty-four (24) hours after the occurrence of such event and shall provide Seller with appropriate evidence of such SLA Failure. If Buyer fails to notify Seller within such twenty-four (24) hour period, Buyer shall not be entitled to any Credits for the period from the day on which such SLA Failure commenced until the day immediately preceding the day on which Buyer notifies Seller of such SLA Failure, provided that the SLA Failure is still in effect. To the extent that Credits in respect of liquidated damages under this Article __ exceed any payment due from Buyer to Seller, Seller shall pay to Buyer such excess amount, as liquidated damages, on the next Payment Date or, if there is no such Payment Date, on the second Business Day after termination of the applicable Term. Except to the extent an SLA Failure constitutes a Product Termination Event, Seller expressly limits its liability to Buyer for any SLA Failure to the liquidated damages set forth in this Article __.

SERVICE LEVEL AGREEMENT – IP TRANSIT

This Service Level Agreement ("SLA") applies to the purchase of IP Transit within North America pursuant to the Confirmation to which this SLA is attached. A failure under this SLA (an "SLA Failure") shall be deemed to occur upon the occurrence of any of the events described below under: (a) Unavailability, (b) Latency or (c) Packet Loss.

Notification:

Buyer shall notify Seller of the occurrence of an SLA Failure in accordance with Article __ of the Master Agreement.

"Demarcation Point" means the point(s) of interconnection of Seller and Buyer designated in connection with a Transaction.

"Latency" means the average ***round trip time*** period (rounded up to the next full millisecond) for the transmission of an IP packet between points of presence on the Provider's network, as reported by such Provider, during a Period. ***Latency measurements shall not include latency attributable to outages or disruptions caused by Buyer or the facilities on Buyer's side of the Demarcation Point*** connecting Buyer-designated termination points to the applicable Demarcation Point.

"Period" means the consecutive individual periods of time (e.g., weekly, monthly) comprising the Term; provided, however, that if the Parties fail to specify a Period, the Period shall be deemed to be a calendar month.

Credits:

The following credits ("Credits") shall apply in accordance with Article __ of the Master Agreement in the event of an SLA Failure during any Period. A Credit due under clause (a) (Unavailability) shall be the exclusive remedy for any applicable period of Unavailability. Credits due under clause (b) (Latency) and clause (c) Packet Loss shall be cumulative.

Unavailability:
During any Period with respect to any Unavailable Product, if the cumulative Unavailability with respect to a Demarcation Point exceeds [_____ (__)] minutes, Buyer shall be entitled to request a Credit equal to the positive difference, if any, between the Replacement Value and the Contract Value applicable to such Period, plus the ***pro rata*** portion of the Contract Value attributable to any Contract Price actually paid by Buyer ***in respect of such Period***.

Latency:
During any Period with respect to a Product, if Latency exceeds [_____ (____)] milliseconds, Buyer shall be entitled to request a Credit equal to ***[ten percent (10%)] of the Contract Price applicable to such Period***. During any Period with respect to a Product, if Latency exceeds [_____(___)] milliseconds, Buyer shall be entitled to request a Credit equal to twenty percent (20%) of the Contract Price applicable to such Period.
Packet Loss:
During any Period with respect to a Product, if Packet Loss exceeds [_____percent (___%)], Buyer shall be entitled to request a Credit equal to [five percent (5%)] of the Contract Price applicable to such Period.
Product Termination Event:

[With respect to IP Transit, a "Product Termination Event" means (i) that IP Transit is Unavailable in excess of [120 hours][__% of the total number of hours] in a Period, or (ii) the non-Claiming Party terminates its obligations with respect to Product(s) affected by Force Majeure in accordance with Article 3.

ARTICLE 3
FORCE MAJEURE

To the extent either Party is prevented by Force Majeure from carrying out, in whole or part, its obligations in respect of a Product, such Party (the "Claiming Party") shall orally notify the other Party of the Force Majeure as soon as practicable after the occurrence thereof and shall provide to the other Party a written description of the details of such Force Majeure within five (5) Business Days after the date of such oral notice. The Claiming Party shall make reasonable efforts to mitigate the effects of such Force Majeure with reasonable dispatch. If the Claiming Party complies with the foregoing procedures, such Claiming Party shall be excused from the performance of its obligations with respect to such Product (other than the obligation to make payments then due or becoming due with respect to performance prior to the Force Majeure). The non-

Claiming Party shall not be required to perform or resume performance of its obligations to the Claiming Party which correspond to the obligations of the Claiming Party excused by Force Majeure. If the Force Majeure continues for a period of thirty (30) days after the date of the Claiming Party's oral notice, the non-Claiming Party shall have the option, upon three (3) days' written notice to the Claiming Party, to terminate its obligations with respect to the affected Product(s) (other than payment obligations for prior performance) pursuant to Section ___.

Example Clause #5: In this example a latency specification is assigned to a single service, called "Fast Ethernet" service, not for all services offered under the master services agreement. Latency is defined as a unidirectional specification. Unlike some of the other examples, latency is not defined as a round-trip specification.

This service level agreement is simplified in two respects. First, latency and interruption of service are combined into a single service level agreement. If the service fails to satisfy either specification, the customer is entitled to a service credit. Second, the service credit consists of one day of monthly recurring charges for every day in which both specifications are not satisfied, after a grace period of thirty minutes.

Service Availability:

Service Provider provides a 99.99% logical wire availability objective on a monthly basis.

Latency Objectives for Service:

The following latency numbers assume Fast Ethernet (100 Mbps) or faster Service Interfaces. Latency is defined as the time interval from the arrival of the first bit at the source Service Interface until the departure of the last bit from the destination Service Interface. For Service, ***the average latency is 1.5 ms and the 95th percentile is 5 ms***. This objective only applies to packets that conform to the Bandwidth Profile on each Service Interface.

Packet-Loss Delivery Objectives for Service:

A packet received on a Service Interface with a valid Frame Check Sequence (the sequence necessary to make sure that the frame is not corrupt) and a valid destination MAC (Media Access Control which determines the end address) address that is not delivered to the proper destination Service Interface is deemed to be a lost packet. The packet loss ratio is defined as the ratio (as a percentage) of the number of lost packets divided by the number of transmitted packets. For Service, the packet loss ratio is .1%. This objective only applies to packets that conform to the Bandwidth Profile on each Service Interface.

Monetary Penalties for Excess Latency/Packet-Loss/Down-Time:

Logical Wire Service is considered down when the objectives for packet ***delay*** (both average and 95th percentile) ***and loss*** are not all met for a continuous period of two (2) minutes and remains down until all objectives are met for a period of two (2) minutes. If Logical Wire Service is down ***for more than thirty (30) minutes in a day*** and it is not during a Scheduled Maintenance Window, the full MRCs for the impacted logical wire for that day will be credited to Customer's account. Customer must notify Service Provider of any such occurrences so that a case can be opened to document the event. (Note: Penalties do not apply to Customer-caused outages or disruptions, or Force Majeure events.)

Example Clause #6: The following clause imposes strict penalties in the event that the latency specification is not maintained. Latency is measured by the Seller at the time that the service is activated, and later at any time when requested by the Customer. If it is found that the latency specification has not been maintained, the Customer is entitled to a service credit of $25% of the applicable monthly recurring charge for the service.

Delay (Latency):

Seller will measure the Delay (Latency) for a wavelength (i) upon activation of service (both for the initial Implementation and Additional Service Activations) for such wavelength, and (ii) ***at Customer's request at any time***. Delay (Latency) will be measured in milliseconds. The one way delay measurement will be calculated as half of the round trip delay. These measurements will be made in 15 minute increments over a period of 24 – 72 hours (the "Measurement Period") between the applicable demarcation points (i.e. the points of interconnection between Customer's network and Seller's network). It is understood and agreed that a wavelength will be unavailable for the transmission of production data during the Measurement Period, and the Measurement Period will be excluded from Outage Time calculations as provided above.

For each wavelength, the Delay (Latency) Service Level shall be <5 milliseconds for >95% of all measurements for such wavelength during the Measurement Period.

Achieving the Delay (Latency) Service Level will be a requirement for successfully completing service activation.

If Customer requests Seller to measure Delay (Latency) for a wavelength after activation of service, and the Delay (Latency) Service Level is not met during the Measurement Period, Customer may assess a ***Service Credit*** with respect to such wavelength equal to ***25% of Seller's total MRC*** for such wavelength for the month in which Delay (Latency) is measured.

Part 2, Chapter 5 - Latency Service Level Agreement

This chapter sets forth an example of a Service Level Agreement that includes a latency specification. For a more detailed discussion of low latency service, please consult the chapter entitled "Low Latency Service."

Latency for a telecommunications service is the time interval or time delay measured between the end points of a service from the transmission of a signal or packet to the receipt of the signal or packet.

Often latency for a service is measured as "round trip latency." Round trip latency is simply a bidirectional measure. It is the amount of time delay between the origination of transmission to the other end point of the service plus the time delay to return to the end point of origin.

Latency is primarily a function of distance. Service on a longer route will necessarily have a greater latency than service on a shorter route. The differences in latency on routes of different lengths may seem very slight and insignificant for most uses. However, for some customer uses even small differences in latency are very important.

Because latency specification is a function of distance, a latency specification is specific to each Service that is being ordered. Therefore, a latency specification is not stated in the MSA or in the service exhibits. The latency specification must be stated in the Service Order. Exhibit A, Section 1, of the form below, would direct a customer to the Service Order for the latency specification that is applicable to each particular service that is being ordered.

Also because latency is a function of distance, there must be agreement on the end points between which latency will be measured. In Exhibit B below, under the title "Service Availability," there is agreement that latency will be measured "between the Carrier Termination Points." The Carrier Termination Points are defined in Exhibit A, Section 1, and in the Service Order.

Carrier Termination Points are the defined end points of the Carrier's network. A Service is typically delivered to end points on the Carrier's network, to the Customer at the Carrier Termination Points. However, the parties may in some instances agree that a Service will be delivered to a location beyond a Carrier Termination Point, an extended demarcation point. In the form below, in such a case latency would still be measured between the Carrier Termination Points. The SLA for an outage or interruption of Service might include the extended segment, unless the extended segment is a third party facility for which the Carrier is not responsible. However, the segment beyond the Carrier Termination Point would not be included in the measurement of latency.

Most service level agreements do not provide for service credits relating to latency. In those circumstances when latency is included in the serviced level agreement, the inclusion of latency can be accomplished simply by adding a latency specification to the definition of Outage or Unavailable. In Exhibit B below, the definition of Unavailable would be modified to include the latency specification. Therefore, a Service would be Unavailable and the Customer would be entitled to a service credit for any period during which there is an outage or interruption of service, and for periods during which the latency of the Service exceeds the latency specification. The failure to satisfy the latency specification is defined in Exhibit B below as a Latency Deficiency.

Carrier: ___________________________

Low Latency Optical Wavelength Service

These Exhibits A, B, and C are exhibits to, incorporated into, and are subject to all of the terms and conditions of that certain Service Order dated ___________, by and between ___________________________ ("**Carrier**") and Customer. These Exhibits are subject to the Master Services Agreement entered into by the Parties dated ______________________________. Any terms not defined herein have the meaning given in the SOF or, if not therein, in the MSA.

Exhibit A
To
Service Order

Service Description

1. SERVICE DESCRIPTION: CARRIER LOW LATENCY OPTICAL WAVELENGTH SERVICE

Carrier Optical Wavelength Services provide point-to-point optical connectivity between Carrier's facilities (the "**Carrier Termination Points**"). The Low Latency optical wavelength service is available at one (1) Gbps, ten (10) Gbps or one hundred (100) Gbps line rates, and is offered as a single path, unprotected service only. Dual entrance facilities are not available for the low latency optical wavelength Service.

The Carrier Termination Points and the **Latency Specification for each Service are set forth in the Service Order.**

2. RESTRICTIONS ON USE OF THE LOW LATENCY OPTICAL WAVELENGTH SERVICE:

The Carrier Low Latency Optical Wavelength Service is subject to restrictions on assignment and use, as set forth in Section ___ of the MSA.

Carrier will have a right to inspect Customer's use of the Service at any time during normal business hours and with at least twenty-four (24) hours prior notice by Carrier, in order to verify Customer's compliance with Section __ of the MSA.

3. COLLOCATION: If collocation space is provided by Carrier in connection with a Service, the terms of the collocation will be addressed in a separate agreement.

4. DEFINITIONS:

Dual Path Optical Wavelength Services
Dual Path Optical Wavelength Services provide two optical circuit paths between the Carrier Demarcation Points. These optical circuit paths will be diverse from each other. These Services are delivered with a four-fiber handoff at the Carrier Demarcation Point.

Single Path Optical Wavelength Services
Single Path Optical Wavelength Services provide a single optical circuit path between the Demarcation Points. These Services are delivered with a two-fiber handoff at the Carrier Demarcation Point.

Exhibit B
To
Service Order

Service Level Agreement (SLA)

1. Service Provisioning. Subject to availability and Carrier's acceptance of Service Orders, Carrier agrees to provide those Services set forth in the Service Order and **Exhibit A**. Each Service will be provisioned pursuant to the terms and conditions of the Service Order, including but not limited to those service level objectives set forth in this **Exhibit B**.

2. Associated Equipment. Unless otherwise agreed to in writing by the Parties, the Customer is responsible for providing, installing, connecting and maintaining all customer premises equipment ("CPE") associated with any Service provisioned under a Service Order. Carrier Service charges will continue to apply regardless of whether the CPE is unavailable or inoperable.

3. SLAs. The Services are subject to the following Service Level Agreement, as applicable to each particular Service as specified. If Carrier does not achieve a Service Level in a particular month, Carrier will issue a credit to Customer as set forth below upon Customer's request. To request a credit, Customer must contact Carrier Customer Service within thirty (30) days following the end of the month for which a credit is requested. Carrier Customer Service may be contacted by calling toll free in the U.S. (800) ___-____ or by electronic mail at customerservice@____________________. In no event shall the total amount of all credits issued to Customer for a particular month exceed the monthly recurring charges ("**MRC**") for the applicable Service.

SERVICE AVAILABILITY

Service availability is measured between the Demarcation Points (including extended Demarcation Points, where applicable), except that **Latency is measured between the Carrier Termination Points identified above.**

Availability for dual path, route-diverse service is defined as the relative amount of time that at least one end-to-end Circuit is available for use. Service availability is calculated separately for each individual Service connection order.

The Service availability objectives exclude periods of Excused Outages.

If Carrier discovers or is notified by Customer that a Service is Unavailable, Carrier will use commercially reasonable efforts to determine the cause of the Unavailability, and will cooperate with the Customer to resolve such deficiency as soon as possible.

Definitions

"**Demarcation Point**" means a point of interconnection of Carrier's facilities equipment and Customer's equipment or facilities at a location agreed by Carrier and Customer; provided that if the Parties do not specify a Demarcation Point for a Circuit, the Demarcation Point shall be deemed to be the Carrier Termination Points listed above.

Service Unavailability

"**Unavailable**" or "**Unavailability**" for wavelength Circuits means the duration of an interruption in transmission measured from the first of ten (10) consecutive severely erred seconds ("**SES**") on the affected Service until the first of ten (10) consecutive non-SES. "**SES**" means the point when the Bit Error Ratio (BER) exceeds 10^{-6} for a period of ten consecutive seconds and ends when the BER drops below 10^{-6} for a period of one second.

For the Low Latency Optical Wavelength Service only, a Service is also Unavailable if it experiences a Latency Deficiency.

For purposes of calculating a credit, Unavailability is to be measured from the time that Carrier receives a verified notification from Customer that a Circuit is Unavailable until the time that Carrier determines that the Service has been restored.

Service Latency

"**Latency**" means the average round trip time period (rounded up to the next full microsecond) for the transmission of an IP packet between the Carrier Termination Points, as reported by Carrier, during a month. Latency measurements shall not include latency attributable to outages or disruptions caused by Customer or the facilities on Customer's side of the Carrier Termination Points.

"Latency Deficiency" shall mean transmission Latency in excess of the Latency Specification.

Service Availability Objectives

Carrier's Service availability objectives vary depending on the type of service being provided.

Credits Structure

The credit structure for Service that is Unavailable is based on monthly billing calculations. For any billing month in which Carrier fails to meet the Service Availability SLA, the following credit structure will be applied to the net Monthly Recurring Charge for the Circuit(s) affected.

Table B-1 – Service Availability Credit

Availability Service Level for Unprotected Low Latency Optical Wavelength Service.

If a Low Latency Optical Wavelength Service becomes Unavailable (as defined above) for reasons other than an Excused Outage, Customer will be entitled to a service credit. The amount of the credit shall be based on the cumulative Unavailability for the affected Service in a given calendar month, as set forth in the following table:

Cumulative Monthly Unavailability	**Service Level Credit for Unprotected Low Latency Optical Wavelength Service (% of MRC)**
Less than 24 hours	No Credit
24 hours or longer but less than 30 hours	___% of the MRC
30 hours or longer but less than 36 hours	__% of the MRC
36 hours or longer but less than 42 hours	__% of the MRC
42 hours or longer but less than 72 hours	__% of the MRC
72 hours or longer but less than 96 hours	__% of the MRC
96 hours or longer but less than 120 hours	__% of the MRC
120 hours or longer	___% of the MRC

The Availability Service Levels and associated credits set forth in this Section shall not apply to Off-Net Service provisioned by Carrier through a third party carrier for the benefit of Customer. Carrier will pass through to Customer any availability service level and associated credit (if applicable) provided to Carrier by the third party carrier for such Off-Net Service.

TIME TO RESTORE:

Time to Restore (TTR) objective is the time required to restore the failure and resume availability of a Service. The time is measured from the moment an outage is reported until the Service is available.

TTR Objectives:

Carrier's TTR objective is to restore a fiber cut within twelve (12) hours, and to resolve Unavailability from any other cause within four (4) hours.

Exhibit C
To
Service Order

Carrier Pricing

Pricing Structure

Monthly Recurring Charges and One-Time Installation Charges: Monthly recurring charges ("MRC") vary according to the type of Service and bandwidth ordered by the Customer. In addition to the MRCs, there may be nonrecurring installation charges ("NRC") associated with each Circuit installed by Carrier, all as set forth in to the Service Order.

One-Time Installation Charges

Installation charges vary depending on whether a Service is terminated in a building or facility, and are set forth in the Service Order.

Carrier's Service demarcation point is typically at a fiber patch panel (FPP).

Inside Wiring

Inside wiring is defined as the wiring within a building or facility from the building Minimum Point of Entry (MPOE) (where the Carrier Circuit enters the building or facility) to the location of the Carrier service demarcation equipment. Installation of inside wiring and associated inside plant work will be assessed an installation fee on an individual case basis (ICB).

Dual Entrance Facilities

Carrier Services are optionally available using dual entrance facilities in which the Carrier Circuit enters a building or facility at two separate MPOEs. Installation of Dual Entrance Facilities will be assessed an installation fee on an individual case basis (ICB).

Building Access Charges

Carrier may encounter building or property owners which require monthly recurring or other charges to allow telecommunication services to be provided to their tenants. Carrier will attempt to minimize these charges, but if Carrier incurs such charges, Carrier will pass these charges on to the Customer, which the Customer agrees to pay.

No Access Charges

A Carrier field technician will arrive at a building or facility to install the Service on an agreed installation date. It is the Customer's responsibility to ensure that the building or facility is available for access, including any necessary permits from the building or facility owner. If Carrier cannot complete the installation because it is not allowed access to a building or facility, Carrier will charge the Customer a "No Access Charge" of $___ per incident.

AGREED AND ACCEPTED BY:

Customer: ______________________________

By: ______________________________

Print Name: ______________________________

Title: ______________________________

Date: ______________________________

Carrier: ______________________________

By: ______________________________

Print Name: ______________________________

Title: ______________________________

Date: ______________________________

Part 2, Chapter 6 - Diversity

Diversity means redundancy and separation. Most lit fiber services are "protected," which means that they are offered over redundant or dual paths. This affords protection in the event of damage to the network over which the services are provided. If a primary path is damaged, the service is instantly routed to the secondary, redundant or "diverse" path.

Diversity may be defined in several different ways. Sometimes a diverse service means merely that a seller agrees to provide a protected service using two diverse paths, such as in the following example, with no specific diversity standards.

Example Clause #1:

> Diversity: The Circuits connecting two (2) locations shall have a physically diverse service and protected path (except with respect to laterals off the backbone and building entrances).

Often a seller will include, or a customer will require, more specific diversity requirements within a service description or within a master services agreement. A service description will sometimes prescribe a minimum separation between the primary and secondary paths.

Example Clause #2: In the following example the Seller and the Customer may collaborate to establish the primary and secondary paths over which a proposed service will be delivered. The Seller in this example has agreed that, once these paths have been agreed by the parties, the route of the service will not be modified.

> Diversity Planning
>
> (a) If requested by Customer, Seller and Customer ***will agree a primary path*** through the Seller network for a specific Wave ordered by Customer (the "Agreed Path"). Details of the additional charge and the Agreed Path will be set out in the Order Form for the Wave. Where an Agreed Path is agreed, Seller undertakes ***not to reroute*** the Wave from the Agreed path except on a temporary basis (1) for the purposes of scheduled maintenance, in which case seven days' notice shall be provided to Customer or (2) emergency maintenance in which case Seller shall provide Customer with as much as notice as is reasonable practical in the circumstances.
>
> (b) Diversity Planning is available only where requested on an individual case basis for specific Waves by Customer and agreed to by Seller. If agreed, the commitment will be subject to the payment of an additional charge, which will be reflected in an increased Monthly Recurring Charge or (in the case of a Wave purchased on a Prepaid Lease basis) in an increased prepayment amount.

If the Seller has accepted such limitations on the Seller's control over relocations, the parties should then consider what the Customer's remedies are if the Seller does not adhere to this standard. Would the Seller then be in breach of this covenant under the contract, or would the Customer's remedies be limited to a right to terminate the agreement or the relevant service order?

Example Clause #3: The next example illustrates that the Customer's concerns about diversity may not be limited to the design of the Provider's network. The Customer may take into account not only the design the network but also Customer's network and the networks of Customer's other vendors. The Customer may have entered into this agreement specifically to obtain route diversity from Customer's other network elements. For this reason, the Provider has agreed not to modify the primary path of the service without Customer's consent.

Diversity:

Provider acknowledges that network diversity is an integral component of Service, and that Customer has entered into this Agreement and will place Service Orders hereunder, in part, ***to create a diverse path for Customer's Network***. In order to confirm that Provider is providing network diversity to Customer, Provider shall provide to Customer, on Customer's request, information regarding the elements, architecture and configuration of Provider's Network, including, but not limited to, (a) the route (including the specific right of way used within the route) and location of all circuits leased by Customer from Provider, and (b) the owner of the network comprising such route(s) if such owner is other than Provider, and the use of such network has been approved by Customer in writing pursuant to Section ___ above. ***Provider shall not change the primary physical path of network elements leased by Customer without Customer's prior approval.***

Example Clause #4: The Customer in the following example has made a similar point. This Customer has previously purchased service from a different vendor. The Customer requires that the route of this service be diverse from the previously acquired service. Therefore, the parties have defined diversity for this service by reference to the route of the previously acquired service. If the Seller modifies the route of this service in a way that reduces diversity with the alternate service, the Customer will have a right to terminate its agreement.

Note that as the network approaches the end points, which are referred to here as "Customer Facilities," diversity requirements change or are no longer applicable at all. The diverse segments will necessarily converge at the end points.

RELOCATIONS OF FIBER

Seller shall have the right to change the location of the fiber path connecting the Customer Facilities. Seller shall provide Customer as much advance notice of a relocation (the "Relocation Notice") as practicable but shall use reasonable efforts to provide at least sixty (60) days' prior notice of any relocation, if reasonably feasible. The Relocation Notice shall include reasonable detail of the proposed new route and Seller's plan for implementing the relocation.

Customer may terminate this Agreement without liability to either party, as of a date specified in a written notice of termination, in the event that any such relocation or proposed relocation fails to satisfy one or more of the following requirements: (i) no segment connecting any two Customer Facilities shall exceed sixty (60) fiber miles; (ii) the western route connecting the _____ Facility and _____ Facility shall cross the _______ River at least five (5) miles west of _________________; and (iii) there shall be ***at least a one (1) mile separation*** of Seller's Western Path and the path of Customer's diverse carrier located along the Interstate ______ corridor (***except within a two (2) mile radius of each Customer Facility***) The path of Customer's diverse carrier is set forth in Exhibit ____ which is attached hereto and incorporated herein by this reference. Seller shall have the right to determine the extent and timing of, and methods to be used for, such relocation of the fiber path. Seller shall keep Customer informed of the progress of any such relocation and coordinate with Customer as required to maintain uninterrupted service.

Example Clause #5: In the next example the Seller has defined diversity requirements differently for several components of its network, for customer facility diversity, for backbone diversity, for building lateral diversity, for building riser diversity, and for diversity for customer points of presence that are not controlled by the seller. Diversity has a different meaning with respect to each of these network components.

As diverse paths converge toward the end points, at some point it becomes impossible to maintain the prescribed separation if both paths will terminate at the same end points. This question arises in this example because the service will terminate at two fiber termination panels at the customer's point of presence (POP diversity).

1. CUSTOMER FACILITY DIVERSITY

A Customer facility is defined as the space within Seller's data centers, running line sites, and collector nodes, including those ***sites owned and maintained by other parties***. A Customer facility includes fiber entrance and termination facilities, risers, cable tray, optical cross-connections, and fiber jumpers.

Full Customer facility diversity requires the following:

- There shall be separate entrance facilities for each fiber route that requires diversity. A minimum of 50 feet of separation is required between entrance facilities. It is not necessary to create separate entrance facilities to segregate metro routes from intercity routes.
- Each route that requires diversity shall have its own fiber entrance cabinet (FEC). FECs between routes shall be placed with a minimum of 50 feet of separation. Multiple cables from a single backbone path may be spliced within a single FEC.
- Inter-facility cables (IFCs) serving diverse routes shall follow different physical paths from the FECs to the fiber termination panels (FTPs). A minimum of 25 feet of separation shall be maintained between paths up to the point where the paths converge to terminate at the optical cross-connections. Common cable tray, conduit, or fiber trough shall not be used to carry the IFCs of separate paths from the FECs to

the optical cross-connections.

- The optical cross-connection blocks of diverse fiber routes as well as optical cross-connection blocks serving distribution risers may be located within a single optical cross-connection frame.
- Diverse routes shall be terminated at different optical cross-connection blocks. Additionally, distribution risers shall be terminated at optical cross-connection blocks isolated from backbone routes. Multiple cables from a single backbone path or riser may be terminated to a single optical cross-connection block.
- Work and protect inter-bay fiber distribution (jumpers) shall follow a different path to the end locations. Work and project jumpers shall not be placed in the same fiber trough path.

Note that this specification is not intended to address other systems within the facility such as power, cooling, etc.

2. BACKBONE DIVERSITY

Backbone is defined as ***core route intended to carry metro or intercity fiber backbone cables, segments, and loops***. Metro backbone originates at a Customer facility and is not built to a specific building or location, but is built to serve multiple potential locations. Intercity backbone inter-connects Customer facilities. Backbone is generally built fully underground in accordance with Customer construction specifications.

Full backbone diversity requires maintaining ***a minimum full city block separation between opposing routes***. While the actual distance between city blocks is variable, in general a ***100 foot minimum*** separation is expected using this methodology.

In the event a third party backbone route travels through a third party facility such as a collector node or termination facility, the standards set forth in the Customer Facility Diversity section of this document shall govern the portion of the route that is within the third party facility.

When using leased duct for backbones, more separation between routes should be maintained if possible if the manhole system is physically interconnected between both backbone routes. This measure is intended to protect against events (such as fires) which could traverse the interconnected systems and impact both routes. Interconnected manhole systems are not uncommon with utility companies such as department of public works, power companies, or metropolitan transit authorities.

In the event aerial construction is used on one backbone route, the route should still remain fully diverse from the underground route as defined above (i.e. one city block separation). However, if a perpendicular crossing between aerial and underground routes is necessary or if aerial and underground routes must parallel to one another, then the aerial and underground routes should maintain a ***minimum of 25 feet of vertical separation***. If this cannot be obtained, the underground route should be supplemented with additional protection such as steel pipe, steel plating, or concrete encasement. Local practices should be used to select a method to further protect the underground route.

In some circumstances, backbone routes may be constructed within the same railroad, utility, or railroad right-of-way. If this is deemed necessary, the routes shall be kept to the extreme edges of the right-of-way rather than within the street or near the tracks. Based upon the distance between routes, additional protection such as concrete encasement or steel plating may be required for one or both of the routes.

3. BUILDING OSP (LATERAL) DIVERSITY

Building OSP (lateral) is defined as the portion of route starting at the backbone and ending at the building entrance. The building entrance may be a building core or a serving manhole.

Full building OSP diversity requires a ***minimum of 50 feet*** of separation between building laterals.

In the event a third party lateral route travels through a third party facility such as a collector node or termination facility, the standards set forth in the Customer Facility Diversity section of this document shall govern the portion of the route that is within the third party facility.

When using leased duct for laterals, more separation between routes should be maintained if possible if the manhole system is physically interconnected between both laterals. This measure is intended to protect against events (such as fires) which could traverse the interconnected systems and impact both routes. Interconnected manhole systems are not uncommon with utility companies such as department of public works, power companies, or metropolitan transit authorities.

In the event aerial construction is used on one lateral route, the route should still remain fully diverse from the underground route as defined above (i.e. 50 feet of separation). However, if a perpendicular crossing between aerial and underground laterals is necessary or if aerial and underground routes must parallel to one another, then the aerial and underground routes should maintain a ***minimum of 25 feet of vertical separation***. If this cannot be obtained, the underground route should be supplemented with additional protection such as steel pipe, steel plating, or concrete encasement. Local practices should be used to select a method to further protect the underground route.

In some circumstances, laterals may be constructed within the same railroad, utility, or railroad right-of-way. If this is deemed necessary, the routes shall be kept to the extreme edges of the right-of-way rather than within the street or near the tracks. Based upon the distance between routes, additional protection such as concrete encasement or steel plating may be required for one of the laterals.

At times, the location of building entrances are restricted by physical conditions, building moratoriums, building management, or the location of existing building entrance facilities. In these cases, additional protection should be applied to one of the routes at any location where the laterals have less than 50 feet of separation. Typically the protection would occur to the "primary" or working lateral. Note that it is not necessary to apply additional protection to both laterals. The additional protection can take the form of steel pipe, steel plating, or concrete encasement as prescribed by local practices and conditions.

4. BUILDING ISP (RISER) DIVERSITY

Building ISP is defined as the portion of route starting at the building entrance and ending at the point-of-presence (POP). It also includes any distribution risers built within the building to serve other locations or POPs.

Full building ISP diversity requires a ***minimum of 25 feet of separation*** between building risers through the course of the riser paths up to the point where the risers converge at the POP. The riser entrances into the POP shall maintain as much separation as possible, preferably entering the POP on opposite sides. Distribution risers shall maintain similar diversity from the risers serving the POP from the OSP laterals.

At times the location of building entrances as well as access to the POP are restricted for various reasons. In these cases, diversity shall be maintained to the extent practical. Additionally, some buildings have limited riser shafts available for vertical riser runs. In these cases, the risers may be non-diverse through the riser shaft, but the risers shall be made fully diverse as soon as conditions allow.

5. POP DIVERSITY

A POP is defined as the ***space where Customer equipment is maintained*** such as distribution POPs, end user buildings, customer demarcation points, or third party/neutral collocation spaces. The POP space includes end riser runs, cable tray, FTPs, and fiber jumpers.

Full POP diversity requires the following:

- Diverse paths of OSP fiber shall follow different physical paths to the FTPs. A ***minimum of 25 feet of separation*** shall be maintained between these paths ***when the POP area allows***. At a minimum, common cable tray or conduit shall not be used to cable the fiber for separate fiber paths.
- Diverse lateral FTPs as well as FTPs serving distribution risers may be located in a single rack or cabinet.
- ***Diverse laterals shall be terminated to separate FTPs***. Additionally, distribution risers shall be terminated at FTPs isolated from lateral fiber terminations. Multiple cables from a single lateral path or riser may be terminated to a single FTP.
- Work and protect inter-bay fiber distribution (jumpers) shall follow a different path to the end locations. Work and project jumpers shall not be placed in the same fiber trough path.

Note this specification is not intended to address other systems within the POP such as power, cooling, etc.

Additionally, POPs may be placed within a third party or neutral collocation environment. As such, diversity may be limited by the standards of the collocation provider. Such situations shall be managed via the proper approvals and project documentation.

Example Clause #6: In the following clause the parties have distinguished between diversity that is prescribed for protected service, and diversity that may not apply to an unprotected service.

Wavelength Access Product Offerings

Customer may choose to purchase any of the following Wavelength Access Products from Supplier.

A. ***Protected*** DWDM Option:

Supplier installs and maintains the DWDM systems providing protection switching and supporting facilities between Customer POP & End-User Premises.

Required:

Support for 2.5Gbps and 10Gbps speeds:

Supplier DWDM platforms will provide protected services between Customer PoP and Supplier facilities in End-User building by utilizing ***diverse route/entrance facilities*** and will provide protection switching on these systems. The Supplier DWDM systems are cross-connected respectively to the Customer NTN Optical Distribution Frame (ODF) and End-User designated ODF (in End-User Premise) using mutually agreed upon cables and testing methods for hand-off of services. The ***hand-offs beyond the Supplier DWDM systems will not be protected*** (i.e. fiber pair lateral to End-User suite from Supplier POP or Supplier Entrance facilities).

B. Managed DWDM Option:

Supplier deploys and maintains dedicated DWDM system in End-User provided space.

Required:

Support for ***Unprotected*** 2.5Gbps and 10Gbps speeds:

DWDM platform at the End-User building will most likely be located in a common space or in Supplier's POP. The DWDM equipment will be Supplier managed and installed. Risers and laterals may be required to reach End-User space, and they are to be provided by the Supplier. The Supplier Demarcation Point will be an ODF (Optical Distribution Frame) or ODP (Optical Distribution Panel) at the End-User floor. End-User applications with significant Wave requirements may require DWDM located in the End-User suite (ICB process).

DWDM systems are cross-connected respectively to the Customer Network To Network (NTN) ODF and the End-User designated ODF (in End-User Premise) using mutually agreed upon cables and testing methods for hand-off of services.

Service is unprotected using fiber pair end-to-end.

Options:

Diversely routed network pathways to End-User building (Outside Plant). This would allow support for multi-wave applications (East-West routes) that require diversity in order to maintain End-User designed network integrity. End-User may require this diversity to manage their network protection scheme.

Diverse entrance in End-User building. Maintain pathway ***diversity upon entry and within building*** (approx. 30+ ft. between routes) with termination at Supplier DWDM platform. Diversity would then also need to be maintained with any riser or lateral cross-connects to the End-User designated demarcation point, wherever available.

Example Clause #7: Sometimes diagrams such as following accompany diversity standards in order to illustrate and explain the diversity standards of a service. Diagrams might be used in conjunction with a description of route diversity, and not as a substitute for express diversity requirements.

Furthermore, diversity may be defined differently for different network components and for different types of services. The seller in this example describes diversity differently depending on whether the service is SONET, wavelength, or Ethernet service.

References in these diagrams to "Seller Network" mean the Seller backbone network, which does not include laterals that branch off from or diverge from the backbone portion of the network. Therefore, diversity for a "Protected SONET Service" with "1:0 Hand-off" in this example means that there are primary and secondary paths on the Seller's backbone network, but when the path of the service leaves the backbone, it follows a single path only. It is sometimes said that this service is "core protected" only. On the other hand, if a service is "Protected SONET Service" with "1+1 Hand-off," the two paths are diverse not only on the Seller backbone, but also on the laterals, until they reach the Demarcation Points of the service.

SELLER SONET SERVICE.

1. DESCRIPTION.

Protected SONET Services

These Protected SONET Services deliver traffic along a primary and secondary path within the Seller network, and include automatic protection switching. Customer can select either a two fiber connection (1:0 hand-off) or a four fiber connection (1+1 hand-off) to connect its equipment to the Seller network. All DS3 services are provided with an electrical hand-off.

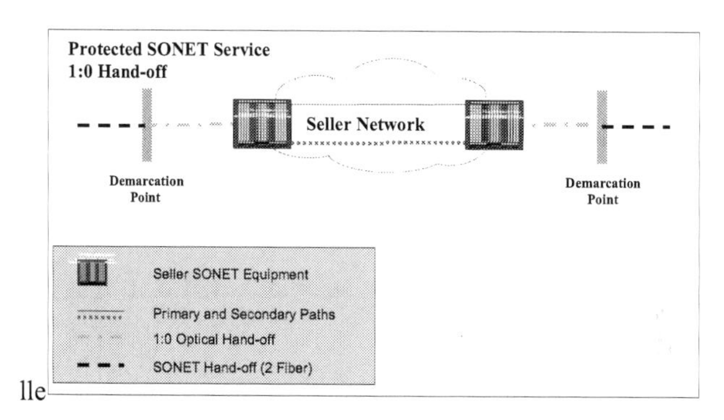

lle

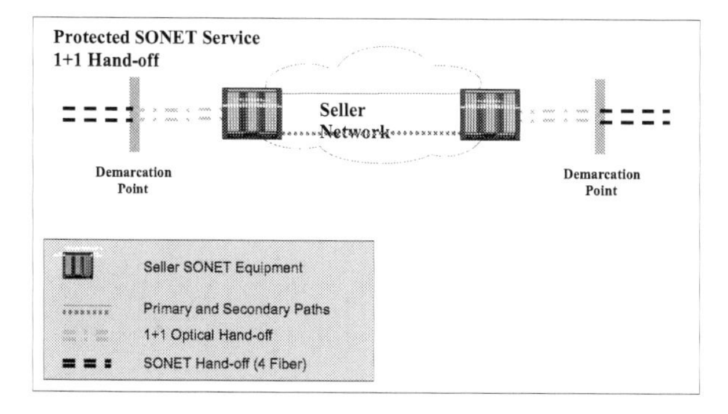

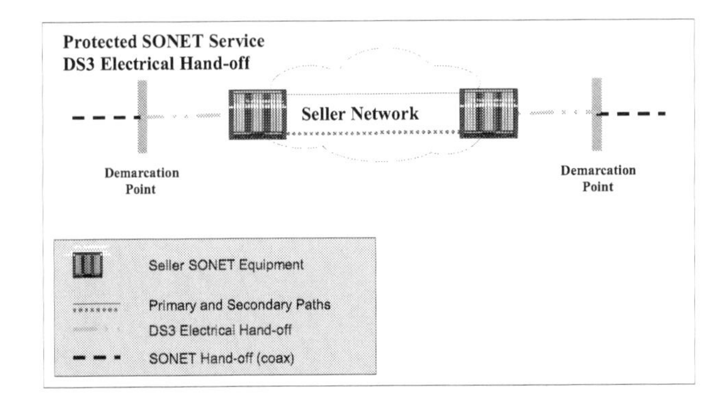

2. TECHNICAL SPECIFICATIONS.

	Protected SONET Services						
Service Characteristics							
Line Rate	44.7 Mbps	155.5 Mbps		622.1 Mbps		2.488 Gbps	
Optical Signal or Digital Signal	DS3	OC-3/OC-3c		OC-12/OC-12c		OC-48/OC-48c	
SDH Equivalent	N/A	STM-1		STM-4/STM-4c		STM-16/STM-16c	
Wavelength	N/A	1310 nm		1310 nm		1310 nm	
Hand-off	Coax	1:0	1+1	1:0	1+1	1:0	1+1(ICB)
Availability*	99.999%	99.9%	99.999%	99.9%	99.999%	99.9%	99.999%

*** See MSA for applicability and exclusions.**

SELLER OPTICAL WAVELENGTH SERVICE.

1. DESCRIPTION.

Dual Path Optical Wavelength Services

These Dual Path Optical Wavelength Services provide two optical circuit paths between the Seller demarcation points. These optical circuit paths will be diverse from each other. These Services are delivered with a four-fiber hand-off at the Seller demarcation point.

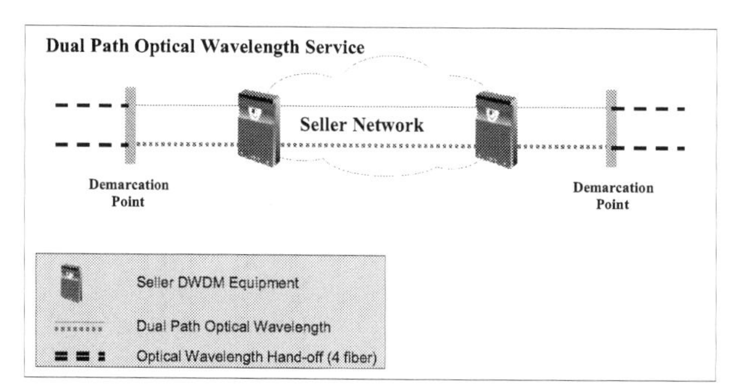

Core Protected Optical Wavelength Services

These Core Protected Optical Wavelength Services provide two optical circuit paths within the Seller network. These optical circuit paths will be diverse from each other. In the event of a failure along the primary path, the traffic will be automatically re-routed to the secondary path. These Services are delivered with a two-fiber hand-off at the Seller demarcation point.

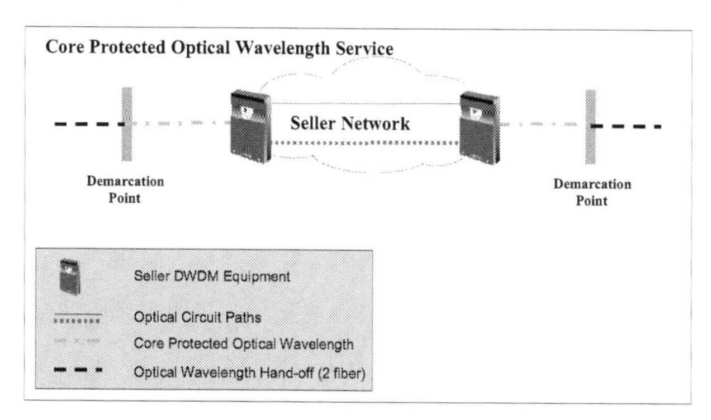

Single Path Optical Wavelength Services
These Single Path Optical Wavelength Services provide a single optical circuit path between the Seller demarcation points. These Services are delivered with a two-fiber hand-off at the Seller demarcation point.

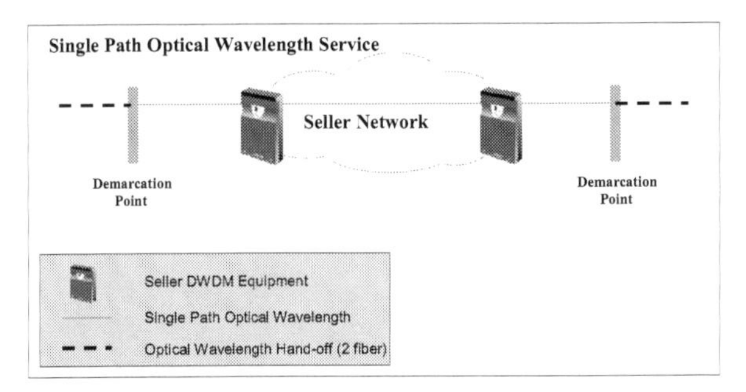

2. TECHNICAL SPECIFICATIONS.

	Dual Path			Core Protected			Single Path		
Service Characteristics									
Line Rate	1.25 Gbps	2.5 Gbps	10 Gbps	1.25 Gbps	2.5 Gbps	10 Gbps	1.25 Gbps	2.5 Gbps	10 Gbps
SONET Equivalent	N/A	OC-48	OC-192	N/A	OC-48	OC-192	N/A	OC-48	OC-192
Comparable SDH	N/A	STM-16	STM-64	N/A	STM-16	STM-64	N/A	STM-16	STM-64
Wavelength	1310 nm	1310 nm	1310 nm	1310 nm	1310 nm	1310 nm	1310 nm	1310 nm	1310 nm
Hand-off	4 fiber	4 fiber	4 fiber	2 fiber	2 fiber	2 fiber	2 fiber	2 fiber	2 fiber
Availability*	99.999%	99.999%	99.999%	99.9%	99.9%	99.9%	99%	99%	99%

*** See MSA for applicability and exclusions.**

SELLER ETHERNET SERVICE.

1. DESCRIPTION.

SERVICES.

Scalable Ethernet Service
This Scalable Ethernet Service delivers IP (Internet Protocol) data packets between the demarcation points using Ethernet frames. Each Scalable Ethernet Service will have a Bandwidth Profile, which indicates its maximum data throughput. Customer may change a Bandwidth Profile provided it does not exceed the physical line rate of the port on which the circuit is provisioned (e.g. Customer could not have a 150 Mbps Bandwidth Profile on a circuit delivered via 10/100BaseT connection).

Fixed Ethernet Service
This Fixed Ethernet Service delivers IP (Internet Protocol) data packets between two demarcation points using Ethernet frames. Each Fixed Ethernet Circuit is offered at the 1 Gbps line rate and may be provided in an unprotected or a protected version. The Bandwidth Profile of this service may not be changed.

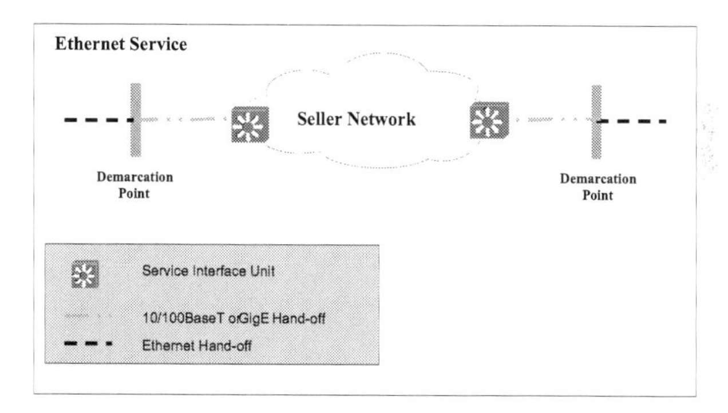

A. INTERFACE OPTIONS.

Service Interface Unit

To provide Ethernet service, Seller must install, at or near the demarcation point, a Service Interface Unit ("SIU"), which it will own and maintain. If Customer is unable to provide a suitable environment for the SIU, Seller may terminate the Service Order without penalty and/or decline to award service credits due to a degradation in Service resulting from Customer's acts or omissions.

STANDARD SIU
The standard interface for Ethernet service is an untagged interface. This interface provisions one Ethernet Service per physical port on the network. Multiple customers may be provided service from one Standard SIU.

DEDICATED SIU
A Dedicated SIU provides Service to only one customer. A Dedicated SIU is required if Customer orders a Service with a Bandwidth Profile of greater than 400 Mbps. If Customer requests a Dedicated SIU for a Service with a Bandwidth Profile of less than 400 Mbps, an additional charge will apply and will be included in the price listed on the SOF. The maximum Bandwidth Profile for each Dedicated SIU is 1 Gbps.

DEDICATED SIU -- Tagged Interfaces

For an additional charge, which will be included on the price stated on the SOF, Customer can receive multiple Services over one Gigabit Ethernet hand-off by selecting a Dedicated SIU with a tagged interface. This will permit Customer to assign its own VLAN tags to each Ethernet Service. The sum of Bandwidth Profiles of the Ethernet Services provisioned to a Dedicated SIU with a tagged interface shall not exceed 1 Gbps.

2. **TECHNICAL SPECIFICATIONS**.

	Scalable Ethernet	**Fixed Ethernet**	
		Protected GigE	**Unprotected GigE**
Service Characteristics			
Bandwidth Profile	10,15,20,30,40,50,60,70,80,90,100, 150,200,300,400,500,600,800,1000 Mbps	1 Gbps	1 Gbps
Hand-offs	10/100BaseT, GigE	GigE	GigE
Scalable	Yes	No	No
Availability*	99.9%	99.9%	99%
Service Quality Objective (packet loss)	.1%	.1%	.1%

*** See MSA for applicability and exclusions.**

3. **SERVICE PROVISIONING.**

BANDWIDTH PROFILE CHANGES.

Customer may request a change in the Bandwidth Profile of a scalable Ethernet service at any time during the Service Term. Seller reserves the right to refuse a Bandwidth Profile change request at Seller's sole discretion.

4. **ETHERNET PRICING.**

Ethernet Service will be billed one month in advance based on the Bandwidth Profile in place at the end of the previous month. The monthly charge is calculated by aggregating daily charges, which are measured at the peak Bandwidth Profile in effect during a midnight-to-midnight day. Following the first month's bill, each subsequent bill will contain the amount for the next month as well as a debit or credit based on the changes in the Bandwidth Profile during the previous month. Disconnect charges will be based on the peak Bandwidth Profile during the Service Term.

If Customer reduces the Bandwidth Profile, there will be a $500 fee for each reduction. If Customer reduces the Bandwidth Profile below the original Bandwidth Profile listed on the Service Order Form, in addition to the $500 fee, the MRC will be changed to equal the new Bandwidth Profile plus 75% of the difference between the rates for the new and the original Bandwidth Profile.

Part 2, Chapter 7 - Undersea Networks

The segments of telecommunications networks that are located under large bodies of water present special challenges. Usually different service levels apply to a network or segment that is undersea. The Service Schedule below was prepared for a Service that includes an "undersea" or "submarine" segment.

One difference between this and other service level agreements is the need to define what is meant by an undersea segment of a network. Where does an undersea segment begin and where does it end? The distinction between terrestrial and undersea segments is important for calculating service outages and chronic outages. An "outage" or "interruption of service" on an undersea segment will usually require a longer period to repair, and therefore will justify a longer interruption of service before a chronic outage may be claimed by the customer.

In the form below, the undersea segment is defined as beginning at the last landing station before the water's edge.

In this form, the schedule of response times uses the concept of a Mean Time to Restore "objective." The use of the term "objective" often implies that a service level standard is not associated with a service level credit. That is, a service level objective is stated for information purposes only. The carrier aspires to meet this specification. But if the carrier fails to meet an "objective," the failure does not give rise to a service credit.

Another term that is pertinent to the maintenance and repair of undersea segments is "Unworkable Weather." The definition of Unworkable Weather is similar in some ways to the concept of force majeure. If the carrier experiences Unworkable Weather, repair efforts on an undersea segment must be suspended until the Unworkable Weather condition has passed. And periods of Unworkable Weather are not included in the calculation of outage or outage credits.

One of the exceptions for service outages or Unavailability is a force majeure event. The same exception may apply to Unworkable Weather or chronic outage. That is, a force majeure event is an exception to the definition of chronic outage, and therefore the customer will not be entitled to claim a service credit or terminate the applicable service for chronic outage if the outage is caused by a force majeure event. This can cause inconvenience to a customer if the outage continues for a long period of time. The solution to this would be either to exclude force majeure from the definition of chronic outage, or to limit the period for which force majeure may apply. Here is an example clause that would allow the customer to claim a service credit if a force majeure event outage continues for eleven days.

> The foregoing notwithstanding, if a Service is not available due to a Force Majeure Event for a period longer than eleven (11) days, beginning on the twelfth (12th) day the Customer shall be entitled to claim a credit against all charges for such Service for the remainder of such Force Majeure Event, unless such unavailability is caused by the Customer, its employee, agent, contractor or its customer. For the avoidance of doubt, the Parties agree and acknowledge (a) with respect to a Submarine Segment, the 11-day period shall commence with the onset of the Force Majeure Event and not with the arrival

of a repair ship at the site; and (b) the credit in this paragraph is not subject to the cap specified elsewhere for Outage Credits.

Service Schedule
Service Level Agreement

Optical Wavelength Service

This Service Schedule is an exhibit to, incorporated into and is subject to all of the terms and conditions of that certain Service Order (the "**Order**") dated ____________ by and between ____________________________ ("**Carrier**") and Customer. This Service Schedule is further subject to the Master Services Agreement entered into by the Parties dated ______________________________. Any terms not defined herein have the meanings assigned to them in the Order or the MSA.

Scope

This Service Schedule will cover the applicable Service(s) ordered by the Customer under this MSA as indicated by the Customer in the Order Form.

Service Descriptions

The Optical Wavelenght Service (the "**Service**") provides point-to-point optical connectivity between two locations over Carrier network.

Definitions

"**Availability**" means the total number of minutes in a month during which there is no Service Outage, divided by the total number of minutes in such month, expressed as a percentage.
"**Excused Outage**" means any Service Outage, unavailability, interruption or degradation of any part of the Service associated with or caused by: (a) Planned Maintenance; (b) any service outage, unavailability, interruption or degradation of service on the Customer's or a third party's communications network; (c) any act or omission by Customer or its employee, affiliate, agent, contractor or invitee; (d) the failure of the Customer and/or any of its employees, agents or contractors to adhere to Carrier's reasonable instructions, including but not limited to instructions to allow Carrier access to network components at the Customer site which is required to complete acceptance testing, trouble shooting, repair or diagnosis; (e) a failure on the part of the Customer and/or any of its employees, agents or contractors to release the circuit for testing; (f) switching time associated with back haul restoration; or (g) a Force Majeure Event.
"**Force Majeure Event**" means an occurrence (other than a failure to comply with payment obligations) caused by any of the following conditions: act of God; fire; tsunami, flood or other severe weather condition; labor strike; sabotage; fiber cut; material shortages or unavailability or other delay in delivery not resulting from the responsible party's failure to timely place orders therefor; power blackouts; lack of or delay in transportation; government codes, ordinances, laws, rules, regulations, permits or restrictions; failure of a governmental entity or other party to grant or recognize a right of way; war or warlike actions (whether or not declared) or civil disorder or riot; landslide; earthquake; fire; explosion; nuclear disaster or radiation leak; insurrection; blockade, terrorism, sabotage, piracy, epidemic, quarantine restriction, strike, lockout or work stoppage (other than of such Party's own employees, agents or subcontractors),

government action or inaction (including delay in issuing permits, through no fault or negligence of Carrier); discovery of hazardous materials, ordinances, nuclear waste or minefields in the performance of the work; Unworkable Weather days; or any other cause beyond the reasonable control of such Party.

"MTTR" or "Mean Time to Restore" means the elapsed time following the opening of a trouble ticket until Service is fully restored to normal operating performance. MTTR is calculated by taking a quarterly average of the time taken to repair all trouble tickets reported for an affected Service. Restoration time for an Undersea Segment begins when the repair ship is on site and excludes Unworkable Weather days.

"**Outage**" or "**Service Outage**" means a period during which there is a break in transmission. For these purposes a break in transmission is signalled by the first of ten (10) consecutively severely errored seconds ("**SESs**"), and the end is signalled by the first of ten (10) consecutive non-SESs. An SES is defined in ITU Standard (G.821).

"**Terrestrial Segment**" means a segment of a Service that transits Supplier's land-based network (with limited water crossings, including, without limitation, bay and channel crossings) and does not transit any part of Carrier's trans-oceanic network. A Terrestrial Segment terminates at the last landing station before the water's edge and does not include the link between a landing station and a beach manhole.

"**Undersea Segment**" means a segment of a Service that transits Supplier's trans-oceanic network between landing stations and does not include any part of a Terrestrial Segment.

"Unprotected" means any Service that does not include a protection scheme that would allow traffic to be switched to a protected path in the event of a fiber cut or material equipment failure.

"**Unworkable Weather**" means winds at Beaufort Force 6 or above for cable ship repair operations on an Undersea Segment and Beaufort Force 5 or above for a remotely operated underwater vehicle (ROV) launch or recovery, and Beaufort Force 6 or above for transit delays. As used herein, Beaufort Force has the meaning given to such term in the American Practical Navigator by Bowditch. The foregoing notwithstanding, the continuation or cessation of operations in weather shall be determined at the sole discretion of the master of the cable repair ship with due regard to the safety of the cable repair ship, equipment, the network and crew. If operations are suspended under such conditions it shall conclusively be deemed to be Unworkable Weather.

> **Alternate Clause**:
> "**Unworkable Weather Day**" means any day on which, in the reasonable opinion of the captain of the repair vessel in consultation with Customer's on board representative, and having due regard to the safety of the crew, vessel and equipment, work on the Undersea Segment should not proceed.

Service Levels Agreement

If a Service becomes experiences an Outage for reasons other than an Excused Outage, Customer will be entitled to a service credit. The amount of the credit shall be based on the cumulative Outage for the affected Service in a given calendar month, as set forth in the following table:

For an Undersea Segment, a Service Outage shall be measured from the time of the home port departure of the repair vessel until the Outage has been remedied, or, if the repair vessel is at another repair location, from the time of making headway from the previous location. Servcie Outage shall not include periods of Unworkable Weather days.

Service	Protection	Availability (%)
100 Gbit/s (OTU4, 100GbE)	Unprotected with back-haul restoration	99.9%
40 Gbit/s (OTU3, 40G Base-R)	Unprotected	99.5%
10 Gbit/s (OTU2, STM64, OC192, WAN Phy, LAN Phy, Fibre Channel)	Unprotected	99.5%

Outage Credits

The Customer shall be entitled to an Outage credit as calculated in accordance with the following table.

Monthly Availability (%)	Outage Credit as a % of MRC
99.5 or Above	0%
99 to 99.5	5%
98.99 – 98	10%
97.99 – 97	15%
96.99 – 96	20%
<95	25%

In no event shall the total amount of Outage Credits per month exceed the total MRC for the affected month .

In order to receive credits, Customer must submit a claim within 180 days following the end of the calendar month in which the fault occurred. The credits will be based on the actual fault time. In the event of any dispute concerning the duration of a fault, Carrier's fault monitoring and clearance records will govern. If the Customer fails to submit a claim within the applicable 180 day period as defined above, the Customer shall irrevocably waive the right to claim any credits for the Services affected by the fault. Notwithstanding anything to the contrary in the Agreement, the claim and award of credits or the right to terminate because of the occurrence of a Chronic Outage shall be Customer's sole and exclusive remedy in the event that the Service is unavailable or fails to meet the specified service levels.

MTTR Objectives

Carrier shall use reasonable commercial endeavours to achive the following MTTR objectives. Service levels with respect to Mean Time To Restore relate to Services with backhaul restoration only and are based on the time it takes Carrier to restore Services from their restoration (secondary) path to their working (primary) path.

Service	MTTR Objective
Terrestrial Fiber	24 Hours
Undersea Network	11 Days

Terrestrial Network	8 Hours

For an Underseach Segment, the calculation of time to repair will begin when the reair ship is onsite to begin the repair and will exclude any periods of Unworkeable Weather.

Chronic Outage Termination

On the occurrence of a Chronic Outage, Customer may cancel a Service and the related Order, without incurring any liability for an Early Termination Charge or other penalty (but subject to Carrier rights with respect to such Service and the related Order, including the right for payment of all accrued but unpaiid Charges) on written notice given to Carrier within ninety (90) days following the occurrence of a Chronic Outage. For purposes of this clause "**Chronic Outage**" means:

(a) Three (3) or more Service Outages of eight (8) hours or more each during any thirty (30) day period; or
(b) A continuous Service Outage that continues for eleven (11) days or more on a Undersea Segment of the System or forty-eight (48) hours or more on a Terrestrial Segment of the System.

If Customer fails to give the notice during such ninety (90) day period, the right to terminate for Chronic Outage shall lapse and shall no longer be exercisable, but Customer's failure to terminate for such Chronic Outage shall not prejudice Customer's right to terminate for any subsequent Chronic Outage.

Undersea Segment. For an Undersea Segment, "**Chronic Outage**" means:

(a) A continuous Service Outage lasting more than eleven (11) consecutive days; or

(b) More than seven (7) Service Outages lasting at least one (1) hour each over any thirty (30) consecutive days period.

Exceptions

An interruption of Service will not constitute a Service Outage or give rise to a right to terminiate when such failure or interruption is attributable to any of the following:

(a) An Excused Outage; or
(b) A disconnection or suspension of the service by Carrier as permitted under the MSA.

Part 3 – THE SERVICE ORDER FORM

A Service Order Form serves two purposes. First, it sets out all of the commercial terms that are specific to a particular transaction. Second, it integrates the commercial terms of this transaction to the service description and the contract terms and conditions. It is most important that each service order form identify the contract terms and conditions that are found in the associated Master Services Agreement and any other contract or other service terms that might pertain to each service.

A service order form can also be used to modify the terms of the related Master Services Agreement or service description. It can be used to create exceptions to the standard contract terms that are found in a Master Services Agreement or in a service description.

The normal order of precedence of the documents in these sample forms is that if there is a perceived conflict between the MSA, a service description, and the SOF, the MSA controls. (See MSA Section 9(i)). It is not always this way. Some are of the opinion that because a Service Order Form is more specific to a particular transaction, the Service Order From should always take precedence and control.

On the other hand, some believe that the MSA more properly reflects established Seller company positions on many questions that may arise in interpreting a contract, and therefore the MSA should not be superseded by an inadvertent inconsistency in a Service Order Form that may have been done in haste.

Irrespective or the order of precedence contained in an MSA, the MSA allows for an amendment of the contract terms and conditions.

> The foregoing notwithstanding, if an SOF or Exhibit specifically cites a provision of this MSA or an exhibit that is to be modified, superseded, or changed, the SOF shall control and take precedence, but only as to the specific provision identified in the SOF. (See MSA Section 9(i))

There is space in this service order form for such modifications in the area just above the signature blocks.

Service Order Number

Service Order

Customer: _______

Fiber Pair Number :

	P O Opportunity #		

Customer Desired Due Date:

Customer Equipment Location

	Location A	*On-net locations only* **Location Z**
Street Address		
Floor		
Cage/Rack/Cabinet		
City, State, Zip		
County		
Network Access Contact Name		
Network Access Contact Phone		
Network Access Contact Email		

Delivery Method

Cross Connect to Customer		
Demarc to Meet Me		
Demarc to Seller		
Enterprise Extension		
Network Extension		
Riser to Customer		
Service Extension		
Other		

Service Type:

			1-Year	2-Year	3-Year Term
Fiber Pair		Unprotected			

Term	*Price*	
Mo-to-Mo	**$___ MRC**	
	$________ NRC	

-

Fiber Pair Number :

	P O Opportunity #		

Customer Desired Due Date:

Customer Equipment Location **Location A** *On-net locations only* **Location Z**

Street Address		
Floor		
Cage/Rack/Cabinet		
City, State, Zip		
County		
Network Access Contact Name		
Network Access Contact Phone		
Network Access Contact Email		

Delivery Method

Cross Connect to Customer		
Demarc to Meet Me		
Demarc to Seller		
Enterprise Extension		
Network Extension		
Riser to Customer		
Service Extension		
Other		

Service Type			**Short-Term**			*Term*	*Price*
Fiber Pair							$___ **MRC**
							$____ **NRC**

Part 4 – OTHER FORMS AND CLAUSES

There are many clauses and contract forms that are not essential to every service contract but are nevertheless used from time to time to address specific questions as they arise. The clauses and contract forms that follow are used to address less common questions.

Part 4, Chapter 1 - Portability

Portability refers to a customer's right to terminate a service and replace it with another. Normally if a customer terminates a service before the normal expiration of its term, the customer would be liable to pay an early termination charge. A typical early termination clause is found in the sample Master Services Agreement in Section 3(b). A right of portability would create an exception to an obligation to pay early termination charges. It would allow a customer to avoid these charges, but only if the conditions set forth in the portability clause are satisfied. The interrelation between early termination charges and portability is also discussed in the chapter dealing with early termination charges.

While the concept of portability is simple, a right to "port" a service may be highly qualified. Often a portability right is exercisable only under strict conditions, and, although a customer would not be required to pay early termination charges, there might be other charges that would apply.

Furthermore, portability may not be available for all services. Portability may be available only for services that meet many conditions prescribed by the seller.

In the first example below there are several conditions to a right of portability. In this example, portability is limited to services of OC-12 capacity or less. And portability would not be available for any service of bandwidth greater than OC-12.

Second, the replacement service must be turned up immediately. The termination of a service must coincide with the commencement of the replacement service. To a Seller, this means that an exercise of portability must not result in an interruption of service related revenue. There can be no gap between the termination of one service and the commencement of the replacement service.

Finally, the replacement service must be an On-Net Service. On-Net Service is defined in the MSA as "as service that originates from and terminates to locations that are both on the Seller network."

Example Clause #1:

> Customer may terminate ***either one or both points of termination for*** **any Service** ***of OC-12 capacity or less*** on _____ days' notice and substitute a new service as a replacement service without incurring the early termination charged described in Section 3(b) of the MSA, provided that Customer ***simultaneously*** replaces the existing Circuit with another Circuit order for ***On-Net Services***, and provided that the following conditions are met:

Comment: At the end of this introductory paragraph of this clause, the seller leads to a list of additional qualifying conditions. The list of qualifying conditions can be very extensive, and might include many of the following.

1. The Circuit must have been installed for at least twelve (12) months;

[or The Circuit must have completed at least three (3) months of service]

2. The Circuit must be replaced ***simultaneously*** by a new Circuit with a ***term that is equal to or greater than the term remaining on the Circuit being terminated***, and otherwise on the terms and conditions of this Agreement.

3. The term of the replacement Circuit must be the longer of: (a) the remainder of the term of the Circuit being replaced, or (b) one year.

4. The new Service Element has a Service Element Term which is equal to or greater than the remainder of the Service Element Term of the Service Element that is being replaced, rounded up to the next highest Service Element Term if the remainder of the Service Element Term falls between 1, 2, 3, or 5 years.

5. The MRC and NRC of the new Circuit are equal to or greater than those of the Circuit that is being replaced.

6. Customer may terminate ***no more than of ten percent (10%)*** of Customer's total active Circuits ordered pursuant to this Service Exhibit, or ***two (2)*** total circuits, whichever is greater, in any ***during a calendar year … during any 12-month period***.

7. The replacement Circuit must be of equal or greater bandwidth.

8. The Circuit being terminated must be On-Net and the replacement Circuit must be On-Net ***where Seller has capacity available***.

9. The portability right may be exercised only one time.

10. There must not be any amount past due for any Service.

In addition, if Customer exercises this portability option, Customer agrees to pay for:

1. All nonrecurring disconnect and installation charges ***due to Seller*** for the terminated Service, including all termination charges for local access services or any other third party services or provided facilities that are affected by the disconnection and replacement of the circuit.

2. All nonrecurring disconnect and installation charges due to Seller as set forth in the attached services schedules or services orders.

3. Any one-time local access services charges and the new monthly recurring local access service charges.

4. All charges for Seller constructed interconnections, if applicable.

5. All other third party charges, including but not limited to cross-connect charges, early termination charges, cancellation charges, one-time charges, and the new monthly recurring charges, if applicable.

"On-Net Services" means Services provided to a building that has Seller's fiber optic cable terminated at the building's Minimum Point of Entry ("MPOE");

Comment: The following several examples show the application of many of these conditions as well as others.

Example Clause #2: Under the following clause portability would be permitted only under very limited circumstances. Customer would be allowed only to convert a one gigabit service to a ten gigabit service. Furthermore, portability is allowed only between the same termination points, the replacement service must be turned up at the same time that the original service is terminated, and the replacement service must have a service term of at least six months. A customer would not be allowed to port services under any other circumstance.

In several of the examples there is an additional qualifying condition that the Seller must have capacity available to accommodate portability. This essentially puts portability to a large degree at the discretion of the Seller. If the Seller determines that capacity is not available on its network, the portability right would not be available.

PORTABILITY: If Customer has purchased ***a one (1) Gbps Service***, Customer may terminate such Service on thirty (30) days' notice without incurring the early termination charges described in Section 3 of the MSA, provided that Customer ***simultaneously replaces the existing Service with a ten (10) Gbps Service between the same termination points***, and provided that the following conditions are met:

1. The Service that is being terminated must be On-Net, and the replacement Service must be to On-Net locations ***where Seller has capacity available***.
2. Customer agrees to pay a nonrecurring charge associated with the new Service in an amount reasonably determined by Seller.
3. There must not be any amount past due for any Service.
4. The Service Term of the replacement Service must be the longer of: (a) the remainder of the Service Term of the Service being replaced, or (b) six (6) months.

In addition, if Customer exercises this portability option, Customer agrees to pay:

1. All nonrecurring disconnect and installation charges for the terminated Service, including all termination charges for local access services or any other third party provided services or facilities that are affected by the disconnection and replacement of the Service;
2. All one-time local access services charges and the new monthly recurring local access service charges;
3. All charges for Seller constructed interconnects, if applicable; and

4. All other third party charges, including but not limited to cross-connect charges, early termination charges, cancellation charges, one-time charges, and the new monthly recurring charges, if applicable.

Example Clause #3: The following clause does not require that a replacement service be turned up immediately on the termination of the service that is being replaced. There may be an interruption of service and an interruption of revenues to the Seller.

In this clause, portability is available only for bandwidths less than a DS-3. Portability is not available for larger bandwidth orders. And portability is available only for services utilizing the same termination points. Portability is available for some but not all on-net services.

There are two time intervals applicable to portability. First, the Customer must notify the Seller of an intention to port within 30 days following termination of service. Second, the replacement service must be installed within 60 days following the replacement order. Note that the replacement service need only be ordered within 30 days following the termination notice. The replacement service need not be in service and generating revenues within 30 days.

EARLY TERMINATION CHARGE.

5.1. Early Termination Charge. Seller is providing the rates under this Exhibit __ based on Customer's agreement to purchase data Services for the Term. If Customer terminates this Exhibit __ under Section ____ of the Agreement or if Customer breaches its obligations and Seller terminates this Exhibit __ under Section ___ of the Agreement, Customer must pay Seller, in addition to all other applicable charges, including, but not limited to, Early Termination Charges for access under Section ____ of the Agreement, an Early Termination Charge equal to $_____. Customer will comply with Section ___ under this Exhibit ___.

5.2. Service Early Termination Charge. In addition to the charge provided for in Subsection ___ of this Exhibit __, if Customer terminates any Service before the expiration of the Order term for that Service, Customer must pay Early Termination Charges described in the Attachments to Exhibit __ and as provided in Section ______.

(a) Access Early Termination Charges.

(1) If Customer terminates Service having access of DS3 or greater bandwidth before the end of the Service Order term, Customer must pay Seller 100% of the remaining access charges as follows:

(A) For Customer-provided access, Customer must pay Seller the EFC and COC MRCs for each month remaining in the Order term; or

(B) For Seller-provided access, Customer must pay Seller the ACF, COC, and local loop MRCs for each month remaining in the Order term.

(2) ***Except for the access Early Termination Charges on bandwidth of DS3 or greater***, Seller will waive the otherwise applicable Early Termination Charges for any Service if Customer orders a replacement Service of the same or greater bandwidth with a term equal to the greater of:

(A) The Order term remaining on the original Service; or
(B) 12 months.

Customer must order the replacement Service ***during the 30 days immediately after termination of the original Service***. The replacement Service ***must be installed within 60 days following the Order for the replacement Service***. Customer must pay the access Early Termination Charges in this Section even if Customer orders a replacement Service ***unless the replacement Service is at the same Service location*** and utilizes the same access facilities as the terminated Service.

Example Clause #4: The following clause was prepared for a contract in which a customer orders multiple services. The Seller has imposed limitations on the Customer's right to port more than one Service. The Customer is not permitted to port more than the three circuits or 15% of all of its circuits, whichever is greater, in any calendar year.

This clause allows for the possibility that there might be an interruption of service revenues. The Customer is permitted to port a Service if the Customer orders a replacement service within 30 days following cancellation of the original circuit.

Portability: On thirty (30) days' notice, Customer may change one or both endpoints of an existing Circuit without incurring early disconnection fees if the following conditions are met:

1. The Circuit must have been installed for at least twelve (12) months;
2. The replacement Circuit must be ordered by Customer ***within thirty (30) days following Customer's cancellation of the original Circuit***, and Customer must request a planned delivery interval that is within Seller's standard installation intervals set forth in this MSA. Such replacement Circuit shall have a term that is equal to or greater than one year.
3. The new Circuit's MRC must be equal to or greater than that associated with the Circuit being replaced.
4. Customer's right to exercise this portability option shall be ***limited to the greater of (i) three (3) Circuits during a calendar year, or (ii) fifteen percent (15%) of all active Circuits under this MSA during a calendar year***.
5. The Circuit being replaced must have been an On-Net Circuit and the new Circuit must be On-Net, and ***sufficient capacity must be available***.

In addition, if Customer exercises this portability option, Customer agrees to pay:

1. All disconnect and installation charges for the terminated Circuit, as determined by Seller and agreed to by Customer.

2. All Third Party Provider charges, including but not limited to cross-connect charges, early termination charges, cancellation charges, one-time charges, and the new monthly recurring charges, if applicable.

Example Clause #5: This portability clause could lead to an interruption of service revenue. In order to invoke this clause, a customer would need only to notify Seller that Customer intends to order a replacement service within 30 days following its notice of termination. The Seller has imposed certain other limitations on the Customer's ability to port a Service.

Early Termination Charge:

A service disconnect involves the disconnecting of a purchased Circuit prior to the end of its specific Initial Circuit Term. Except as otherwise provided in Exhibit __, disconnection of a circuit requires 30 days' prior written notice to Seller. Billing for the Circuit will continue up to the end of service date. For disconnections not the result of Seller's failure to provision the Service as required hereunder, Seller will assess an early disconnection fee applicable to all Circuits disconnected prior to the completion of each Circuit's specific Initial Circuit Term. If the disconnection occurs in the first year of the Initial Circuit Term, the disconnection fee will be equal to seventy-five percent (75%) of the amount determined according to the following formula, and if the disconnection occurs in any year beyond the first year of the Initial Circuit Term, the disconnection fee will be equal to fifty percent (50%) of: (x) the number of months remaining in the applicable Initial Circuit Term multiplied by (y) the MRC associated with the disconnected Circuit(s).

Portability.

(a) Notwithstanding the foregoing, after a given ***circuit has been in effect for a period of at least twelve (12) calendar months***, Customer has the option to terminate such circuit and order a new circuit as a replacement without incurring the early termination charge described in Section __ above. To exercise this "Portability Option," Customer must, simultaneous with its termination notice, give written notice to Seller that Customer intends to exercise this portability option ("Portability Notice"). The Portability Notice shall reference the specific circuit to be terminated and Customer's ***intention to order a replacement circuit within thirty (30) calendar days***.

(b) The replacement circuit(s) shall be for a ***minimum term of one (1) year*** and must ***generate revenue equal to or greater than the revenue remaining on the disconnected circuit***. (For example, Customer has one circuit with a monthly recurring charge of $10,000 with a five (5) year term that has ten (10) months remaining in the term. The remaining revenue is equal to $100,000 ($10,000 times 10 months). The replacement circuit must have a term of at least one (1) year and generate revenue equal to or greater than remaining revenue of $100,000). If the replacement circuit is not ordered within the thirty (30) calendar day time period, Customer shall be liable for the early termination charge set forth above. Notwithstanding the foregoing, ***no more than ten percent (10%) of all existing circuits*** purchased under this Agreement may be replaced as provided in this Section ***in any one-year period*** commencing on the Effective Date.

(c) If Customer exercises this portability option, Customer agrees to pay for:

(1) All nonrecurring disconnect and installation charges due to Seller for the replacement Service;

(2) All charges for Seller custom constructed interconnects, if applicable; and

(3) All third party provider charges, including but not limited to cross-connect charges, early termination charges, cancellation charges, one-time charges, and the new monthly recurring charges, if applicable.

(4) Exercise of this Portability Option is ***restricted to Seller Network Service*** only, for Circuits that terminate within On-Net Buildings.

Example Clause #6: The following clause limits portability to services that are in bandwidth of OC-12 or less. Furthermore, the replacement service must terminate at the same end points. In addition, the contract sets forth an example of the application of the portability clause.

Customer may order Service Provider to disconnect any Service Element by giving Service Provider twenty-four (24) hours' advance written notice of the disconnection. If Customer orders Service Provider to disconnect a Service Element prior to the expiration of its Service Element Term, Service Provider shall permit Customer to do so without liability for an Early Disconnect Charge (as described below) if:

(a) Customer orders such disconnection in accordance with a provision of this Agreement that permits Customer to discontinue an affected Service Element without liability;

(b) Customer elects a portability option whereby Customer orders a new Service Element under the guidelines as stipulated below; or

(c) The Service Element is replaced by another ***higher capacity*** Service Element (*e.g.*, a DS-3 is replaced with an OC-3 Service Element) ***to the same Premises***, *provided that* Customer accepts a new Service Element Term for the higher capacity Service Element.

If Customer elects the portability option as described in subsection (b) above, Customer shall order a new Service Element under the following guidelines to replace the Service Element that is being terminated:

(1) New Service Element's ***Monthly Recurring Rates and Charges are equal to or greater than those associated with the Service Element being replaced***;

(2) New Service Element has a Service Element ***Term which is equal to or greater than the remainder of the Service Element Term of the Service Element being replaced***, rounded up to the next highest Service Element Term if the remainder of the Service Element Term falls between 1, 2, 3, or 5 years (see example below);

(3) The Service Element being replaced ***has been installed for a minimum of twelve (12) months***; and

(4) ***No more than ten percent (10%)*** of all existing Service Elements may be ported in any Annual Period commencing on the Effective Date of this Agreement. If Customer orders such disconnection for its convenience, rather than in accordance with a provision of this Agreement that permits Customer to discontinue an affected Service Element without liability, Service Provider may, as its sole and exclusive remedy, charge Customer the following disconnect charges ("Early Disconnect Charge"). Early Disconnect Charges will be incurred in accordance with the following guidelines:

If a Service Element is terminated: (i) in the first year of its Service Element Term, Customer shall pay seventy-five percent (75%) of its remaining Service Element Term Monthly Recurring Rates and Charges for the remainder of its Service Element Term; (ii) in the second year of its Service Element Term, Customer shall pay fifty percent (50%) of its remaining Service Element Term Monthly Recurring Rates and Charges for the remainder of its Service Element Term; (iii) in the third year of its Service Element Term, Customer shall pay thirty-five percent (35%) of its remaining Service Element Term Monthly Recurring Rates and Charges for the remainder of its Service Element Term; (iv) in the fourth year of its Service Element Term, Customer shall pay twenty-five percent (25%) of its remaining Service Element Term Monthly Recurring Rates and Charges for the remainder of its Service Element Term; and.(v)) in the fifth year of its Service Element Term, Customer shall pay ten percent (10%) of its remaining Service Element Term Monthly Recurring Rates and Charges for the remainder of its Service Element Term.

<u>An example for Section __above:</u>

Customer elects to terminate ("port") a 10 Mbps Service Element in its seventeenth month of being in service and it has an initial 3 year Service Element Term affiliated with it. The new 10 Mbps Service Element (or higher bandwidth) that shall replace the existing 10 Mbps Service Element shall have a Service Element Term that is a minimum of 2 years (36 months [original Service Element Term] – 17 months [time in service] = 19 months, then rounded to the next highest Service Element Term which is a 2 year term).

Early Disconnect Charges will be incurred in accordance with the following guidelines: Early Disconnect Charges shall be calculated as the difference between what the Customer's recurring charges would have been, as stated in Exhibit C, if Customer had originally signed up for a shorter Service Element Term (dependent on the number of months a Service Element has been in Service from the Start of Service Date to the date of termination rounded to the closest circuit term rate available) and the recurring charges actually paid to Service Provider by Customer prior to the disconnection of the Service Element based on the original Service Element Term the Customer had signed up for as stated in the initial LWOF. For example, if the Customer signs up for a 10 Mbps Service Element for a three (3) year term, the Customer would be billed at a monthly recurring rate of $688 per month ($588 bandwidth MRC + $100 SIU MRC). If the Customer terminates this 10 Mbps Service Element, without cause, nine (9) months into the three (3) year term, the Customer would then be billed for the one (1) year period at the rate based on the one (1) year term rate, which is $718 per month ($618 bandwidth MRC + $100 MRC). So for the $688 rate, the Customer has paid $6192 ($688 x 9 months) for the first year. If the Customer had initially signed up for a one-year term, the Customer would have paid $8616 in recurring charges ($718 x 12 months). So for this particular 10 Mbps Service Element, the termination liability charges to the Customer would be $2424 ($8616 - $6192 = $2424).

Example Clause #7: In this example portability is limited to services of OC-12 or less. A replacement service must be ordered before the effective date of the cancellation, although this clause does not require that the replacement service be activated and in service before the first service terminates.

Portability. Customer may cancel an existing On-Net Circuit with ***up to OC-12 capacity*** (a "Cancelled Circuit") without incurring any cancellation or early termination fees or penalties, so long as:

(f) Cancelled Circuit has been installed and in use for at least sixty (60) calendar days;
(g) Customer replaces the Cancelled Circuit with another On-Net circuit having ***equal (or greater)***: (i) ***IXC MRCs, and (ii) term commitment*** (the " Replacement Circuit");
(h) the Replacement Circuit is ***ordered by Customer before the effective date of the cancellation*** of the Cancelled Circuit (the effective date of cancellation being the "Cancellation Date");
(i) the total number of Cancelled Circuits ***does not exceed ten percent (10%)*** of their total active Circuits ***in any annual period***; and
(j) Customer is not otherwise in breach of this Agreement.

If the number of Cancelled Circuits exceeds ten percent (10%) of their total active Circuits in any annual period, or if Customer fails to meet any of the other above criterion, Customer shall pay all applicable shortfall charges and early termination fees for each additional Cancelled Circuit.

Example Clause #8:

PORTABILITY

Customer shall have the right, in its sole discretion and without charge to Customer, to cancel any ASR up to three (3) days after Customer receives a Design Layout Record (hereinafter "DLR") from Carrier. If the Circuit has been activated, Customer may terminate the Circuit prior to the expiration of an agreed term by providing Carrier thirty (30) days prior written notice of its intent to cancel such Circuit. No later than sixty (60) days after the effective date of such termination, Customer shall pay to Carrier, as liquidated damages and not as a penalty, a termination charge equal to:

a. One hundred percent (100%) of the monthly recurring charge that would have been incurred for the Circuit for months 1 - 6 of the agreed term, plus
b. Seventy-five percent (75%) of the monthly recurring charge that would have been incurred for the Circuit for months 7 – 12 of the agreed term, plus
c. Fifty percent (50%) of the monthly recurring charge that would have been incurred for the Circuit for the remaining months of the agreed term.

The foregoing shall be Carrier's sole and exclusive remedy in the event that Customer cancels a Circuit order prior to the expiration of an agreed term. Notwithstanding the foregoing, Customer shall not be liable for any termination charge if Customer ***submits an ASR for a Circuit or Circuits requested to be installed within ninety (90) days of termination*** having monthly recurring charges which are at least equal to the monthly recurring charges for the cancelled Circuit(s).

Example Clause #9: This clause allows a customer to port a service in order to take advantage of lower charges that are part of Seller's tariff filings for services to other customers.

Customer may, at its option, terminate this Agreement without penalty if: **(a)** Customer replaces this Agreement with an agreement entered into between ***Seller and a Similarly Situated Customer*** for comparable volumes and services, as set forth in ***Seller's tariffs*** filed with the Federal Communications Commission (the "FCC") or a state regulatory commission, and with a Term at least equal to that remaining on this Agreement at the time of termination or **(b)** if the introduction of a new technology decreases rates for a Similarly Situated Customer and Customer replaces this Agreement with an agreement entered into between Seller and a Similarly Situated Customer for comparable volumes and services, as ***set forth in Seller's tariffs*** filed with the Federal Communications Commission (the "FCC") or a state regulatory commission, and with a Term at least equal to that remaining on this Agreement at the time of termination. For purposes of this section, a "Similarly Situated Customer" is any customer of Seller which has executed a contract for toll free inbound and outbound intrastate, interstate, and international telecommunications services originating or terminating over dedicated access facilities with a ***term of at least three (3) years and a Revenue Level Commitment that is plus or minus five (5) percent of that contained in this Agreement***. It shall be Customer's responsibility to request termination of the Agreement pursuant to this Section ___.

Example Clause #10: The following portability clause could result in an interruption of service and service revenues. A Customer that wishes to invoke this portability clause must give notice of its intent to port a service within 30 days following the date that Customer cancels a service, and must actually order the replacement service within 60 days following termination of the old service.

<u>Circuit Portability</u>.

(a) Notwithstanding the above, Customer may disconnect an existing On-Net Circuit with ***up to and including OC-12 capacity*** for which no special construction is or was required (a "Disconnected Circuit") without incurring any cancellation or early termination fees or penalties, so long as all of the following are met: (i) the Disconnected Circuit has been installed and in use for at least sixty (60) calendar days; (ii) Customer replaces the Disconnected Circuit with another On-Net Circuit having: (x) equal or greater On-Net Circuit MRCs and (y) an equal or greater (but in no event less than 12 month) term commitment (collectively, the "Replacement Circuit"); (iii) the Replacement Circuit is ordered by Customer within sixty (60) calendar days of the date that Seller disconnects the Disconnected Circuit; (iv) the total number of Disconnected Circuits does not exceed ten percent (10%) of Customer's total active circuits ordered pursuant to this Service Exhibit or five (5) total circuits, whichever is greater, in any twelve-month period; (v) Seller determines, in its sole discretion, ***that adequate capacity is available*** on the Seller Domestic Network for the Replacement Circuit; and (vi) Customer is not otherwise in breach of the Agreement. If Customer fails to meet any of the above criteria for any particular Disconnected Circuit, Customer shall pay all applicable shortfall charges and early termination fees for such Disconnected Circuit. Customer shall also be liable for any third party expenses which are charged to Seller by a third party as a result of any Disconnected Circuit.

(b) To exercise this portability option, Customer must, ***within 30 days of the date that Customer cancels the Disconnected Circuit, notify Seller in writing of its intent to replace a Disconnected Circuit with a Replacement Circuit***. Seller shall credit Customer all applicable shortfall charges and early termination fees charged to Customer's account, ***provided that Customer orders the Replacement Circuit within sixty (60) days of the Disconnected Date***. Notwithstanding the foregoing, in no event shall Customer be credited back for any On-Net Circuit installation NRCs or any third party expenses that were charged for any Disconnected Circuits which Customer cancels pursuant to this Section.

<u>Example Clause #11</u>: The following clause would not allow for an interruption of service revenues. A replacement service must be activated at the same time that the first service terminates.

On sixty (60) days prior written notice, Customer may change one or both of the endpoints of an existing Service that has been installed and billing for at least twelve (12) months during the Service Term without incurring early termination fees if the following conditions are met:

(a) The existing Service must be replaced ***simultaneously*** by a new Service with a term that is equal to or greater than the original term of the existing Service; and
(b) The new Service's MRC must be equal to or greater than those associated with the existing Service, and a new NRC may apply, and
(c) The new Service must be On-Net, unless mutually agreed; and
(d) No more than a total of two (2) or ten percent (10%) of all Services installed

under this MSA, whichever is greater, may be ported during a calendar year. Within five (5) business days of receiving Customer's written notice of its intent to change endpoints, Seller will provide to Customer a Service Order Form that will include the new endpoints and the new NRC and new MRC for the new Service. If Customer wishes to port the Service, Customer will execute and return the Service Order Form to Seller within five (5) business days of receipt. If Customer does not return the Service Order Form, the existing Service will continue uninterrupted. If Customer does not port the Service, and chooses to terminate the existing Service, then applicable early termination fees will apply as described herein.

Example Clause #12: The following clause is even more explicit in a requirement that there can be no interruption of service or service revenues. The term of the original service must continue with the Replacement Circuit.

As in some of the earlier portability clauses, in order for a customer to exercise a right to port a service, the Provider must determine that there is sufficient network capacity available for the Replacement Circuit "in its sole discretion." The effect of such a provision is to put great discretion in the Provider to determine whether a customer will be permitted to port a service.

Upon thirty (30) days prior written notice, Customer may terminate an existing On-Net circuit during the initial Service Order period (a "Terminated Circuit") without incurring any early termination charges so long as: (i) Customer replaces the Terminated Circuit with another On-Net circuit having an equal (or greater) monthly recurring charge as the Terminated Circuit (the "Replacement Circuit"); (ii) ***the Replacement Circuit is ordered by Customer under a binding Service Order within thirty (30) calendar days of the notice of intended termination*** of the Terminated Circuit and the initial Service Order period of the Terminated Circuit ***continues with the Replacement Circuit***, provided the remaining Service Order Period is at least (12) twelve months, and if it is less than (12) twelve months, it will be deemed to be (12) twelve months; and (iii) there is sufficient ***network capacity available*** for the Replacement Circuit ***in Provider's sole discretion***. After Customer submits its request for a Replacement Circuit that is acceptable to Provider, Provider will send Customer a new Service Order that will list the new endpoint locations and new MRC and NRC. If the new Service Order is acceptable to Customer, Customer will execute and return to Provider and provisioning of the Replacement Service will commence. If no Replacement Service is ordered pursuant to this paragraph, Customer shall pay all applicable early termination charges for the Terminated Circuit as set forth herein.

Upon thirty (30) days prior written notice, Customer may terminate an existing Off-Net circuit during the initial Service Order period Customer will be liable for any termination charges passed through by the Third Party provider, if any. Customer and Provider will work together to mitigate such charges.

Upon thirty (30) days prior written notice, Customer may terminate an existing On-Net Services - Facilities Construction Required circuit during the initial Service Order period. Customer will incur an early termination charge as set forth in Section ___. In addition,

Customer will reimburse to Provider actual, documented costs for construction of such On-Net Services – Facilities Construction Required. Customer and Provider will work together to mitigate such charges.

Part 4, Chapter 2 - On-Net, Off-Net, and Near-Net Services

A carrier may provide lit fiber services using its own network entirely, or may use the network of another carrier in whole or in part. As we have noted in other chapters, outage credits, service activation intervals, and other service terms may vary depending on whether a service is offered on-net or off-net.

The distinction between on-services and off-net services can sometimes be confusing because a carrier will often lease facilities from another carrier and then integrate those facilities into its own network. For example, if a carrier leases dark fibers from another carrier and then integrates those fibers with its owned dark fibers, the entire route, including the leased fibers, are usually treated as the carrier's network. Therefore, if a service is intended to be treated as partially or entirely off-net for any reason, it is important to make that clear in the service contract.

Note that in Example Clause #3 below the carrier did not make that distinction. It is not clear whether "entirely on Seller's network" includes any facilities that have been leased by the Seller and integrated into the Seller's network. This question can be addressed simply by specifying in the applicable service contract that the Service or a segment of the Service is off-net. Otherwise it will probably be presumed that the entire Service is on-net.

Example Clause #1: In the first example, On-Net and Off-Net are defined by reference to the termination points for the service. If both end points are on the Seller's network, the service is an On-Net service. If either or both of the termination points are not on the Seller's network, the service is an Off-Net service.

> "On-Net" means service that originates from and terminates ***to locations*** that are both ***on the Seller network***.
>
> "Off-Net" means a service that originates from or terminates ***to a location*** that is ***not on the Seller network***.

Example Clause #2: The following example adopts a different concept to define On-Net and Off-Net services. On-Net and Off-Net are defined by reference to the provider that owns the services. Therefore, if the Seller is reselling to Customer services of another carrier, these would be considered Off-Net Services.

> "On-Net" means any telecommunications ***capacity and related ancillary Services owned, operated and provided by Provider***.
>
> "Off-Net" means any telecommunications capacity and related ancillary ***services provided by a third party vendor ordered by Provider on Customer's behalf***.

Example Clause #3: If services are ordered from Seller but provided by means of the facilities of another carrier, Off-Net Services are obtained by the Seller in its own name, rather than ordered "on Customer's behalf."

"On-Net" or "On-Net Service" shall mean Service that is provisioned entirely on Seller's Network.

"Off-Net" or "Off-Net Services" shall mean Service that is provisioned in part or wholly on a third party's network and where ***Seller is the customer of record*** with such third party. In the instances where Seller is able to provide Off-Net Service, the pricing of such Service shall be on an individual case basis.

Example Clause #4: The following clause defines On-Net by reference to locations that are connected to the Carrier's network.

On-Net. "On-Net" shall refer to ***Premises which are served by the Carrier Network*** and is also referred to as "Type I." Service provided to an On-Net location utilizes solely the Carrier Network.

Off-Net. "Off-Net" shall refer to ***Premises which are not served by the Carrier Network*** but are in an On-Net city. Service provided to an Off-Net location utilizes the network of both the Carrier and a competitor of Carrier's choice.

Example Clause #5: In the following example, On-Net and Off-Net are defined by reference to the locations that are connected to Provider's network. An On-Net Service is one for which both termination points are served by Service Provider's Network. This clause raises an additional feature. If Customer wishes to purchase a service to a location that is not on Provider's network, the Provider may provision that service by means of another carrier that already has facilities to the new termination point, or Seller may extend its own network to the new location on an "individual case basis."

On-Net Services. "On-Net Services" shall mean those Services which connect ***two (2) locations*** served by Service Provider's Network. On-Net Services are provided entirely by Service Provider.

Off-Net Services. "Off-Net Services" are those where ***one or both locations*** to be connected are ***not served by Service Provider's Network***. Off-Net Services have a portion of the Services provided by another local access provider. In the instances where Service Provider is able to provide Off-Net Services, the terms, conditions and pricing of such Service will be provided on an individual case basis ("ICB").

Example Clause #6: The following clause sets forth a very narrow definition of On-Net Services. A Service is On-Net only if it can be terminated at a collocation space to which the Seller has already established a connection and the Customer's network has already been terminated to this location. Any other Service would be treated as an Off-Net service.

On-Net. "On-Net" shall mean Services which (a) have fiber-optic cable, at a minimum, ***terminated from Service Provider's collocation space*** on a particular floor in a particular building to the Service Provider's demarcation point; ***and*** (b) the Customer and/or End User is also located in this same collocation space on the same floor in the same building, and shall have the ability to Cross-Connect to Service Provider's Network or Service Provider otherwise has the right to Cross-Connect without the necessity of incurring Construction Costs ("On-Net Location"). Service Provider shall provide On-Net Services to Customer in accordance with Section ___.

<u>Off-Net</u>. "Off-Net" shall mean Services which involve a ***building/facility that is not connected*** to a Service Provider Distribution Network ("Off-Net Location"). In the case of an Off-Net Service, Service Provider shall decide, at its sole option, whether or not to elect to build facilities to an Off-Net Location in order to provide Customer and/or End Users with Services. A Site Assessment is required for Off-Net Services. Under no circumstances shall Off-Net Services designate that an underlying Access Service Provider be utilized to provide such Service.

<u>Example Clause #7</u>: The following clause introduces an additional concept, "Near-Net" building. A Near-Net building is one that is not On-Net, but adjacent to which the Supplier has built network.

Further, rather than relying on a definition of On-Net buildings, this clause provides for a list of buildings that the parties agree are On-Net or Near-Net; all other buildings are Off-Net.

"*On-Net Building*" means a building into which Supplier has installed a fiber patch panel or fiber optic cable connecting from Supplier's Local Access Grid to either a Supplier-owned fiber patch panel or to the minimum point of entry or other point of interconnection in the building as agreed by the Parties pursuant to Section __ (Activities at Seller Serving Office).

"*Off-Net Building*" means a building that is not passed by or adjacent to Supplier's Local Access Grid and for which physical construction (*i.e.,* building fiber from a non-adjacent point) will be required. Supplier's On-Net Buildings and Near-Net Buildings as of the Effective Date are listed on Schedule __ (On-Net Buildings) and Schedule __ (Near-Net Buildings) of this Agreement.

"*Near-Net Building*" means ***a building passed*** by or adjacent to Supplier's Local Access Grid, where Supplier has installed fiber to the property line or a manhole or handhole just outside the property line for such building or for adjoining property(ies).

<u>Example Clause #8</u>: The following clause utilizes the concept of "Near-Net." However, the Seller reserves the right to determine, in is sole discretion, which building are Near-Net.

"Near-Net Services" means Services provided to a location that does not have On-Net Services available to it but is *near* the Seller Network, ***as determined in Seller's sole discretion***.

"Network" means either Seller or Customer's proprietary leased or owned network, as the context of the provision requires or as contemplated herein;

"Off-Net Services" means Services provided to a location that does not have either Near-Net or On-Net Services available to it. All Off-Net Services are offered on an individual case basis;

"Seller POP" means a location on the Seller Network designated in the Seller POP list as may be amended from time to time;

"On-Net Services" means Services provided to a location that is connected to the Seller Network, as determined in Seller's sole discretion.

Example Clause #9: The following is another example of a clause that utilizes a very narrow definition of "On-Net." A service is On-Net only if both the Service Provider and the customer (or end user) are already in the same collocation space.
This clause employs two definitions for "Near Net Services." One definition of Near-Net applies to services provided within "collocation type buildings." In this type of service, "Near-Net" means that both the Service Provider and the customer are in the same building but not in the same collocation space. The second definition applies to "enterprise type buildings." In this type of service "Near-Net" means that the Service Provider's network is adjacent to the building in which the customer is located.

On-Net Services. "On-Net Services" shall mean Services which (a) have fiber-optic cable, at a minimum, terminated from Service Provider's collocation space on a particular floor in a particular building to the Service Provider's Point of Demarcation; and (b) the ***Customer and/or End User is also located in this same collocation space*** as Service Provider, on the same floor in the same building, ***and shall have the ability to Cross-Connect to Service Provider's Network or Service Provider otherwise has the right to Cross-Connect without the necessity of incurring Construction Costs***. Service Provider shall provide On-Net Services to Customer in accordance with Section ___. The agreed On-Net Service locations are set forth in Exhibit ___ that is attached hereto and incorporated herein by this reference. Service Provider shall, from time to time, in its discretion, amend Exhibit __ to set forth buildings that are connected to the Seller Network. Service Provider shall provide Customer with an updated electronic version of Exhibit __ on a quarterly basis.

Off-Net Services. "Off-Net Services" means Services provided to a location that does not have either Near-Net or On-Net Services available to it. All Off-Net Services are offered on an individual case basis.

Near-Net Services. "Near-Net Services" ***for collocation type buildings*** means Services that (a) have fiber-optic cable or Category 5 cable, at a minimum, terminated from Service Provider's collocation space on a particular floor in a particular building to the Service Provider's Point of Demarcation; and (b) ***the Customer and/or End User is located in the same building but not located in this same collocation space*** as Service Provider, on the same floor in this building; and (c) the Service Provider would have to incur Construction Costs in order for Customer to connect to Service Provider's Network ("Near-Net Location"). "Near-Net Services" ***for enterprise type buildings*** means Services that (a) Service Provider has fiber optic cable located ***adjacent to such building***; and (b) the Service Provider would have to incur Construction Costs in order for Customer and/or End User to connect to Service Provider's Network ("Near-Net Location"). A Site Assessment is required for Near-Net Services. Service Provider shall provide Near-Net Services to Customer in accordance with Section ___. The agreed Near-Net Service locations are set forth in Exhibit __ that is attached hereto and incorporated herein by this reference.

Example Clause #10: The following clause introduces an additional concept, that of a "Minimum Point of Entry" or "MPOE." A Minimum Point of Entry is a location at a point of entry to a building, where the Seller's network first penetrates a building. In this clause, services provided to a building are considered On-Net if Seller's network extends to the building's MPOE.

"Near-Net Services" shall mean Services which involve a building located adjacent to an Seller distribution Network, but not considered an On-Net building.

"Network" means either Seller or Customer's proprietary leased or owned network, as the context of the provision requires or as contemplated herein.

"Off-Net Services" shall mean Services which involve a building/facility that does not have either Near-Net or On-Net Services available to it. All Off-Net Services are offered on an individual case basis.

"On-Net Services" shall mean Services which involve a Near-Net building, as defined above, which has fiber-optic cable, at a minimum, terminated from its Minimum Point of Entry ("MPOE") to the serving Seller hub.

Example Clause #11: In contrast to the previous example, in the following clause if the Seller's network extends only to the MPOE, the building is treated as an Off-Net building.

Where Private Line Service is being terminated Off-Net at the Customer Premises through an Off-Net Local Loop to be provisioned by Seller on behalf of Customer, the charges set forth in the Customer Order for such Private Line (Off-Net) Service assumes that such Private Line ***(Off-Net) Service will be terminated at a pre-established demarcation point or minimum point of entry (MPOE)*** in the building within which the Customer Premises is located, as determined by the local access provider. Seller may charge Customer additional non-recurring charges and/or monthly recurring charges not otherwise set forth in the Customer Order for such Private Line (Off-Net) Service where the local access provider determines that it is necessary to extend the demarcation point or MPOE through the provision of additional infrastructure, cabling, electronics or other materials necessary to reach the Customer Premises. Seller will notify Customer of any additional non-recurring charges and/or monthly recurring charges as soon as practicable after Seller is notified by the local access provider of the amount of such charges.

<u>Example Clause #12</u>: The following clause uses terms that are similar to the several of the previous such clauses. On-Net is defined very narrowly. A Service is On-Net only if both the Service Provider and the Customer are located within the same collocation space.

<u>Definitions</u>

<u>On-Net Services</u>. On-Net Services shall mean Services which (a) have fiber-optic cable, at a minimum, ***terminated from Service Provider's collocation space*** on a particular floor in a particular building to the Service Provider's Point of Demarcation; ***and*** (b) the ***Customer and/or End User is also located in this same collocation space*** on the same floor in the same building, and shall have the ability to Cross-Connect to Service Provider's Network or Service Provider otherwise has the right to Cross-Connect without the necessity of incurring Construction Costs. Service Provider shall provide On-Net Services to Customer in accordance with Section ____.

<u>Near-Net Services</u>. "Near-Net Services" shall mean Services that involve ***a building located adjacent to a Service Provider distribution Network, but not considered an On-Net*** building. As it pertains to Ethernet Services, Near-Net Services shall mean Ethernet Services that (a) have fiber-optic cable or Category 5 cable, at a minimum, terminated from Service Provider's collocation space on a particular floor in a particular building to the Service Provider's Point of Demarcation; and (b) the Customer and/or End User is located in the same building but not located in this same collocation space on the same floor in this building; and (c) Service Provider would have to incur Construction Costs in order for Customer to connect to Service Provider's Network ("Near-Net Location"). A Site Assessment is required for Near-Net Services. Service Provider shall provide Near-Net Services to Customer in accordance with Section ___.

<u>Off-Net</u>
An "Off-Net" building is a building/facility that is not adjacent to a Service Provider distribution Network.

<u>Example Clause #13</u>: The following example sets forth simple definitions of On-Net, Near-Net, and Off-Net. However, this clause incorporates detailed installation intervals with these definitions. There are three categories of installation intervals, one of each of the definitions. And this clause allows for service credits if the Service Provider fails to meet an installation interval commitment.

<u>Definitions</u>

<u>On-Net</u>
An "On-Net" building is a "Near-Net" building (as defined below) which has fiber-optic cable, at a minimum, terminated from its Minimum Point of Entry ("MPOE") to the serving Service Provider hub.

<u>Near-Net</u>

A "Near-Net" building shall be defined as a building located adjacent to a Service Provider distribution Network, but not considered an On-Net building.

Off-Net
An "Off-Net" building is a building/facility that: is not adjacent to a Service Provider distribution Network.

Installation Interval Objectives: Service Provider's ***Installation Interval Objectives for all buildings/facilities are shown in the tables below***. These intervals begin with the receipt of a complete and accurate ASR. The interval is complete when Service Provider has finished its Service activation process including the Service Provider acceptance testing to be done jointly with Customer, at Customer's option. The installation intervals are exclusive of any acceptance testing that the End Users may require.

Table 5 – On-Net Service Delivery Interval

	Seller Building Installation Interval Objectives				
	Add/Change Orders		Inside Plant Construction Required*		
	Additional Service to Existing End User[1]	Change Bandwidth of Existing End User[2]	Basic Inside Plant Work[3]	Extensive Inside Plant Riser Work[4]	
Interval Components					
Site Assessment	N/A	N/A	N/A	N/A	
Firm Order Commitment (FOC)	2 Day	2 Day	5 Day	5 Day	
Installation	N/A	N/A	5 Days	20 Days	
Service Provider Turn up **(Per agreed upon testing procedures)**	5 Days	5 Days	5 Days	5 Days	
Total	**7 Days**	**7 Days**	**15 Days**	**30 Days**	

* Where inside plant construction is required, Service Provider may [based on the complexity of the build] adjust service delivery intervals prior to the issuance of an FOC date. [1]Add Orders - Additional service in the same subscriber location of a building/facility that already has one or more of Service Provider's Services terminating in said building/facility and the Customer wishes to terminate a new service of the same or different type in the same building/facility. [2] Change Orders – changing bandwidth of an existing Service occurs when a Customer already has one of Service Provider's Services terminating in an applicable building/facility, but it wishes to change the bandwidth of the Service terminating in that building/facility without changing the type of Service. [3] Basic Inside Plant Only Work --- Fiber has been or can easily be run from the Point of

Demarcation to the telecommunications closet serving the floor or area where the termination is located. Fiber must then be run from this telecommunications closet to the actual termination point. [4] Extensive Inside Plant Riser Work --- Service Provider will need to perform significant work to either construct riser space or to pull fiber through existing, congested riser space in order to connect the Point of Demarcation to the telecommunications closet serving the floor or area where the termination is located. Fiber must also be run from this telecommunications closet to the actual Point of Demarcation.

Table 6 – Near-Net Service Delivery Intervals

	Conduit Lateral/No Fiber[1]	New Lateral Build Required[2]	Extensive Lateral Build Required[3]
Interval Components			
Site Assessment	4 Days	4 Days	4 Days
Firm Order Commitment (FOC)	1 Day	1 Day	1 Day
Installation	35 Days	45 Days	60 Days
Service Provider Turn up **(Per agreed upon testing procedures)**	5 Days	5 Days	5 Days
Total	**45 Days**	**55 Days**	**70 Days**

[1] Conduit Lateral to Building (No Fiber) - Service Provider has a conduit system adjacent to the building/facility and Service Provider has conduit available from the property line to the building/facility. Fiber must be pulled from the Service Provider system, through the conduit lateral, to the building. [2] New Lateral Build Required - Service Provider has a conduit system adjacent to the building/facility, but Service Provider does not have conduit available from the property line to the building/facility. [3] Extensive Lateral Build Required - Service Provider has a conduit system adjacent to the building/facility, but significant work may be required to build conduit to the building/facility Point of Demarcation.

Table 7 – Off-Net Service Delivery Intervals

	Off-Net Building Installation Interval Objectives	
	Building in Service Provider Jurisdiction1	Building Not in Service Provider Jurisdiction2
Interval Components		
Site Assessment	ICB	ICB
Firm Order Commitment (FOC)	10 days	ICB
Installation	ICB	ICB
Service Provider Turn up **(Per agreed upon testing procedures)**	ICB	ICB

Total	ICB	ICB

ICB – Individual Case Basis [1] Building in Service Provider Jurisdiction -The building/facility requiring connectivity is located within a jurisdiction where Service Provider offers Service, but the building/facility is not adjacent to existing Service Provider Network facilities. [2] Building Not in Service Provider Jurisdiction - The building/facility requiring service is not located within a jurisdiction where Service Provider offers connectivity.

Conditions

The above tables represent Service Provider's Installation Interval Objectives. Service Provider's ability to meet these intervals is contingent on many variables, including the Customer's completion of prescribed tasks as well as specific assumptions regarding the installation process. These conditions include:

- City/State/Region has no jurisdiction over private property (requires no permitting in this environment).
- As required, jurisdiction permit cycle takes no longer than 30 calendar days.
- Complete and signed Tenant Request for Building Entrance and Owner/Management Agency contact information received, when requested by Service Provider for Customer to provide when no access agreement is currently in place and Service Provider is having issues with the building owner and therefore soliciting Customer's assistance with this matter.
- Service provider has obtained building access rights from the building owner.
- Maximum distance from an end building to a Service Provider hub will be no longer than 15 route miles.
- Rack space, accessibility, and power (if necessary) are available in the building/facility to be installed at the time the order is received.
- Optical handoff at the building/facility to be installed is assumed.
- All intervals indicated are in calendar days.
- Excluding Services that are being ported in accordance with Section ___ in this Agreement, Add/Change orders must terminate in the same location/facility as the original ASR.

The FOC represents Service Provider's committed date for the Installation Interval SLA. Actual Service delivery intervals committed at FOC may be different from standard intervals if any above conditions for installation are not met.

Credit Structure

For any Service installation in which Service Provider fails to meet the Installation Interval SLA Objectives, the following credit structure will be applied to the Nonrecurring Installation Charges (NRCs) associated with the Service Elements(s) affected:

Table 8 – Service Delivery Credit Structure

Days Exceeding Installation Interval Commitment	Credit Structure (% Of NRC Credited)
4 or more Business Days	100%

The Service credit structures provided for in this Appendix No. __ represent the Customer's exclusive remedies for any failure of the Services to conform to the specifications/provisioning intervals set forth herein. The Service credits, if any, accruing hereunder shall be applied against: (a) the MRCs invoiced to the Customer for the affected Services, and shall not exceed, in the aggregate, the total MRCs invoiced during that month for the affected Services; and/or (b) the nonrecurring charges invoiced the Customer for the affected Services, and shall not exceed, in the aggregate, the total NRCs invoiced during that month for the affected Services.

Part 4, Chapter 3 - Most Favored Treatment

Although clauses that extend most favored treatment to a customer are not part of a seller's standard forms or service offerings, it is not uncommon to find that for certain customers a seller is willing to extend favorable treatment. This is sometimes referred to as a "most favored nation" status or an "MFN" clause.

Exactly what favored treatment will consist of in a given instance will be the subject of careful negotiations. Sometimes most favored treatment will mean that a customer will be entitled to the very best pricing for a particular service. Other times most favored treatment will encompass other terms as well. In any event, an MFN clause will always present many questions for the parties to consider.

If most favored treatment extends to all of the terms of a service, this can present challenges when making comparisons between different service offerings. Whether terms are more favorable can be very subjective. And perhaps a seller and a customer will agree to modify many terms of a standard service offering, some in a manner more favorable to the seller, and some in a manner more favorable to the customer. When comparing the contract terms for purposes of an MFN clause, should all of the changes be considered together? If so, would a customer benefiting from an MFN clause be entitled to the superior terms only if they also accept the less desirable terms? Or would the customer benefiting from an MFN clause be entitled to choose selectively among the more favorable terms?

Does an MFN clause apply only to services of comparable terms that are part of a final contract entered into with another customer? Or does the MFN clause apply to all terms offered to another prospective customer irrespective of whether such offer leads to a final contract.

Also, what are a preferred customer's remedies if it is later found that a term offered to another customer is arguably more favorable? Does an MFN clause expose the seller to a claim arising perhaps several years later?

If most favored treatment applies only to pricing, the questions are less complicated. But important questions remain. What services are comparable? Will an MFN clause apply to services sold to all other customers? Or will service offerings made to certain customers, such as large volume purchasers, be excluded from comparison? Here is an example of one attempt at a definition.

> Vendor Warranty. Vendor warrants that the rates and charges are either the result of competition or set by tariff or regulation. Vendor warrants that the rates charged for the Services are the lowest prices charged by Vendor to ***buyers of the same class*** as Customer under ***conditions similar to those specified in this Agreement***, and comply with applicable government regulations in effect at the time of quotation or sale.

A clause such as the foregoing is problematic. It does not explain what other customers are of the same "class." And it does not explain what other "conditions similar" might be pertinent for purposes of an MFN comparison. Because of these omissions, a clause such as this may lead to confusion.

Example Clause #1: In the following clause the parties have limited the application of the MFN clause to pricing only. And the application of the MFN clause is limited to one year, not the entire term of the contract. Furthermore, the type of service offering that may be used for an MFN comparison is highly qualified. Customer's service may be compared only to other services of the same or shorter term, the same route mileage, the same bandwidth, the same end points, and the same latency specification. An MFN clause such as this would be less likely to lead to confusion or disputes.

> ***For a period of one (1) year*** following the Agreement Effective Date, Supplier will treat Customer as a Most Favored Customer. "Most Favored Customer" means that if, at any time within one (1) year following the Agreement Effective Date, Supplier provides a Comparable Service to any other customer or customers, and the average monthly charge for a Comparable Service is less than the average monthly charge to Customer for the same Service, Supplier shall be entitled to receive a Service credit calculated and payable as described below.
>
> For purposes of this Section ____, "Comparable Service" means a Service provided to other customers that has the ***same or a shorter term, the same route mileage, the same bandwidth or number of dark fiber pairs, the same end points, and the same latency specification*** as the Services provided to Customer.
>
> The amount of the credit shall be the difference between:
>
> (1) The average monthly charge (inclusive of lease fees and fees for collocation and O&M, but exclusive of service and other credits) (the "AMC") for any single Comparable Service multiplied by the number of months of the term of the Comparable Service; and
>
> (2) The AMC for the Service provided to Cusotmer multiplied by the number of months of the term of the Comparable Service.
>
> If more than one Comparable Service would give rise to a credit under this Section, Customer would be entitled to a credit for only the one Comparable Service that would result in the largest credit.
>
> Twelve (12) months after the date of Customer Acceptance, Supplier agrees to review in good faith the pricing for all agreements entered into for Comparable Services during the previous year. Supplier agrees complete the review within thirty (30) days. The credit would be issued to Customer in equal monthly installments beginning with the next subsequent monthly invoice and monthly invoices thereafter until the credit is satisfied. A credit under this Section will be in addition to any Outage Credits to which Customer may be entitled.

Example Clause #2: In contrast with the previous example, the following clause is much broader and more general. This MFN clause pertains not only to pricing, but to all terms and conditions that are "more favorable."

This clause also raises a question about a Supplier's potential liability under an MFN clause. The Supplier has agreed to offer better terms to this Customer whenever Supplier offers better terms to any other customer. If Supplier offers better terms to another customer, this contract will be "deemed amended" to include the more favorable terms. If the Supplier fails to amend the contract in the manner described, that would likely constitute a breach of the contract. The Supplier would be liable for contract damages running from the date that this contract would be "deemed amended."

A similar result would arise from any MFN clause in which an adjustment of the pricing or other terms would be effective automatically and without any other action of the parties. A Supplier might be unaware of a liability that is accruing for an extended period of time. This concern is magnified by the uncertainty surrounding terms such as "more favorable terms and conditions" and "similarly situated customer."

> Supplier will treat Customer as Supplier's most favored customer ("MFC"). "MFC" means a customer(s) of Supplier who receives ***terms and conditions*** (including pricing terms) that are ***more favorable*** than those received by any other similarly situated Supplier customer. If Supplier offers more favorable terms and conditions to any other customer than are offered to Customer under this Agreement and/or under any Schedule or any pricing/discount arrangement between Customer and Supplier, ***then Supplier will concurrently extend those terms and conditions to Customer***, and this Agreement and all Schedules and/or any pricing/discount arrangement, at Customer's option, ***will be deemed amended*** to provide those terms and conditions to Customer. Any amounts charged to Customer in excess of prices and fees offered by Supplier to any other ***similarly situated customer*** will promptly be refunded or credited to Customer by Supplier at Customer's option.

Example Clause #3: The following clause was written very simply. However, this clause also illustrates uncertainty that can result from lack of clarity. The parties failed to explain how to recognize a "similarly situated customer" or a "similar services." Is a similarly situated customer one that purchases the same service and the same bandwidth? Or is a similarly situated customer one that has made a total order commitment similar to this Customer?

What is a similar service? Is that only an identical service? Or is a service that resembles in some respects but not others as similar service?

> The rates, charges, volume commitments (if any), and early termination charges specified in the applicable Service Order or this Agreement shall not be increased during the Term. Furthermore, Vendor agrees to review with Customer in good faith the rates and charges for the Services, and to reduce the rates, including the rates on the installed circuits on a prospective basis, if the rates are either higher than the rates charged by Vendor to other

similarly situated customers as Customer ***purchasing similar services*** in the same service territories.

__**Example Clause #4**__: In the following clause, the Provider introduces new pricing by making an announcement of revised pricing. Revised pricing will be applied immediately to all of Customer's Service Orders.

PRICE AND PRICE REVISIONS

The prices for all Service ordered by Customer are specified in Exhibit __, unless this Section __ applies.

Prices shall not increase during the Term of this Agreement except as set forth in Exhibit __. Any price decrease or discount increase shall be ***effective immediately*** upon ***announcement by Provider*** and shall apply to all Service Orders that have not been invoiced for payment by Provider. In addition, Customer shall receive ***credit or refund, at Customer's option, within thirty (30) days, for the difference between the price paid by Customer and the reduced price for all affected Services***.

Under special circumstances, Customer may purchase Service pursuant to the terms of this Agreement at prices and discounts quoted by Provider that are more favorable than those prices and discounts set forth in Exhibit __. Such special circumstances include, but are not limited to, (a) competitive allowances and (b) Provider promotional offers.

Quotations provided by Provider shall be valid for ninety (90) days from their date unless otherwise mutually agreed.

__**Example Clause #5**__: In the following clause, the Provider and the Customer have adopted a process for a Competitive Review of Provider's pricing and other terms and conditions. Changes to service offerings are implemented once Provider and Customer read agreement on new pricing and terms of service.

MORE FAVORABLE PRICING

If Provider offers more favorable ***prices, terms, conditions, warranties, or other benefits*** to ***any other commercial customer***, for ***like volumes and quantities***, during the term of this agreement, then, at the option of customer, this Agreement and any Service Order affected thereby will be ***modified to include such more favorable prices, terms, conditions, warranties or benefits*** ("More Favorable Pricing").

COMPETITIVE REVIEW

The parties shall review in good faith the prices for Services as set forth in Exhibit __ to this Section ("__Competitive Review__"), including any review of More Favorable Pricing terms, biannually or as agreed to by the parties.

Customer shall submit the initial Competitive Review request to Provider no sooner than six (6) months from the Effective Date. Subsequent Competitive Reviews may be requested by Customer six (6) months following the initial or any subsequent Competitive Review.

The parties agree to complete each Competitive Review within thirty (30) days from Provider's receipt of the Competitive Review request.

The More Favorable Pricing provisions of Section ___ of this Agreement will apply to prices for the same Services and in the same geographic markets as the Services Customer ordered (as described in Section __).

Adjustments to ***prices on a going forward basis***, if any, as a result of a Competitive Review must be specifically agreed to in writing by the parties and will go into effect the next month after the Competitive Review is completed.

Example Clause #6: The following clause resembles a portability clause in that if the Seller offers more favorable pricing to another customer, this Customer's recourse is merely to terminate its contract and at the same time replace it with a new order that contains more favorable terms. The more favorable terms would be either available in Seller's tariffs, which are published by Seller and available to anyone, or would result from improvements in technology.

The right to substitute a new agreement on more favorable terms would be limited to those terms offered by Seller to a Similarly Situated Customer. And "Similarly Situated Customer" is defined in a very limited way.

> ***Customer may***, at its option, ***terminate this Agreement*** without penalty if (a) ***Customer replaces this Agreement*** with an agreement entered into between Seller and a Similarly Situated Customer for comparable volumes and services, as set forth in Seller's tariffs filed with the Federal Communications Commission (the "FCC") or a state regulatory commission, and with a Term at least equal to that remaining on this Agreement at the time of termination ***or*** (b) if the introduction of a ***new technology decreases rates for a Similarly Situated Customer*** and Customer replaces this Agreement with an agreement entered into between Seller and a Similarly Situated Customer for comparable volumes and services, **as set forth in Seller's tariffs filed with the FCC** (the "FCC") or a state regulatory commission, and with a Term at least equal to that remaining on this Agreement at the time of termination. For purposes of this section, a "***Similarly Situated Customer***" is any customer of Seller which has executed a contract for toll free inbound and outbound intrastate, interstate, and international telecommunications services originating or terminating over dedicated access facilities with a ***term of at least three (3) years*** and a ***Revenue Level Commitment that is plus or minus five (5) percent of that contained in this Agreement***. It shall be Customer's responsibility to request termination of the Agreement pursuant to this Section ____.

<u>Example Clause #7</u>: The following clause sets forth a very broad and general standard for comparison of pricing and other service terms. In addition, the Vendor's Chief Financial Officer must certify each calendar quarter that Customer has been given the most favorable pricing and other terms of service.

> <u>Most Favored Customer</u>. Vendor shall not offer or provide services that are comparable to Services described in the attached Schedules and provided to any Vendor's non-affiliate customer which purchases such services from Vendor (a) at ***prices that are lower*** than those charged to Customer pursuant to this Agreement, ***or*** (b) pursuant to ***terms and conditions more favorable*** than the terms and conditions in this Agreement. In the event that Vendor fails to comply with its obligations as described in this Section, then pricing under this Agreement and related SOFs shall be adjusted such that they are equal to or less than corresponding prices offered to any other non-Affiliate customer (1) ***on a going-forward basis*** for all future applicable purchases under this Agreement, ***and*** (2) ***retroactively*** to all applicable purchases previously made back to the date such prices were first offered to such other non-affiliate customer, and the resulting adjustment in charges shall be provided to Customer, at Customer's option, as either a credit for future purchases or in the form of a lump-sum payment. Upon the request of Customer during the term of this Agreement (which shall not be made more frequently than once per a calendar quarter), ***Vendor's Chief Financial Officer*** ("Supplier's CFO") ***will certify in writing*** whether Vendor has satisfied its obligations set forth in this Section.

<u>Example Clause #8</u>: The following clause sets forth pricing adjustments under two circumstances. First, if Carrier offers a lower price to any other customer that purchases similar services in the same or similar terms, this Customer is entitled to that lower price.

Second, if after the first three years of the term of the service agreement Customer receives an offer from another telecommunications provider on more favorable pricing, Carrier must offer the same lower price to this Customer. If Carrier fails to do so, the Customer may terminate this service contract.

> Section ___ Reductions in Price.
>
> (a) Carrier ***represents and warrants*** to Customer that the unit ***prices*** listed on <u>Exhibit __</u> are and shall be equal to or lower than the aggregate ***prices or rates charged***, after discount, ***to any of its other customers for similar products or services in like or comparable (or less) quantities*** under ***similar*** (or less favorable to the customer) ***terms and conditions***. To the extent that Carrier offers rates to other comparable (or smaller-volume) customers that are more favorable to such customers than the rates set forth in <u>Exhibit __</u>, Carrier shall be obligated to provide Telecommunications Services to Customer based upon such ***favored rates*** (which shall include a retroactive adjustment in the rates from the date such rates were provided to such other customers). For purposes of this <u>Section __</u>, agreements involving leases or indefeasible rights of use granted respecting dark fibers shall not be considered comparable to this Agreement. Customer shall have the right, no more frequently than ***once annually, to request an audit*** of any and all information and documents possessed by Seller the inspection of which is

reasonably required to confirm the accuracy of the warranty contained in this section. All audits shall be performed by an independent accounting firm selected by Customer (and reasonably acceptable to Carrier) on no less than seven (7) business days' advance written notice, and the costs of retaining such accounting firm shall be paid by Customer. After the audit, the independent accounting firm shall be instructed to provide Customer with a written report stating whether any adjustment in the rates charged to Customer is warranted under this Section ___, and, if so, what rates are required to satisfy the requirements of this Section ____ and the date upon which such rates should be applicable.

(b) Carrier agrees that, if at any time after the third anniversary of the Agreement Date, ***Customer secures a firm bid*** for the delivery of telecommunications services which:

(1) is for telecommunications services of equal capacity and geographic scope as the Telecommunications Services then being delivered to Customer hereunder;

(2) has been provided by a reputable provider of such services; and

(3) contains a rate which is less than the rate set forth in Exhibit __, then

Carrier shall adjust the per unit price set forth in Exhibit __ to match the rates offered by such reputable provider. In the event that Carrier does not, within fifteen (15) days of receipt of a request from Customer, adjust its rates to match the rates offered by such reputable provider, and Customer is able to secure a contract from such reputable provider for services equal to or lower than the rates contained in the firm bid, then ***Customer shall be permitted to terminate this Agreement*** in its entirety without any further liability to Carrier, and without payment of any Termination Charge.

Example Clause #9: In the following clause, the Customer is entitled to the benefit of more favorable pricing calculated as the lesser of a discounted value of Seller's list prices, or the most favorable pricing that Seller offers to any other customer.

RATES AND CHARGES

(a) Rates (the "Rates") and charges for Services (collectively, the "Charges") shall be set forth in the Exhibits and shall be calculated in accordance with this Agreement. The Rates and Charges shall be expressed in U.S. dollars, unless otherwise specifically provided for in a particular Service Exhibit.

(b) ***Seller may periodically introduce new list pricing*** for its carrier services, which it intends to communicate to Customer from time to time through its revised pricing schedules. At the time of ordering. Customer will receive either the ***lower of***: (x) the ***prices quoted in Exhibit*** ___ , which is inclusive of all applicable discounts, ***or*** (y) the ***discounted list prices*** determined by the following formulae: Seller will provide its Services to Customer with a discount of: (i) fifteen percent (15%) off its then current carrier list prices for Services having speeds of less than 150 Mbps; and (ii) twenty-five percent (25%) off its then current carrier list prices for all other Services. However, under no circumstances will the net price paid by Customer for such Services available under this discount formulae be (i) more than ten percent (10%) lower than the price generally available based on Seller's discount plan for its largest volume carriers for such Service. These referenced discounts will be in addition to any discounts provided for by multi-year term contractual commitments to which Customer would be entitled. For example: if the highest volume discount plan available to Seller's largest volume carriers provides for a 11% discount off a circuit having a list price of $1000, resulting in a net circuit price of $890, then Customer's price for the same circuit, applying the discount formulae above, would result in a $750 price, assuming a 25% discount. Since this amount is more than ten percent (10%) lower than the $890 price made applying Seller's largest volume discount plan, the price to be made available under this discount formulae to Customer for this same circuit shall be revised upward to $801, which is 90% of $890. If the $801 price was, in fact, higher than the Exhibit ___ pricing, then the Exhibit ___ pricing would apply. Notwithstanding either above pricing methodology, Seller agrees to meet or beat market place competitive pricing under similar situations if the negotiated prices herein and the Service Exhibits become non-competitive. The determination of whether Seller's prices are non-competitive shall be reviewed every six months.

(c) ***Seller will provide its Services to Customer at prices and terms that are no less favorable than that which it charges its other customers who purchase comparable volumes and Services from Seller***.

Example Clause #10: The following clause is from a bilateral services agreement entered into between two telecommunications carriers. They agreed to sell services to each other on favorable terms.

Either Party may periodically introduce new ***list pricing*** for its carrier services, which it intends to communicate to the other Party from time to time through its revised pricing schedules. At the time of ordering. Telco A will receive either ***the lower of***: (x) ***the prices quoted*** in Exhibit __ (Access Service Agreement), which is inclusive of all applicable discounts, ***or*** (y) ***the discounted list prices determined by the following formulae***: Telco B will provide its Services to Telco A with a discount of: (i) fifteen percent (15%) off its then current carrier list prices for Services having speeds of less than 150 Mbps; and (ii) twenty-five percent (25%) off its then current carrier list prices for all other Services. However, under no circumstances will the net price paid by Telco A for such Services available under this discount formulae be (i) more than ten percent (10%) lower than the price generally available based on Telco B's discount plan for its largest volume carriers for such Service. These referenced discounts will be in addition to any discounts provided for by multi-year term contractual commitments to which Telco A would be entitled. For example: if the highest volume discount plan available to an Telco B's largest volume carriers provides for a 11% discount off a circuit having a list price of $1000, resulting in a net circuit price of $890, then Telco A's price for the same circuit, applying the discount formulae above, would result in a $750 price, assuming a 25% discount. Since this amount is more than ten percent (10%) lower than the $890 price made applying Telco B's largest volume discount plan, the price to be made available under this discount formulae to Telco A for this same circuit shall be revised upward to $801, which is 90% of $890. If the $801 price was, in fact, higher than the Exhibit 1 pricing, then the Exhibit 1 pricing would apply. Notwithstanding either above pricing methodology, Telco B agrees to meet or beat market place competitive pricing under similar situations if the negotiated prices herein and the Service Exhibits become non-competitive. The determination of whether Telco B's prices are non-competitive shall be reviewed every six months.

Telco A will provide its Services to Telco B at prices and terms that are no less favorable than that which it charges its other customers who purchase comparable volumes and Services from Telco A.

Part 4, Chapter 4 - Minimum Purchase Obligation

A minimum purchase obligation is a contractual revenue commitment from a customer to purchase services from seller of no less than an agreed value within a stated period of time. If a customer fails to order services of an agreed revenue value during that time, the seller may charge customer for the deficiency. Such arrangements are also sometimes referred to as "take or pay" agreements.

A customer might make such a concession in return for favorable terms extended by the seller, such as favorable pricing that would apply to the services purchased to satisfy the minimum purchase obligation, or for a reciprocal commitment for seller to purchase services from the customer.

Example Clause #1: The first example sets forth a simple minimum purchase obligation. The obligation is stated a minimum of new services that customer must order each month. If the customer fails to purchase the agreed minimum during any month, the Seller may charge a "shortfall liability" in an amount equal to the difference between the minimum purchase obligation and the amount revenues of new orders actually agreed to by customer.

> MINIMUM MONTHLY COMMITMENT
>
> A. Minimum Monthly Commitment ("MMC") is the amount of Services that Customer commits to purchase ***during each month*** of the Term. Customer's MMC is stated in Attachment __.
>
> B. MMC Shortfall Liability. If Customer fails to meet its MMC, unless caused by Seller's material failure to perform under this Agreement, Customer will pay Seller, in addition to other applicable charges, ***the difference between the MMC and Customer's actual MMC Contributory Service Usage Charges*** for each month in which Customer does not achieve the MMC.

Example Clause #2: In the next example the minimum purchase obligation is stated on a quarterly purchase obligation and as a total purchase obligation for the full term of the contract.

The Seller has committed to a network development schedule. Customer's purchase obligation is conditioned on Seller meeting certain milestones in that schedule. If Seller fails to deploy its network in accordance with the schedule, Customer would not be excused entirely from the minimum purchase obligation, but there would be an appropriate reduction in the minimum purchase obligation.

Customer may order from any of Seller's service offerings. Taxes and certain other charges are excluded from the calculation of "Contributory Charges."

> MINIMUM PURCHASE OBLIGATIONS; PREFERRED VENDOR STATUS. Each Exhibit attached hereto sets forth the revenue, term and utilization commitments, if any, associated with the specific Services provisioned pursuant to the Exhibits hereto.

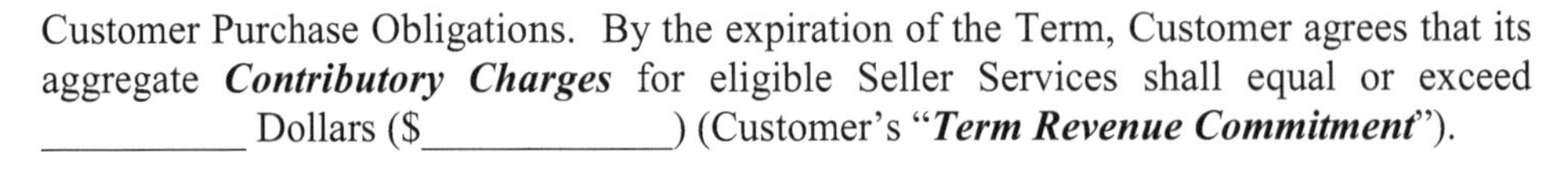

Customer Purchase Obligations. By the expiration of the Term, Customer agrees that its aggregate ***Contributory Charges*** for eligible Seller Services shall equal or exceed __________ Dollars ($____________) (Customer's "***Term Revenue Commitment***").

(a) Until Customer's Term Revenue Commitment is achieved, Customer agrees to satisfy on a quarterly basis the quarterly revenue commitment(s) set forth in Exhibit _ (Customer's "***Quarterly Revenue Commitment(s)***"). Subject to Section ___(c) below, ***Seller shall be authorized to assess a shortfall charge at the expiration of each quarter*** of the Term beginning with the first (1st) quarter of 20__ equal to the difference between its Quarterly Revenue Commitment applicable to said quarter and Customer's aggregate, actual Contributory Charges invoiced during the quarter just ended.

(b) If Customer does not satisfy its applicable Term Revenue Commitment by the end of the Term, then, subject to Section ___(c) below, Customer agrees, in addition to all other charges due under the Agreement, to pay a shortfall charge equal to the difference between its applicable Term Revenue Commitment and Customer's aggregate, actual Contributory Charges invoiced under the Agreement during the Term for eligible Seller Services. Only revenue accrued and booked during the Term shall contribute to Customer's applicable Term Revenue Commitment.

(c) Seller and Customer agree that ***Customer's Term Revenue Commitment and Quarterly Revenue Commitment are based, in part, upon Seller's projected Building Schedule***, as more fully described in Exhibit __. ***Should Seller fail to meet this Building Schedule, Customer shall be entitled to a reduction in any shortfall*** charge due under Section ___(b) above equal to the amount by which Customer demonstrates it failed to satisfy the Term Revenue Commitment as a result of Seller's inability to meet the Building Schedule. Additionally, if during any calendar quarter Seller has failed to pass the number of buildings set forth in Exhibit __ for such quarter, then the Quarterly Revenue Commitment for such quarter shall be reduced. The Quarterly Revenue Commitment for that quarter shall equal the Quarterly Revenue Commitment for such quarter multiplied by a fraction, the numerator of which is the total number of buildings actually passed, and the denominator is the number of buildings that would have been passed by the end of such quarter according to the Building Schedule.

$$QRC = QRC \times \frac{\text{\# of buildings actually passed}}{\text{\# of buildings according to Building Schedule}}$$

(d) For purposes of this Section, eligible "Seller Services" shall include all of Seller's Service offerings made available generally to its customers, and shall include, at a minimum, transport services.

"Contributory Charges" means recurring charges, usage charges, one-time purchase charges in the case of dark and lit fiber sales, and other qualifying charges applicable to the eligible Services accruing to a Party's account under this Agreement, before application of all eligible discounts and excluding all: (a) taxes; (b) surcharges; (c) fees;

(d) amounts owing for or related to, credits, uncollectable charges, pass-through charges, installation charges; and (e) any other charges and/or amounts expressly excluded in the applicable Exhibits.

Example Clause #3: In the following clause, there is a single minimum purchase obligation applies to the entire five-year term of the agreement. There are no intermediate revenue objectives that must be satisfied.

Also in this clause, the Carrier has committed to make services available between certain city pairs in accordance with "Milestone Dates," either on Carrier's own network, or by means of services acquired by another carrier and made available to this Customer. Specific values of projected revenues are projected over each of the city pairs. If the Carrier fails to make such resources available so that Customer may order services between the agreed city pairs, the Contract Revenue Commitment is reduced according to the revenues projected for the related city pairs.

> Section ___ Contract Revenue Commitment. Customer hereby commits to Carrier that, except as provided otherwise in this Section ___ or through application of the remedies for Service Outages or Carrier Defaults as set forth in Articles _ and _, the aggregate amount paid by Customer for all Telecommunications Services ***ordered from Carrier during the initial 5-year Agreement Term*** shall equal not less than $____ million ($__________) (the "Contract Revenue Commitment"). In the event that Customer does not satisfy the Contract Revenue Commitment, then Customer shall, no later than thirty (30) days ***following expiration of the initial 5-year Contract Term***, pay to Carrier, in certified funds, the Termination Charge (as calculated under Section ___ of this Agreement). In the event that Customer elects to terminate this Agreement with respect to any Segments, City Pair Connections or Telecommunications Services due to: (a) ***failure of Carrier to meet the Milestone Date*** under Section ____; (b) events described in Article __; or (c) failure or inability to deliver forecasted Telecommunications Services as provided in Section _____, then the Contract Revenue Commitment shall be reduced by the MRC for the remainder of the Agreement term for the Telecommunications Services that were to have been delivered over such Segments or such City Pair Connections, but the prices set forth in Exhibit __ shall remain unchanged with respect to the balance of Telecommunications Services delivered to Customer.
>
> The following payments shall be included when determining whether the Contract Revenue Commitment has been satisfied: payments of the MRC for on-net and off-net Telecommunications Services, and any other payments made in consideration for the delivery of telecommunications services of any kind (including payment for switched services), regardless of whether such services are included in the definition of Telecommunications Services under this Agreement. ***Non-recurring charges*** for circuit changes, installation or reconfiguration of the Telecommunications Services ***shall not be included*** when determining whether the Contract Revenue Commitment has been satisfied.

Section ___ Schedule of Performance.

(a) Carrier shall provide Customer with ***OC-12 capacity*** which has passed the Acceptance Tests specified in Sections ___ and ___ on or before the dates specified in Exhibit A (each a "Milestone Date"), with respect to each of the following (each a "City Pair Connection"):

[To be completed.]

(b) In addition, Carrier shall provide Telecommunications Services to Customer (through leased circuits not located within the Network) consisting of ***one OC-3 and one OC-3 circuit***, commencing on ________ and continuing through the applicable Milestone Date, for the following City Pair Connections:

[To be completed.]

Example Clause #4: In the example that follows, the underutilization charge is not 100% of the deficiency. In delivering services to Customer, the Seller will incur costs in provisioning services. Therefore, if the Customer fails to order the full amount of the revenue commitment, the Seller would suffer a loss of revenues, but the Seller would also not be required to incur the costs of provisioning. The parties have agreed that the Underutilization Charge will be 25% of the amount of the shortfall, not the full amount.

Here the Customer has agreed to an annual revenue commitment, but the shortfall, if any, is determined monthly.

MINIMUM ANNUAL VOLUME COMMITMENT ("AVC").

Customer agrees to pay Seller no less than ____________ thousand dollars ($______) in Total Service Charges (as hereinafter defined) ***during each Contract Year***. A "Contract Year" shall mean each consecutive twelve-month period of the Initial Term commencing on the Effective Date. During ***each monthly billing period*** of the Extended Term, Customer's Total Service Charges must equal or exceed one-twelfth (1/12) of the AVC. "Total Service Charges" shall mean all charges, after application of all discounts and credits, incurred by Customer for Services provided under this Agreement, specifically excluding (a) taxes, tax-like charges and tax-related surcharges; (b) charges for equipment and collocation (unless otherwise expressly stated herein); (c) charges incurred for goods or services where Seller or Seller affiliate acts as agent for Customer in its acquisition of goods or services; (d) nonrecurring charges; (e) "Governmental Charges" as defined below; and (f) other charges expressly excluded by this Agreement.

UNDERUTILIZATION CHARGES. If, in any Contract Year during the Initial Term, Customer's Total Service Charges do not meet or exceed the AVC, then Customer shall pay: (a) all accrued but unpaid usage and other charges incurred under this Agreement; and (b) an ***"Underutilization Charge" in an amount equal to twenty-five percent (25%)*** of the difference between the AVC and Customer's Total Service Charges during such

Contract Year. If, in any monthly billing period during the Extended Term, Customer's Total Service Charges do not meet or exceed one-twelfth (1/12) of the AVC then Customer shall pay:

(a) All accrued but unpaid usage and other charges incurred under this Agreement, and

(b) An "***Underutilization Charge***" equal to the difference between one-twelfth (1/12) of the AVC and Customer's Total Service Charges during such monthly billing period.

Example Clause #5: The following clause was entered as an amendment to an agreement that had been in place between the Seller and the Customer. The minimum purchase obligation pertains only to new services that are ordered after this amendment has been entered into, "New Circuits," and not to "Existing Circuits."

The Minimum Revenue Commitment is established for a twelve-month period beginning with the date of this Amendment. That is less than the full term of the contract. The revenue shortfall is collated on a monthly basis. However, if Customer is required to pay a "Shortfall" for any month, Customer may apply the amount of the Shortfall as a credit for other services ordered at any time during the term of the contract. At the end of the term of the contract any credit not utilized by Customer will expire. Customer would not be entitled to a refund of any unused credit.

Minimum Purchase Obligations. This Section sets forth the minimum revenue and term commitments associated with the New Circuits:

Existing Circuits. Exhibit _____ sets forth a schedule of certain Circuits to be provided by Seller under this Agreement, consisting of:

(a) ______ circuits that were being provided to Customer by Seller before the date of this Amendment (the "Existing Circuits").

(b) _______ intra-building circuits for which Seller will charge no MRC (the "Intra-Building Circuits").

(c) ________ inter-building OC-48 Circuits (the "OC-48s"), for which the MRC shall be $_____, and the term shall be five years.

Customer Purchase Obligations. Customer agrees that for each month of the twelve-month period beginning on ____________ and ending on ___________ (the "Minimum Commitment Term"), Customer's aggregate MRCs for New Circuits shall equal or exceed _________________________ ($______) (the "Monthly Revenue Commitment"). Customer's total Monthly Revenue Commitment for the Minimum Commitment Term shall be ____ ($_______) (the "Term Revenue Commitment").

(a) If for any month during the Minimum Commitment Term Customer's actual ***MRCs for New Circuits are less than the Monthly Revenue Commitment***, then, subject to Subsection (c) below, ***Seller shall assess a shortfall charge for such month (a "Shortfall")***, which Customer agrees to pay promptly, equal to the difference between Customer's Monthly Revenue Commitment and Customer's aggregate actual MRCs for New Circuits for such month.

(b) If by the expiration of the Minimum Commitment Term Customer has not satisfied its Term Revenue Commitment, then Customer agrees to pay, in addition to all other charges due under the Agreement, a Shortfall charge equal to the difference between the Term Revenue Commitment and Customer's aggregate actual MRCs paid for New Circuits under the Agreement during the Minimum Commitment Term. Only MRCs that accrue during the Minimum Commitment Term shall be credited against Customer's Term Revenue Commitment.

(c) ***To the extent that for any month during the Minimum Commitment Term Customer pays a Shortfall, Customer shall be entitled to a credit equal to the Shortfall which may be utilized by Customer at any time during the remainder of the Minimum Commitment Term***. At the expiration of the Minimum Commitment Term, to the extent that Customer has not utilized the Shortfall credit, the Shortfall credit shall expire and terminate, and shall be of no further effect.

(d) For purposes of this Section, "***New Circuits***" means all Circuits ordered by Customer from Seller under this Agreement for which Customer pays MRCs, other than the Existing Circuits.

Example Clause #6: The following example sets forth a total contract minimum charge and a monthly minimum charge. The shortfall charge is 100% of the difference between the Contract Minimum and the total value of services ordered.

- Contract Minimum Commitment. ***During the term*** of the Agreement, Customer agrees to purchase Services from Service Provider in an amount which must equal or exceed ________ Dollars ($____) (the "***Contract Minimum***"). If at the expiration of the Agreement, Customer's purchase of Services are less than the Contract Minimum, then Customer will pay: (1) all accrued but unpaid Service charges and other charges incurred by Customer; and (2) ***a shortfall charge*** (which Customer hereby agrees is reasonable) equal to the difference between the Contract Minimum and Customer's total procurement of Service Provider's Services (including any applicable cancellation charges), excluding taxes, tax related surcharges, and tax-like surcharges, during the Agreement's term.

- Monthly Minimum Commitment. ***During each month*** of the Agreement, Customer agrees to purchase Services from Service Provider in an amount which must equal or exceed __________ Dollars ($_____) per month (the "***Monthly Minimum***"). If at the end of each such month, Customer's purchase of Services are less than the Monthly Minimum, then Customer will pay: (1) all accrued but unpaid Service charges and other charges incurred by Customer; and (2) a ***shortfall charge*** (which Customer hereby agrees is reasonable) ***equal to the difference*** between the Monthly Minimum and Customer's total procurement of Service Provider's Services (including any applicable cancellation charges), excluding taxes, tax related surcharges, and tax-like surcharges during that month.

Example Clause #7: In the clause that follows, the minimum purchase obligation is calculated monthly. There is no separate total contract purchase obligation.

The minimum purchase obligation does not begin immediately. Rather, there are a certain number of initial months of the term of the contract during which there is no minimum monthly purchase obligation.

Customer shall be liable for the following monthly minimum charge(s) commencing with Customer's Billing Cycle that starts in the _______ (__) month following the Service Commencement Date (the "Minimum Charge").

MINIMUM PERIOD	MINIMUM CHARGE
______________ month and each month thereafter	$_______________

If during the Minimum Period Customer's net charges (after any available discounts hereunder) for the Services are less than the Minimum Charge, Customer shall pay the shortfall. Calculation of the Minimum Charge includes weekly international billing amounts (if applicable) invoiced after the start of a monthly Billing Cycle and prior to the end of that same Billing Cycle. (Ex: January 19th through February 18th Billing Cycle invoices will include all weekly invoiced amounts dated after January 19th and prior to February 18th.) Taxes, governmental assessments and surcharges, non-recurring charges, operator assistance charges and local loop and third party and regulatory pass-through charges are not included when calculating the Minimum Charge.

Customer agrees that any make-up to applicable Minimum Charges, shortfall charges and surcharges for which it is liable under this Agreement are based on agreed upon minimum commitments on its part and corresponding pricing concessions on Seller's part, and are not penalties or consequential or other damages under Section ___.

Example Clause #8: The following clause is somewhat more complicated in that each party to the contract is both a seller and a buyer to the other company. And both companies have committed to minimum purchase obligations.

However, the two minimum purchase obligations are not entirely reciprocal. The two are defined differently and in different amounts. "Telco B's" purchase obligation is much larger. As a result, the obligations that apply to Telco B are much more flexible than those that apply to Telco A. Telco B may defer a part of its purchase obligation. If Telco B oversubscribes in one year, Telco B may apply the excess to its obligation for subsequent years or claim a refund of shortfall or underutilization charges paid in previous years.

The agreement defines categories of Eligible Charges. Only certain kinds of charges are to be included in the calculation of service revenues in determining whether a minimum purchase obligation has been satisfied. For example, revenues from the lease of dark fibers may or may not be included.

Each party's minimum purchase commitment is conditioned are the other party's satisfaction of performance objectives. If either party as seller fails to complete its network development according to schedule, or for any other reason fails to make services available to the other party as purchaser, the purchaser's minimum purchase obligation is either reduced or postponed.

The parties have assigned an assumed value to constructing network adjacent to commercial buildings. This assumption is based on an expectation of the revenues that will be earned from each building's customers. This assumption be sound or may be the product of "irrational exuberance."

TERM REVENUE COMMITMENTS

(a) ***Telco A will purchase from Telco B, on a "take or pay" basis***, an aggregate of no less than ten million dollars (US$10,000,000) in Services or IRUs (Telco A's "Term Revenue Commitment") by no later than _______________. These Services or IRUs shall include Telco B's Service offerings made available generally to its customers including, but not limited to, dark or lit long haul fiber (purchased on an indefeasible right of use basis or otherwise), leased transport, and collocation (the Telco B "Services"). The quantity and type of Services or IRUs selected for purchase by Telco A in satisfying Telco A's Term Revenue Commitment shall be based on Telco A's then current business requirements determined at its sole discretion. Purchases by Telco A's Affiliates shall count towards its Annual and Term Revenue Commitments.

(b) During the Initial Term (defined in Section ____ below) of the Agreement, ***Telco B will purchase from Telco A***, on a "take or pay" basis, an aggregate of no less than fifty million dollars (US$50,000,000) in Services or IRUs (Telco B's "Term Revenue Commitment"). These Services or IRUs shall include Telco A's Service offerings made available generally to its customers including, but not limited to, leased transport Services (the Telco A "Services"). The quantity and type of Services or IRUs selected for purchase by Telco B in satisfying Telco B's Term Revenue Commitment shall be based on Telco B's then current business requirements determined at its sole discretion. As part of Telco B's Term Revenue Commitment, Telco B also shall meet the annual purchase commitments which are set forth in Telco A's Building Schedule attached hereto as Exhibit ___ and is further described in Section __(d) herein ("Telco B Annual Revenue Commitment"). During the first twelve (12) month take or pay period of this Agreement, ***Telco B may defer up to fifteen percent (15%) of any shortfalls in its Annual Revenue Commitment to the twelve (12) month period following the end of the Initial Term of this Agreement***, prior to any extensions, for the purpose of making up the shortfalls. Purchases by Telco B's Affiliates shall count towards its Annual and Term Revenue Commitments.

(c) If either Party fails to satisfy its respective revenue commitments during the applicable term, (which may have been extended in accordance with Section __(b), __(e), __(c) or __***(d)*** herein), such Party shall pay a shortfall charge equal to the difference between its applicable revenue commitments (which may have been reduced in accordance with Sections __(b), __(e) or _____ herein) and its aggregate Eligible Charges (as defined in Section __(h)) received from the other Party. Payment of a shortfall charge by Telco B for failing to achieve its Annual Revenue Commitment shall apply towards meeting its Term Revenue Commitment. ***If Telco B exceeds its purchase requirements during any year, its excess purchases, at Telco B's sole discretion, may be carried forward and credited towards Telco B's Annual Revenue Commitment for the following year(s), or may be applied to make up for any shortfalls in prior years,*** thus resulting in a recovery of the shortfall charges (through the issuance of Telco A Service credits only) previously paid to Telco A. In either case, the excess purchases shall apply towards Telco B's Term Revenue Commitment. Within forty-five (45) days following the end of each twelve (12) month take or pay period of the Agreement, the Parties shall determine in good faith whether any shortfall charges are due. If the Parties are unable to agree upon this determination, then the dispute resolution provisions in Section ___ below may be invoked by either Party. Each twelve (12) month take or pay period of the Agreement shall commence on __________ and end on ________ of the applicable year (each an "Annual Period").

(d) Telco B's ability to meet its Annual Revenue Commitment and Term Revenue Commitment depends upon Telco A's ability to timely make Services available to Telco B at specific strategic locations. As such, Telco A shall exert commercially reasonable efforts to make such Services available to Telco B in the priority markets ("Priority Markets") described in the construction and deliverable schedule attached hereto as Exhibit __ ("Telco A's Building Schedule"). Telco A shall be deemed to have delivered a "Passed Building" when: (1) the building is either On-Net or Near-Net, as described in Exhibit __ (The Access Service Agreement), attached hereto; and (2) the building is at least 10,000 sq. ft. of useable commercial space for occupancy or 50,000 sq. ft. of usable, multi-tenant residential space for occupancy. Notwithstanding the foregoing, a "Passed Building" shall not include: (1) buildings used primarily for retail, storage, or agricultural purposes unless such building has 100 or more employees; or (2) any building that is unlikely, applying reasonable objective factors, to use high capacity telecom services in the foreseeable future.

(e) If Telco A fails to deliver at least eighty percent (80%) of the required number of Passed Buildings (the "Minimum Passed Buildings") for any of the Priority Markets by the projected quarterly deliverable date(s) set forth in Telco A's Building Schedule attached hereto as Exhibit __ (the projected quarterly deliverable date is the last day of each applicable quarter), then for each Priority Market where the Minimum Passed Building has not been achieved in a quarter, Telco B's Annual Revenue Commitment and Term Revenue Commitment shall be modified pursuant to the methodology described below and illustrated in Exhibit __ attached hereto. For ramping purposes, revenue targets associated with delivered Passed Buildings shown in Table __ reflect a ninety (90) day delay from their delivery by Telco A before they are considered in calculating the applicable Annual Revenue Commitment and Term Revenue Commitment.

(1) Table __ of Exhibit __ shows the Minimum Passed Buildings requirements. ***The value per Passed Building is $_____***. Quarterly revenue projections used in computing the Annual Revenue Commitments and Term Revenue Commitment are set forth therein. The Annual Revenue Commitments are the sum of the revenue targets for each of the Priority Markets during the applicable Annual Period. Eligible Charges incurred from any market will contribute to meeting the overall Annual Revenue Commitments and Term Revenue Commitment. The first Annual Revenue Commitment is to be met by _________ with additional Annual Revenue Commitments to be met by each anniversary thereafter. Table __ assumes the full application of the Section __(b) Annual Revenue Commitment reduction applicable to the first Annual Period. Customer's actual first Annual Revenue Commitment is ____% of the amount set forth in Table __ (the "Actual First Annual Revenue Commitment"). If Telco B does not satisfy the Actual First Annual Revenue Commitment, it shall be entitled to defer the actual shortfall amount (up to the fifteen percent (15%) authorized by section __(b)) until the first quarter following the expiration of the Initial Term.

(2) If Telco A fails to timely deliver the Minimum Passed Buildings for any Priority Market by the projected quarterly deliverable date set forth in Telco A's Building Schedule, but delivers within ninety (90) days thereafter, the affected Priority Markets' Minimum Passed Buildings requirement will be shifted by one quarter. This will be equivalent to substituting the row for the delayed Priority Market in Table __ into Table __ for the delayed quarter and all succeeding quarters. The Annual Revenue Commitment for the period will be calculated using the adjusted quarterly targets. The Initial Term shall be extended by ninety (90) days, and Telco B shall have this additional ninety (90) days to satisfy the deferred portion of the Annual Revenue Commitment and an equal amount of the Term Revenue Commitment, as reflected in Table __ of Exhibit __. The first table would continue to apply for all Priority Markets that are delivered by its projected quarterly deliverable date.

(3) If Telco A fails to timely deliver the Minimum Passed Buildings for any Priority Market by the projected quarterly deliverable date set forth in Telco A's Building Schedule, but delivers within ninety (90) days of the revised Section _(e)(2) due date above, the affected Priority Markets' Minimum Passed Buildings requirements will be shifted by one additional quarter. The revenue targets for each of the affected Priority Markets will be further reduced by ____. This will be equivalent to substituting the row for the delayed Priority Market in Table __ into Table __ for the delayed quarter and all succeeding quarters. The Initial Term shall be extended by a further ninety (90) days, and Telco B shall have this additional ninety (90) days to satisfy the deferred portion of the Annual Revenue Commitment and an equal amount of the Term Revenue Commitment, as reflected in Table __ of Exhibit __. This shift and reduction apply only to those Priority Markets affected by a delay in delivery of this length.

(4) If Telco A fails to timely deliver the Minimum Passed Buildings requirement for any Priority Market by the projected quarterly deliverable date set forth in Telco A's Building Schedule, as revised by the operation of section __(e)(3) above, but delivers within ninety (90) days of the revised __(e)(3) due date, the affected Priority Markets' Minimum Passed Buildings requirements will be shifted by one additional quarter. The revenue targets for each of the affected Priority Markets will be reduced by a further ___. This will be equivalent to substituting the row for the delayed Priority Market in Table ___ into Table __ for the delayed quarter and all succeeding quarters. The Initial Term shall be extended by a further ninety (90) days, and Telco B shall have this additional ninety (90) days to satisfy the deferred portion of the Annual Revenue Commitment and an equal amount of the Term Revenue Commitment, as reflected in Table ___ of Exhibit __. This shift and reduction applies only to those Priority Markets affected by a delay in delivery of this length.

(5) If Telco A fails to timely deliver the Minimum Passed Buildings for any Priority Market within the timeframes set forth in ***Section __(e)(4)***, Telco B's Annual Revenue Commitment and Term Revenue Commitment shall be reduced by a further ___ (for a total of one-hundred percent (100%)) of the total expected revenues for the affected Priority Market, regardless of whether all or part of the required number of Passed Buildings are delivered thereafter. This will be equivalent to removing the row for the delayed Priority Market for the delayed quarter and all succeeding quarters. Nonetheless, the Services that Telco B purchases from Telco A in these markets shall count towards Telco B's reduced Annual Revenue Commitment and Term Revenue Commitment. This reduction applies only to those Priority Markets affected by a delay in delivery of this length.

(6) If Telco B enters into an agreement to obtain substitute services from another vendor to cover for Telco A's failure to timely deliver on any accepted service orders, then in addition to any service installation remedies set forth in Appendix No __ to the Access Service Agreement, Telco B shall be entitled to a reduction in its Annual and Term Revenue Commitments by an amount equal to the amount Telco B pays to the substitute vendor for the replacement services. Telco A shall have no obligation to pay for the cost of such replacement services, nor shall Telco B be required to accept late delivery from Telco A unless Telco A pays any and all early termination charges or penalties incurred by Telco B in terminating substituting services.

(f) Telco B shall treat Telco A and its Affiliates, if such Affiliates are not competitors similarly situated to Telco B, as a preferred provider in the markets where Telco A can provide Services at competitive prices and terms and within comparable provisioning intervals. Telco B shall not be obligated to utilize Telco A as a preferred provider if: (1) Telco B's customer has expressed a preference to use another provider; (2) Telco B has, as of the Effective Date, a conflicting contractual commitment to another provider; (3) Telco B is using its own facilities or those of its Affiliates to provide the Services; or (4) generally accepted economic factors make it commercially not feasible for Telco B to use Telco A. If Telco A acquires an Affiliate after the Effective Date, Telco A shall notify Telco B in writing that it seeks to have such Affiliate treated by Telco B as a preferred provider hereunder. Upon such notice, Telco B shall have thirty (30) days from receipt of such notification to perform due diligence on this Affiliate and give its consent, which consent may be reasonably withheld, to treat such Affiliate as a preferred provider hereunder in accordance with the provisions of this section.

(g) In satisfying each Party's respective revenue commitments, only Eligible Charges shall apply. As used herein, "***Eligible Charges***" are defined as recurring charges, usage charges, one-time purchase charges in the case of dark and lit fiber sales, non-recurring charges, shortfall payments made under this Agreement, and other qualifying charges applicable to the Services accruing to a Party's account under this Agreement, and excluding all: (1) taxes; (2) surcharges; (3) fees; (4) amounts owing for or related to, credits, uncollectable charges, pass-through charges, installation charges; and (5) any other charges and/or amounts expressly excluded in the applicable Exhibits.

Example Clause #9: The final example utilizes many of the passages from previous clauses, but sets out a minimum purchased obligation in the form that could be used as an Addendum or Amendment to an agreement. After a Master Services Agreement has been agreed, the parties might agree to adopt a minimum purchase obligation for future services.

This form also includes an option clause which may be included or omitted. The optional clause would permit the Customer to utilize shortfall payments at a later time as credits against payment obligations for future services.

Addendum

TO
MASTER SERVICE AGREEMENT

This Addendum to Master Service Agreement (the "Addendum") is entered into this ____ day of ______________________, 2____ ("Effective Date"), between ___________________________________ (together with its Affiliates, "Customer") and ________________ ("Seller"). This Addendum modifies the existing agreements between Customer and Seller with respect to Service ordered by Customer from Seller pursuant to that certain Master Service Agreement that was entered into by the parties as of ______________ (the "MSA"). In the event of any inconsistency between the terms of this Addendum and any other agreements between Customer and Seller, the terms of this Addendum shall control.

ARTICLE 1
Minimum Purchase Obligations.

This Section sets forth the minimum revenue and term commitments associated with the New Circuits:

Existing _____Circuits. Exhibit _____ sets forth a schedule of certain Circuits that Seller has agreed to provide to Customer, consisting of:

(a) ***Fourteen (14) circuits*** that were being provided to Customer by ___________ (the "Existing Circuits."

(b) ***Four (4)*** inter-building OC-48 Circuits (the "OC-48s"), for which the MRC shall be $______, and the term shall be five years.

Customer Purchase Obligations. Customer agrees that for each month of the twelve-month period beginning on _______ and ending on _________ (the "Minimum Commitment Term"), Customer's aggregate MRCs for New Circuits shall equal or exceed _________________ ($_______) (the "Monthly Revenue Commitment"). Customer's total Monthly Revenue Commitment for the Minimum Commitment Term shall be ___________________________ ($_______) (the "Term Revenue Commitment").

(a) If for any month during the Minimum Commitment Term Customer's actual MRCs for New Circuits are less than the Monthly Revenue Commitment, then, subject to Subsection (c) below, Seller shall assess a shortfall charge for such month (a "Shortfall"), which Customer agrees to pay promptly, equal to the difference between Customer's Monthly Revenue Commitment and Customer's aggregate actual MRCs for New Circuits for such month.

(c) If by the expiration of the Minimum Commitment Term Customer has not satisfied its Term Revenue Commitment, then Customer agrees to pay, in addition to all other charges due under the Agreement, a Shortfall charge equal to the difference between the Term Revenue Commitment and Customer's aggregate actual MRCs paid for New Circuits under the Agreement during the Minimum Commitment Term. Only MRCs that accrue during the Minimum Commitment Term shall be credited against Customer's Term Revenue Commitment.

(d) Customer agrees that any minimum charge shortfall and any early termination fees for which it may be liable under this Agreement are based on agreed upon minimum commitments on its part and corresponding rate concessions on ________'s part, and are not penalties or consequential or other damages under Section ___ hereof.

Optional clause:
To the extent that for any month during the Minimum Commitment Term Customer pays a Shortfall, Customer shall be entitled to a credit equal to the Shortfall which may be utilized by Customer at any time during the remainder of the Minimum Commitment Term for any MRCs in excess of the Term Revenue Commitment. At the expiration of the Minimum Commitment Term, to the extent that Customer has not utilized the Shortfall credit, the Shortfall credit shall expire and terminate, and shall be of no further effect.

For purposes of this Section, "New Circuits" means all Circuits ordered by Customer from Seller under this Agreement for which Customer pays MRCs, other than the Existing Circuits.

ARTICLE 2
Confirmation

Customer and Seller confirm and ratify in all respects the terms and conditions of the MSA, as amended by this Addendum.

Seller: ______________________	Customer: ______________________
By: ______________________	By: ______________________
Name: ______________________	Name: ______________________
Title: ______________________	Title ______________________
Date: ______________________	Date: ______________________

Part 4, Chapter 5 - Preferred Provider

A "preferred provider" or "preferred vendor" is entitled to special treatment by its customer. A customer agrees to look to its preferred provider to obtain services before considering purchasing those services from another seller.

Such a clause may apply only to certain types of services, or may be limited by geographic region. Furthermore, a customer's obligation to purchase services from a preferred provider is usually qualified by certain conditions. Most often a preferred provider's pricing must be equal to or better than a competing offer.

Because a preferred provider clause attempts to address future service needs that may be uncertain, a customer will often attempt to qualify such an obligation in a manner to allow customer to avoid the obligation under certain circumstances. If a commitment is highly qualified, it might be questionable whether customer has made a binding commitment at all.

Example Clause #1: In the first example, the Customer has agreed to purchase all services offered by the Seller (the "Preferred Supplier") so long as the Seller's price is equal to or better than a competitor's advertised price for the same or similar service.

> Customer agrees to provide "Preferred Supplier" status to Seller for the term of this Master Agreement. For the purposes of this and all Agreements between Seller and Customer, "Preferred Supplier" shall mean that ***Customer will extend to Seller a first right of refusal*** for Seller to provide those services that (a) Seller provides during the normal course of Seller's business: (b) for which Customer formally requests a pricing quotation from Seller: (c) which Customer wishes to secure, and (d) for which Seller's ***quoted price to Customer is the same or less than an advertised competitor's price*** for same or similar services as documented and noticed to Seller by Customer in writing.

Example Clause #2: In the following clause, the Seller and the Customer have each agreed that the other party will be its preferred provider for certain services. The two companies are regional telecom companies. The Seller will be Customer's preferred provider for services outside of Customer's region, and Customer will be Seller's preferred provider for particular services that are offered by Customer.

> Seller will be the "preferred provider" for Customer's 3rd party exchange access service requirements (SONET, Ethernet, Wavelength) for territory ***outside of Customer's Region. Within Customer's Region, Seller will be the "preferred provider"*** for Customer's exchange access service requirements (SONET, Ethernet, Wavelength) where Customer is not the provider of record for such exchange access service.
>
> Customer will be the "preferred provider" for Seller's third party transport and hosting services.
>
> For purposes of this Section, the ***"preferred provider" means that a party shall select the other party*** to this Agreement for provision of services ***unless***: (1) a Party can show that

the "preferred provider" does not have a Service offering available for purchase that is comparable to that being offered by an alternative third party (e.g., features, functionality, and reliability of the alternative service); or (2) the "preferred provider" cannot provision its Service on a comparable timeliness basis with that of the third party.

Example Clause #3: In the following clause Seller would be the preferred provider only within a defined geographic area. Furthermore, Customer's commitment to use Seller as its preferred provider is qualified. Customer is not required to purchase services from Seller if Seller's pricing is not the best price, or if certain other conditions are not satisfied.

And Customer has agreed only to exercise "reasonable efforts" to use Seller as a preferred provider. Even if the other conditions are satisfied, Customer has not agreed simply to purchase Seller's services. This can give rise to a misunderstanding if the Seller's understanding of "reasonable" is different from Customer's.

> Customer shall treat Seller as a preferred provider for the provision of telecommunications services that originate and terminate ***within the metropolitan*** ____________________ area, including dark fiber and other services. Customer will use ***reasonable efforts*** to use Seller if: (a) ***Seller's price quote or pricing represents the best price for Customer***; (b) Customer does not have a conflicting contractual commitment; (c) Customer is confident that Seller will meet Customer's installation interval requirements, (d) Customer's customer has not expressed a preference to use another provider, and (e) Customer is confident that Seller will meet Customer's diversity and other performance specifications.

Example Clause #4: The following clause is mutual. Each party is the preferred provider for the other party.

Here again, use of the term "competitive in all respects" and "comparable timeliness" can give rise to a misunderstanding. It would be less confusing if the parties were either to limit the comparison to pricing only, or require the seller to match a competing offer in price and all material terms.

> For purposes of this Agreement, "preferred provider" means that a party shall select the other party to this Agreement for provision of the specified services if (a) such Party has the prescribed service offering available for purchase that is ***competitive in all respects*** with that being offered by a competing service (e.g., features, functionality, and reliability of the alternative service); (b) the "preferred provider" can provision its service on a ***comparable timeliness*** basis with that of a competing service. The foregoing notwithstanding, the selection of such Party shall not be required if the end user customer expresses a preference for a different service provider, or the promotion or referral to such Party as service provider would conflict with a commitment to another service provider.

<u>Example Clause #5</u>: The preferred vendor obligations of the following clause are limited to service offerings at a particular location, a collocation site. Service offerings that Seller might offer at other locations would not be subject to the preferred provider clause.

The preferred provider comparison applies to pricing only. Other terms and conditions of a competing service offering are not considered.

This Seller wishes to control local access services offered from this collocation site. If the Seller matches the price of a competing local access service at this location, Customer must purchase the service from Seller.

The phrase "competitive with the rates" can be subject to different interpretations. However, the parties have addressed this possibility by requiring the Customer to submit a written pricing proposal to Seller. If Seller matches the competing pricing proposal, Customer would be obligated to purchase the service from Seller.

> <u>Seller to be Preferred Vendor</u>. The parties hereto agree that Seller shall be deemed the preferred vendor of voice and data telecommunications services to Customer ***at the site of the collocation***. This means that Customer is required, unless the parties agree otherwise in writing, to purchase its voice and data telecommunications services ***for use at the collocation site*** from Seller ***if the rate offered is competitive with the rates for comparable services*** provided by other carriers. If Customer challenges the competitiveness of the rate offered by Seller, and Customer has a written quote from another carrier for telecommunications services at a rate below the rate of Seller, then Customer shall deliver a copy of that written quote to Seller. Seller shall then have five (5) business days to ***match that offer***. If Seller chooses not to match the offer, then Customer shall be free to purchase those telecommunications services from any other provider.
>
> *Local Access Service Description ("Local Access Service").* Unless Seller chooses not to match a competitive offer, as provided above, or unless the parties agree otherwise in writing, Seller shall obtain Local Access Services for Customer, which are defined as the telecommunications facilities connecting a Customer-designated termination point to a Seller Point of Presence ("POP"). Customer shall request all Local Access Services in writing to Seller. Customer shall be responsible for all charges, including, without limitation, monthly charges, usage charges, installation charges, non-recurring charges, or applicable termination/cancellation liabilities as set forth in the Agreement and this Addendum.
>
> *Coordination.* In obtaining Local Access Services, Seller shall be responsible for the provisioning and the initial testing of an interconnection between the interexchange Service set forth in a Service Order and the Local Access Services. Seller will coordinate the installation of the Local Access Services with the interexchange Service being provided by Seller. Notwithstanding the foregoing, nothing contained in this Addendum shall be deemed to limit or otherwise affect the Limitation of Liability provision or the Force Majeure provision of the Master Services Agreement.

Customer Obtained Local Access. If Seller chooses not to match a competitive offer, as provided above, or if the parties agree otherwise in writing, then Customer may order its own local access services. In such event, Customer shall be billed directly by the provider of such services; Seller shall not be responsible for the payment or billing of any such charges. If Customer orders its own local access services, Customer shall be responsible for ensuring that such services are turned up at the same time as the Services being provided by Seller. In the event the Customer-ordered local access services are not ready at such time as the Services being provided by Seller, Seller shall nevertheless have the right to begin billing Customer for such Services as of the Effective Date, and Customer shall be liable for payment for such Services as of such date. In sites where Seller owns entrance facilities, Customer must obtain a letter of agency from Seller and will be charged applicable entrance facility charges as quoted.

Example Clause #6: The following is another example of a preferred provider clause in which each party has agreed to treat the other party as a preferred provider. Each party has agreed to treat the other as a preferred provider so long as the service offering of the selling party is "competitive in all respects." Since the phrase "competitive in all respects" is subject to interpretation, it can lead to differences of opinion about whether, in a given case, a party is obligated to purchase from the other party. The Seller might interpret this clause as meaning that its service offering need not be the lowest price, but merely "competitive," for the Customer to be obligated to purchase.

For purposes of this Agreement, "preferred provider" means that a party shall select the other party to this Agreement for provision of the specified services if (a) such Party has the prescribed service offering available for purchase that is ***competitive in all respects*** with that being offered by a competing service (e.g., features, functionality, and reliability of the alternative service); (b) the "preferred provider" ***can provision its service on a comparable timeliness basis*** with that of a competing service. The foregoing notwithstanding, the selection of such Party shall not be required if the end user customer expresses a preference for a different service provider, or the promotion or referral to such Party as service provider would conflict with a commitment to another service provider.

Example Clause #7: In the following example, the Customer is obligated to purchase services from Seller that are offered at "competitive prices and terms." As in several of the previous examples, the following clause uses the term "competitive prices and terms," which might be interpreted to mean that the Seller need not offer the lowest prices and most favorable terms, but merely a price that is "competitive."

Customer shall treat Seller as a preferred provider in the markets where Seller can provide Services at ***competitive prices and terms*** and within comparable provisioning intervals. Customer shall not be obligated to utilize Seller as a preferred provider if: (1) Customer's customer has expressed a preference to use another provider; (2) Customer has, as of the Amendment Effective Date, a conflicting contractual commitment to another provider including; (3) Customer is using its own facilities or those of its

Affiliates to provide the Services; or (4) generally accepted economic factors make it commercially infeasible for Customer to use Seller. Seller agrees to provide the Services in accordance with the terms of the Agreement and this Amendment.

Example Clause #8: In the following clause, the Customer has not made a firm commitment to purchase services from Seller. The Customer has agreed merely to use "reasonable efforts" to purchase from Seller if certain conditions are satisfied.

> Customer shall treat Seller as a preferred supplier where Seller can provide Services at ***competitive prices and terms*** and within comparable ***provisioning intervals***. ***Customer shall not be obligated or liable to use Seller for any requested service***. However, ***Customer will use reasonable efforts to use Seller*** if: (a) Seller's price quote or pricing outlined in this agreement represents the best price for Customer, (b) Customer does not have a conflicting contractual commitment to another provider, (c) ***Customer is confident*** that Seller will meet its installation intervals, and (d) Customer's customer has not expressed a preference to use another provider.

Example Clause #9: In this example the Customer has not made a firm commitment to purchase services from Seller. Before Customer will agree to purchase services it must be satisfied "in its reasonable discretion" that Seller's offering is superior.

> This Agreement is non-exclusive. Nothing in this Agreement shall prevent Customer or Service Provider from entering into similar arrangements with, or otherwise providing similar services to, any other person or entity. Notwithstanding the above, Customer agrees to use ***commercially reasonable efforts to make Service Provider a Preferred Provider*** (defined below) for Services, to the extent that (a) Customer and/or its Affiliates do not provide the Services itself, (b) Customer's End User has not expressed a preference for Customer to use a different service provider; (c) generally accepted ***economic*** factors make it commercially infeasible for Customer to use Service Provider; or (d) Customer has, as of the Effective Date, a conflicting contractual commitment to another provider. "Preferred Provider" means that Customer grants Service Provider the right of simultaneous offer on Customer's needs for Services, provided, however, that Customer shall not be obligated to accept the Service Provider offer unless ***Customer concludes in its reasonable discretion that the Service Provider offer is superior*** to those from other providers. Preferred Provider status is dependent on Service Provider's demonstrated performance basis which shall include, without limitation; price, availability, SLA features and functions, provisioning intervals, provisioning, billing, customer care and systems support.

Example Clause #10: In this example, the Customer's commitment is highly qualified. Before Customer is obligated to purchase form Seller in any given instance the parties must agree on the point of interconnection and many other terms, and even then the Customer retains discretion in whether to choose Seller's services.

Furthermore, the preferred provider commitment runs for two years only, and applies only to services that are offered at Seller's On-Net Buildings.

Preferred Provider:

Customer will purchase dedicated services of the type covered by this Agreement (*"Similar Access Services"*) from Supplier, as opposed to other suppliers of Similar Access Services (*"Other Suppliers"*), ***in certain On-Net Buildings*** under the specific circumstances described below. (The foregoing is referred to as "designating Supplier as primary provider.") ***Primary provider status is not available if the Parties cannot agree on a point of interconnection*** for a particular building. ***Customer will also not be obligated*** to designate Supplier as primary provider with respect to Ethernet Service or Optical Wavelength Service (each, as defined in Appendix ___, Service Description) ***until Customer and Supplier have implemented the service availability, provisioning, ordering, and maintenance systems and processes necessary for Customer to order for such Services*** from Supplier. Customer may, in its sole discretion, also purchase Services from Supplier under other circumstances. Supplier's designation as primary provider in an On-Net Building will last for ***two (2) years*** after Supplier has first begun providing any Service to Customer in such On-Net Building, unless such designation is earlier revoked in accordance with clause (4) below or unless the Agreement earlier expires or is terminated.

(1) If Customer requests Supplier to extend its Local Access Grid into a commercial-type building or a ***Carrier Hotel*** (as defined in Section ___), and Supplier had not previously independently determined to do so, Customer will designate Supplier as its primary provider with respect to Similar Access Services in such building once it has become an On-Net Building, subject to clauses (2) - (4) below. Evidence of such an independent determination by Supplier may include (without limitation) communications between the Parties, communications between Supplier and third parties or public statements by Supplier.

(2) Notwithstanding Supplier's designation as primary provider for an On-Net Building proposed to be included on any OC-X ring, Supplier will not be considered Customer's primary provider for Similar Access Services at OC-X Transmission Rates ("OC-X Ring Services," except under the following circumstances. If a Customer End User or potential End User located in a commercial-type building (but not a Carrier Hotel) requests an Customer-Provided OC-X Service, Customer may request bids for such Similar Access Services from Supplier and Other Suppliers. If Customer determines (in its sole discretion) that Supplier's is the winning bid or declines to seek other bids, then Customer will designate Supplier as its primary provider with respect to OC-X Ring Services as well as other Similar Access Services in such commercial-type building once it has become an On-Net Building, subject to clauses (3) - (4) below.

(3) Notwithstanding Supplier's designation as primary provider in an On-Net Building, Customer may at any time (a) provide Similar Access Services to End Users in such On-Net Building over access facilities owned or leased by Customer or its Affiliates, or (b) obtain any Similar Access Services from any Other Supplier (1) from which it had ordered Similar Access Services in the On-Net Building prior to Supplier's

designation as primary provider therefore, or (2) with which it had entered into an agreement for Similar Access Services (whether or not explicitly covering such On-Net Building) prior to placing its first order for Similar Access Services from Supplier for any End User in an On-Net Building.

(4) Customer need not designate Supplier as primary provider in any On-Net Building, and may revoke such designation in all On-Net Buildings previously granted, if Supplier does not comply with Section 19.B. Additionally, Customer need not designate Supplier as primary provider in a given On-Net Building, and may revoke such designation if previously granted, if:

(A) Customer has determined, in its sole judgment and in good faith, that Supplier's performance in such On-Net Building does not meet the standards articulated in the Requirements Document or this Agreement;

(B) Supplier's provisioning intervals are not better than those of Other Suppliers of Similar Access Services in such On-Net Building;

(C) Supplier's pricing is not competitive with Other Suppliers' pricing for Similar Access Services in such On-Net Building.

A Customer Serving Office in a "carrier hotel" type interconnection facility managed by a third party is referred to herein as a "Carrier Hotel."

The Parties will meet not less frequently than every six (6) months during the term of this Agreement to review Supplier's progress of its build-out, pricing levels and Supplier's performance under this Agreement.

Part 4, Chapter 6 - Acceptable Use Policy

The terms and conditions of a Master Services Agreement govern a customer's use of the services. In addition, a seller will sometimes adopt a policy on the proper and appropriate use of the services. This policy is called an "Acceptable Use Policy." All of seller's customers are required to abide by the seller's Acceptable Use Policy.

Sometimes an AUP is included as an exhibit to a Master Services Agreement. Often an AUP is not actually included in an MSA, but there is a reference in the contract to an Internet link where the Acceptable Use Policy may be found.

Example Clause #1: The first two clauses are very simple, either of which might be included in a Master Services Agreement. The Customer agrees to be bound by the Seller's AUP which is found at Seller's website. The first example is set forth as covenant; the second is a representation and warranty.

> ACCEPTABLE USE POLICY. Customer agrees to be bound by the Seller's Acceptable Use Policy which may be found at (www.____________.com)
>
> Reseller represents and warrants that it and its Customers will comply with Seller' Acceptable Use Policy that is available at www.____.com. Reseller understands that this policy may be revised from time to time in order to keep pace with changing technology and terminology.

Example Clause #2: The following clause would also be incorporated into an MSA. The Seller reserves the right to modify an AUP from time to time. The Customer would be bound by changes made in the future to the AUP.

> Acceptable Use Policy. ***Customer's use of Service shall at all times comply*** with Seller's then-current Acceptable Use Policy and Privacy Policy, ***as amended by Seller from time to time*** and which are available through Seller's web site at www.________.com. Seller will notify Customer of complaints received by Seller regarding each incident of alleged violation of Seller's Acceptable Use Policy by Customer or third parties that have gained access to the Service through Customer. Customer agrees that it will promptly investigate all such complaints and take all necessary actions to remedy any actual violations of Seller's Acceptable Use Policy. Seller may identify to the complaining party that Customer, or a third party that gained access to the Service through Customer, is investigating the complaint and may provide the complaining party with the necessary information to contact Customer directly to resolve the complaint. Customer shall identify a representative for the purposes of receiving such communications. Seller reserves the right to install and use, or to have Customer install and use, any appropriate devices to prevent violations of its Acceptable Use Policy, including devices designed to filter or terminate access to Service.

Example Clause #3: The following clause would allow the Seller to modify the Acceptable Use Police from time to time. Furthermore, if the Customer is found to have violated the AUP, such a breach of the contract would put Customer in default of the contract without the ability to cure the default.

> Customer shall comply with Seller's Acceptable Use and Security Policies (collectively the "Policy") which Policy Seller may modify at any time. The current, complete Policy is available for review at www.________________.com. (Seller may change the web site address via electronic notice to Customer). A failure to comply with the Policy shall constitute ***a material breach of this Agreement not capable of remedy***.

Example Clause #4: Under this clause the Customer is responsible for controlling its customers who might misuse the service. Customer must agree to terminate services to any of its customers that misuse the service.

This clause places particular emphasis on the threats to Supplier that are created by spamming, and requires the Customer to respond promptly.

> Use of Service. Customer agrees that it and its Affiliates will not (a) use Service(s) for any purpose other than that for which it is intended or ***in violation of any law or regulation or in aid of any unlawful act***; (b) use Service(s) so as to interfere with the use of Supplier network by other customers or authorized users of Supplier; (c) use Service(s) for transmission of any unsolicited distribution lists or other ***unsolicited electronic mailing*** ("Spamming"); and/or (d) unless otherwise specified herein, use Service(s) for distribution of any communication, including but not limited to unsolicited electronic mail, that includes Supplier's logos, trademarks, service marks, carrier identification codes (CICs), hosted IP addresses, or any variation thereof. Customer further agrees that it will not, and will use its best efforts to ensure that any End User will not, violate Supplier's Acceptable Use Policy, which may be read at www.________________.com. In the event of violation of this section by Customer's End User, Supplier shall notify Customer, and upon such notification ***Customer hereby agrees to immediately terminate said End User's account***. In the event Customer does not immediately terminate said End User's account, Supplier reserves the right to suspend Services or terminate this Agreement. Supplier will use best efforts to notify Customer of service suspension. Notwithstanding anything to the contrary herein, if in Supplier's sole judgment, Customer, Customer's Affiliates or Customer's End User violates this section, and such violation or failure to comply poses an immediate threat of harm to or destruction of Supplier's network, violates existing law or regulation, or puts Supplier's network at risk with its providers of network services or other customers, Supplier shall have the right to immediately take any and all steps reasonably necessary to remove such threat, including but not limited to suspension or termination of Services immediately and without notice.
>
> Customer and Customer's Affiliates shall indemnify and hold harmless Supplier for Spamming or illegal activities, including but not limited to acts using a third party mail server, caused by Customer's End Users or Affiliates using Supplier's network. All requests for the use of third-party mail servers shall be subject to Supplier's sole approval. Supplier shall not (a) have any obligation or liability to Customer or to any third party for any unlawful or improper use of Services by an End User or Affiliate; nor

(b) have any duty or obligation to exercise control over the use of, or the content or information passing through, the Service.

Example Clause #5: The following clause is similar to others except that it adds an indemnification requirement.

RESTRICTIONS ON USE. Customer agrees that it shall not sell the Services as a whole to others, and that it (and others with access through Customer to the Services) will abide by Seller' Acceptable Use Policy at www.________.com, as periodically revised.

INDEMNIFICATION. Customer agrees to indemnify Seller from any and all third party claims of damages, liability, costs and expenses (including reasonable attorney's fees and expenses) arising from any violation of Seller' Acceptable Use Policy, regardless of whether done with intent or knowledge.

Example Clause #6: In the following example the Seller has combined the AUP with other liability provisions. In addition, the Customer is obligated to impose the AUP on its customers.

Use of Services. Customer will be liable for damages to Seller-provided equipment located on Customer's or End User's premises except reasonable wear and tear or damage caused by Seller. Upon expiration or termination of the Order, Customer must return to Seller any equipment and other Seller-owned property provided to Customer or End Users. Customer will not make any illegal use of the Services and ***will conform to Seller's acceptable use policy ("AUP")*** which may be modified from time to time and is available on the Seller web site at www.________.net/abuse.html. ***Customer will require its End Users to comply with the AUP***. Customer will not use, or permit others to use, Services for any purpose other than its intended use or alter, tamper with, adjust, or repair Services.

Example Clause #7: In this example the AUP is integrated with other default provisions. The Seller has made a distinction between kinds of violations of the AUP that do not affect Seller's operations, and those that do.

Default and Remedies. A "Default" shall occur if (a) Customer fails to make payment as required under this Agreement and such failure remains uncorrected for ten (10) calendar days after written notice from Seller; or (b) either party fails to perform or observe any material term or obligation contained in this Agreement, and any such failure remains uncorrected for thirty (30) calendar days after written notice from the non-defaulting party. If Customer uses the Services for any unlawful purpose or in any unlawful manner, Seller shall have the right immediately to suspend and/or terminate any or all Services hereunder without notice to Customer. ***If Customer violates Seller's Acceptable Use Policy ("AUP") posted on Seller's AUP website at http://www.__________.com/customers/policy/uses.html, which is incorporated herein by reference, and if such activity is affecting Seller's network, or other parties on Seller's network, Seller shall have the right immediately to suspend and/or terminate any or all Services hereunder without notice to Customer.*** Customer affirms that it has reviewed and assented to the AUP. For any ***violations of Seller's AUP that are not affecting either Seller's network or third parties on Seller's network, Customer shall, upon three (3) business days' notice, have the opportunity to cure*** such violation prior to suspension or termination. In the event of a Customer Default for any reason, Seller may in addition to its right available to it at law or in equity: (a) suspend Services to Customer; (b) cease accepting or processing orders for Services; (c) withhold delivery of Call Detail Records (if applicable); and/or (e) except in the event of an AUP violation, terminate this Agreement. If this Agreement is terminated because of a Customer Default, such termination shall not affect or reduce Customer's minimum monthly commitments required under this Agreement, if applicable; and, all Early Termination Charges shall apply.

In the event of any inconsistency between or among a Service Order, a Service Schedule, the provisions contained herein, Seller's AUP website, and any applicable tariff, the following order of precedence shall prevail (from highest priority to lowest): the applicable tariff, if any, a Service Order, a Service Schedule, Seller's AUP website and the provisions contained herein.

Example Clause #8: Next are several examples of sample Acceptable Use Policies. All of these policies have certain elements in common. They all require the Customer to comply will all laws in its use of the service, especially laws pertaining to spamming. The policies differ in with respect to remedies that Seller may invoke for a failure of Customer or its users to comply with the AUP, such as how promptly Seller may suspend or terminate service.

SELLER Acceptable Use Policy

Introduction

SELLER is at all times committed to complying with the laws and regulations governing use of the Internet, e-mail transmission and text messaging and preserving for all of its Customers the ability to use SELLER's network and the Internet without interference or harassment from other users. The SELLER AUP ("AUP") is designed to help achieve these goals.

By using IP Service(s), as defined below, Customer(s) agrees to comply with this Acceptable Use Policy and to remain responsible for its users. SELLER reserves the right to change or modify the terms of the AUP at any time, effective when posted on SELLER's web site at ___________. Customer's use of the IP Service(s) after changes to the AUP are posted shall constitute acceptance of any changed or additional terms.

Scope of the AUP

The AUP applies to the SELLER services that provide (or include) access to the Internet, including hosting services (software applications and hardware), or are provided over the Internet or wireless data networks (collectively "IP Services").

Prohibited Activities

General Prohibitions: SELLER prohibits use of the IP Services in any way that is unlawful, harmful to or interferes with use of SELLER's network or systems, or the network of any other provider, interferes with the use or enjoyment of services received by others, infringes on intellectual property rights, results in the publication of threatening or offensive material, or constitutes Spam/E-mail/Usenet abuse, a security risk or a violation of privacy.

Failure to adhere to the rules, guidelines or agreements applicable to search engines, subscription Web services, chat areas, bulletin boards, Web pages, USENET, applications, or other services that are accessed via a link from the SELLER-branded website or from a website that contains SELLER-branded content is a violation of this AUP.

Unlawful Activities: IP Services shall not be used in connection with any criminal, civil or administrative violation of any applicable local, state, provincial, federal, national or international law, treaty, court order, ordinance, regulation or administrative rule.

Violation of Intellectual Property Rights: IP Service(s) shall not be used to publish, submit/receive upload/download, post, use, copy or otherwise reproduce, transmit, re-transmit, distribute or store any content/material or to engage in any activity that infringes, misappropriates or otherwise violates the intellectual property rights or privacy or publicity rights of SELLER or any individual, group or entity, including but not limited to any rights protected by any copyright, patent, trademark laws, trade secret, trade dress, right of privacy, right of publicity, moral rights or other intellectual property right now known or later recognized by statute, judicial decision or regulation.

Threatening Material or Content: IP Services shall not be used to host, post, transmit, or re-transmit any content or material (or to create a domain name or operate from a domain name), that harasses, or threatens the health or safety of others. In addition, for those IP Services that utilize SELLER provided web hosting, SELLER reserves the right to decline to provide such services if the content is determined by SELLER to be obscene, indecent, hateful, malicious, racist, defamatory, fraudulent, libelous, treasonous, excessively violent or promoting the use of violence or otherwise harmful to others.

Inappropriate Interaction with Minors: SELLER complies with all applicable laws pertaining to the protection of minors, including when appropriate, reporting cases of child exploitation to the National Center for Missing and Exploited Children. For more information about online safety, visit www.ncmec.org or __________.

Child Pornography: IP Services shall not be used to publish, submit/receive, upload/download, post, use, copy or otherwise produce, transmit, distribute or store child pornography. Suspected violations of this prohibition may be reported to SELLER at the following e-mail address: ______________. SELLER will report any discovered violation of this prohibition to the National Center for Missing and Exploited Children and take steps to remove child pornography (or otherwise block access to the content determined to contain child pornography) from its servers.

Spam/E-mail/Usenet Abuse: Violation of the CAN-SPAM Act of 2003, or any other applicable law regulating e-mail services, constitutes a violation of this AUP.

Spam/E-mail or Usenet abuse is prohibited using IP Services. Examples of Spam/E-mail or Usenet abuse include but are not limited to the following activities:

- sending multiple unsolicited electronic mail messages or "mail-bombing" - to one or more recipient;
- sending unsolicited commercial e-mail, or unsolicited electronic messages directed primarily at the advertising or promotion of products or services;
- sending unsolicited electronic messages with petitions for signatures or requests for charitable donations, or sending any chain mail related materials;
- sending bulk electronic messages without identifying, within the message, a reasonable means of opting out from receiving additional messages from the sender;
- s ending electronic messages, files or other transmissions that exceed contracted for capacity or that create the potential for disruption of the SELLER network or of the networks with which SELLER interconnects, by virtue of quantity, size or otherwise;
- using another site's mail server to relay mail without the express permission of that site;
- using another computer without authorization to send multiple e-mail messages or to retransmit e-mail messages for the purpose of misleading recipients as to the origin or to conduct any of the activities prohibited by this AUP;
- using IP addresses that the Customer does not have a right to use;
- collecting the responses from unsolicited electronic messages;
- maintaining a site that is advertised via unsolicited electronic messages, regardless of the origin of the unsolicited electronic messages;
- sending messages that are harassing or malicious, or otherwise could reasonably be predicted to interfere with another party's quiet enjoyment of the IP Services or the Internet (e.g., through language, frequency, size or otherwise);
- using distribution lists containing addresses that include those who have opted out;
- sending electronic messages that do not accurately identify the sender, the sender's return address, the e-mail address of origin, or other information contained in the subject line or header;
- falsifying packet header, sender, or user information whether in whole or in part to mask the identity of the sender, originator or point of origin;
- using redirect links in unsolicited commercial e-mail to advertise a website or service;
- posting a message to more than ten (10) online forums or newsgroups, that could reasonably be expected to generate complaints;
- intercepting, redirecting or otherwise interfering or attempting to interfere with e-mail intended for third parties;

- knowingly deleting any author attributions, legal notices or proprietary designations or labels in a file that the user mails or sends;
- using, distributing, advertising, transmitting, or otherwise making available any software program, product, or service that is designed to violate this AUP or the AUP of any other Internet Service Provider, including, but not limited to, the facilitation of the means to spam.

Security Violations

Customers are responsible for ensuring and maintaining security of their systems and the machines that connect to and use IP Service(s), including implementation of necessary patches and operating system updates.

IP Services may not be used to interfere with, gain unauthorized access to, or otherwise violate the security of SELLER's (or another party's) server, network, network access, personal computer or control devices, software or data, or other system, or to attempt to do any of the foregoing. Examples of system or network security violations include but are not limited to:

- unauthorized monitoring, scanning or probing of network or system or any other action aimed at the unauthorized interception of data or harvesting of e-mail addresses;
- hacking, attacking, gaining access to, breaching, circumventing or testing the vulnerability of the user authentication or security of any host, network, server, personal computer, network access and control devices, software or data without express authorization of the owner of the system or network;
- impersonating others or secretly or deceptively obtaining personal information of third parties (i.e., phishing, etc.);
- using any program, file, script, command or transmission of any message or content of any kind, designed to interfere with a terminal session, the access to or use of the Internet or any other means of communication;
- distributing or using tools designed to compromise security (including but not limited to SNMP tools), including cracking tools, password guessing programs, packet sniffers or network probing tools (except in the case of authorized legitimate network security operations);
- knowingly uploading or distributing files that contain viruses, spyware, Trojan horses, worms, time bombs, cancel bots, corrupted files, root kits or any other software or programs that may damage the operation of another's computer, network system or other property, or be used to engage in modem or system hijacking;
- engaging in the transmission of pirated software;
- with respect to dial-up accounts, using any software or device designed to defeat system time-out limits or to allow Customer's account to stay logged on while Customer is not actively using the IP Services or using such account for the purpose of operating a server of any type;
- using manual or automated means to avoid any use limitations placed on the IP Services;

- providing guidance, information or assistance with respect to causing damage or security breach to SELLER's network or systems, or to the network of any other IP Service provider;
- failure to take reasonable security precautions to help prevent violation(s) of this AUP.

Customer Responsibilities

Customers remain solely and fully responsible for the content of any material posted, hosted, downloaded/uploaded, created, accessed or transmitted using the IP Services. SELLER has no responsibility for any material created on the SELLER's network or accessible using IP Services, including content provided on third-party websites linked to the SELLER network. Such third-party website links are provided as Internet navigation tools for informational purposes only, and do not constitute in any way an endorsement by SELLER of the content(s) of such sites.

Customers are responsible for taking prompt corrective action(s) to remedy a violation of AUP and to help prevent similar future violations.

AUP Enforcement and Notice

Customer's failure to observe the guidelines set forth in this AUP may result in SELLER taking actions anywhere from a warning to a suspension or termination of Customer's IP Services. When feasible, SELLER may provide Customer with a notice of an AUP violation via e-mail or otherwise allowing the Customer to promptly correct such violation.

SELLER reserves the right, however, to act immediately and without notice to suspend or terminate affected IP Services in response to a court order or government notice that certain conduct must be stopped or when SELLER reasonably determines, that the conduct may: (1) expose SELLER to sanctions, prosecution, civil action or any other liability, (2) cause harm to or interfere with the integrity or normal operations of SELLER's network or networks with which SELLER is interconnected, (3) interfere with another SELLER Customer's use of IP Services or the Internet (4) violate any applicable law, rule or regulation, or (5) otherwise present an imminent risk of harm to SELLER or SELLER Customers.

SELLER has no obligation to monitor content of any materials distributed or accessed using the IP Services. However, SELLER may monitor content of any such materials as necessary to comply with applicable laws, regulations or other governmental or judicial requests; or to protect the SELLER network and its customers.

Incident Reporting

Any complaints (other than claims of copyright or trademark infringement) regarding violation of this AUP by an SELLER Customer (or its user) should be directed to

__________. Where possible, include details that would assist SELLER in investigating and resolving such complaint (e.g. expanded headers, IP address(s), a copy of the offending transmission and any log files).

Copyright complaints: If you believe that your work has been copied and posted, stored or transmitted using the IP Services in a way that constitutes copyright infringement, please submit a notification pursuant to the Digital Millennium Copyright Act ("DMCA") in accordance with the process detailed at ________________________ and directed to the designated agent listed below:

SELLER's Designated Copyright Agent for notice of claims of copyright or trademark infringement on the sites can be reached as follows:

____

____

Contact Information: Any notification that SELLER sends to its Customers pursuant to this AUP will be sent via e-mail to the e-mail address on file with SELLER, or may be in writing to Customer's address of record. It is Customer's responsibility to promptly notify SELLER of any change of contact information.

Effective Date: ______________

Example Clause #9: There are two distinguishing features to the following Acceptable Use Policy. The Seller has the right to edit or remove the content of communications that Seller believes are in violation of the AUP. Furthermore, the Seller has the right to suspend or terminate service without notice to the Customer, if Seller believes that Customer has violated the AUP.

Acceptable Use Policy

Introduction

This acceptable use policy (the "Policy") defines acceptable practices relating to the use of Seller's services (the "Service") by customers of Seller ("Customers") and by users that have gained access to the Service through Customer accounts ("Users"). By using the Service, you acknowledge that you and your Users are responsible for compliance with the Policy. You are responsible for violations of this Policy by any User that accesses the Services through your account, which includes without limitation Users for whom you provide services as a reseller of Seller's Services. While it is not our intent to control or monitor online experience or the content of online communications, ***we may edit or remove content*** that we deem to be in violation of the Policy or that we otherwise deem unlawful, harmful or offensive. The Policy applies to all aspects of the Service. "Seller" means Seller and all of its affiliates (including direct and indirect subsidiaries and parents). "Seller Network" includes, without limitation, Seller's constructed or leased transmission network, including all equipment, systems, facilities, services and products incorporated or used in such transmission network. As used in this Policy, "you" refers to Customers, and any reference to "Users" is intended to encompass, as applicable, both Customers and their Users.

This Policy is designed to assist in protecting the Seller Network, the Service, our Users and the Internet community as a whole from improper and/or illegal activity over the Internet, to improve Service and to improve Service offerings. In situations where data communications are carried across networks of other Internet Service Providers (ISPs), Users of the Seller Network must also conform to the applicable acceptable use policies of such other ISPs.
Rights of Seller

Suspension or Termination of Service

If Users engage in conduct or a pattern of conduct, including without limitation repeated violations by a User whereby correction of individual violations does not in Seller's sole discretion correct a pattern of the same or similar violations, while using the Service that violates the Policy, or is otherwise illegal or improper, Seller reserves the right to suspend and/or terminate the Service or the User's access to the Service. Seller will generally attempt to notify you of any activity in violation of the Policy and request that the User cease such activity; however, in cases where the operation of the Seller Network is threatened or cases involving unsolicited commercial email/SPAM, a pattern of violations, mail relaying, alteration of your source IP address information, denial of service attacks, illegal activities, suspected fraud in connection with the use of Service, harassment or copyright infringement, ***we reserve the right to suspend or terminate your Service or the User's access to the Service without notification***. In addition, we may take any other appropriate action against you or a User for violations of the Policy, including repeated violations wherein correction of individual violations does not in Seller's sole discretion correct a pattern of the same or similar violations. Seller reserves the right to avail itself of the safe harbor provisions of the Digital Millennium Copyright Act. We do not make any promise, nor do we have any obligation, to monitor or police activity occurring using the Service and will have no liability to any party, including you, for any violation of the Policy.

Cooperation with Investigations

Seller will cooperate with appropriate law enforcement agencies and other parties involved in investigating claims of illegal or inappropriate activity. Seller reserves the right to disclose Customer information to the extent authorized by federal or state law. As a reseller Customer of Seller's Services (if applicable), by using and accepting Services, you consent to our disclosure to any law enforcement agency, without the need for subpoena, of your identity as the service provider of record (including basic contact information), as applicable, for any User about whom Seller is contacted by the law enforcement agency. In instances involving child pornography, Seller complies with all applicable federal and state laws including providing notice to the National Center for the Missing and Exploited Children or other designated agencies.

Modifications to Policy

Seller reserves the right to modify this Policy at any time without notice. We will attempt to notify Customers of any such modifications either via e-mail or by posting a revised version of the Policy on our Web site. Any such modifications shall be effective and applied prospectively from the date of posting.

Filters and Service Information

We reserve the right to install and use, or to have you install and use, any appropriate devices to prevent violations of this Policy, including devices designed to filter or terminate access to the Service. By accepting and using the Service, you consent to allowing us to collect service information and routing information in the normal course of our business, and to use such information for general business purposes.

Prohibited Uses

Illegal Activity

The Service shall not be used for any unlawful activities or in connection with any criminal or civil violation and the Services shall in all cases be used in compliance with applicable law. Use of the Service for transmission, distribution, retrieval, or storage of any information, data or other material in violation of any applicable law or regulation (including, where applicable, any tariff or treaty) is prohibited. This includes, without limitation, the use or transmission of any data or material protected by copyright, trademark, trade secret, patent or other intellectual property right without proper authorization and the transmission of any material that constitutes an illegal threat, violates export control laws, or is obscene, defamatory or otherwise unlawful.

Unauthorized Access/Interference

A User may not attempt to gain unauthorized access to, or attempt to interfere with or compromise the normal functioning, operation or security of, any portion of the Seller Network. A User may not use the Service to engage in any activities that may interfere with the ability of others to access or use the Service or the Internet. A User may not use the Service to monitor any data, information or communications on any network or system without authorization. A User may not attempt to gain unauthorized access to the user accounts or passwords of other Users.

Unsolicited Commercial Email/Spamming/Mailbombing

A User may not use the Service to transmit unsolicited commercial e-mail messages or deliberately send excessively large attachments to one recipient. Any unsolicited commercial e-mail messages or a series of unsolicited commercial e-mail messages or large attachments sent to one recipient is prohibited. In addition, "spamming" or "mailbombing" is prohibited. Use of the service of another provider to send unsolicited commercial email, spam or mailbombs, to promote a site hosted on or connected to the Seller Network, is similarly prohibited. Likewise, a User may not use the Service to collect responses from mass unsolicited e-mail messages. Seller may in its sole discretion rely upon information obtained from anti-spamming organizations (including for example and without limitation spamhaus.org, spamcop.net, sorbs.net, and abuse.net) as evidence that a User is an active "spam operation" for purposes of taking remedial action under this Policy.

Spoofing/Fraud

Users are prohibited from intentionally or negligently injecting false data into the Internet, for instance in the form of bad routing information (including but not limited to the announcing of networks owned by someone else or reserved by the Internet Assigned Numbers Authority) or incorrect DNS information.

A User may not attempt to send e-mail messages or transmit any electronic communications using a name or address of someone other than the User for purposes of deception. Any attempt to impersonate someone else by altering a source IP address information or by using forged headers or other identifying information is prohibited. Any attempt to fraudulently conceal, forge or otherwise falsify a User's identity in connection with use of the Service is prohibited.

USENET Postings

All postings to USENET groups must comply with that group's charter and other policies. Users are prohibited from cross posting to unrelated news groups or to any news groups where the post does not meet that group's charter. Continued posting of off-topic messages, including commercial messages (unless specifically invited) is prohibited. Disrupting newsgroups with materials, postings or activities that are (as determined by Seller in its sole discretion) frivolous, unlawful, obscene, threatening, abusive, libelous, hateful, excessive or repetitious is prohibited, unless such materials or activities are expressly allowed or encouraged under the newsgroup's name, FAQ or charter.

Complaints

Seller receives complaints directly from Internet users, through Internet organizations and through other parties. Seller shall not be required to determine the validity of complaints received, or of information obtained from anti-spamming organizations, before taking action under this AUP. A complaint from the recipient of commercial email, whether received directly or through an anti-spamming organization, shall be evidence that the message was unsolicited. Seller has no obligation to forward the complaint to the User or to identify the complaining parties.

The following activities are also prohibited:

- Intentionally transmitting files containing a computer virus or corrupted data.

- If we have specified bandwidth limitations for your user account, use of the Service shall not be in excess of those limitations. If a User is accessing the Service via a dial-up connection, we may terminate the user session if the User is connected for more than 7 days in order to protect Seller Network resources and maintain Service availability for others.
- Attempting to circumvent or alter the processes or procedures to measure time, bandwidth utilization, or other methods to document use of Seller's services.
- Advertising, transmitting, or otherwise making available any software, program, product, or service that is designed to violate this AUP, which includes the facilitation of the means to deliver unsolicited commercial email.
- Any activity that disrupts, degrades, harms or threatens to harm the Seller Network or the Service.
- Any use of another party's electronic mail server to relay email without express permission from such other party is prohibited.
- Any other inappropriate activity or abuse of the Service (as determined by us in our sole discretion), whether or not specifically listed in this Policy, may result in suspension or termination of the User's access to or use of the Service.

This listing of prohibited activities is not exhaustive and Seller reserves the right to determine that any conduct that is or could be harmful to the Seller Network, Seller's Customers or Internet users is in violation of this Policy and to exercise any or all of the remedies contained in this Policy.

Responsibilities of Customers

Users are entirely responsible for maintaining the confidentiality of password and account information, as well as the security of their network. You agree immediately to notify Seller of any unauthorized use of your account or any other breach of security known to you. If you become aware of any violation of this Policy by any person, including Users that have accessed the Service through your account, you are required to notify us.

Impending Security Event Notification

All Users of the Seller Network are responsible for notifying Seller immediately if they become aware of an impending event that may negatively affect the Seller Network. This includes extortion threats that involve threat of "denial of service" attacks, unauthorized access, or other security events.

Configuration

All Users of the Seller Network are responsible for configuring their own systems to provide the maximum possible accountability. Seller shall not be liable for any damage caused by such system configurations regardless of whether such configurations have been authorized or requested by Seller. For example, Users should ensure that there are clear "path" lines in news headers so that the originator of a post may be identified. Users should also configure their Mail Transport Agents (MTA) to authenticate (by look-up on the name or similar procedures) any system that connects to perform a mail exchange, and should generally present header data as clearly possible. As another example, Users should maintain logs of dynamically assigned IP addresses. Users of the Seller Network are responsible for educating themselves and configuring their systems with at least basic security. Should systems at a User's site be violated, the User is responsible for reporting the violation and then fixing the exploited system. For instance, should a site be abused to distribute unlicensed software due to a poorly configured FTP (File Transfer Protocol) Server, the User is responsible for re-configuring the system to stop the abuse.

Complaints

In most cases, we will notify our Customer(s) of complaints received by us regarding an alleged violation of this Policy. You agree to promptly investigate all such complaints and take all necessary actions to remedy any violations of this Policy. We may inform the complainant that you are investigating the complaint and may provide the complainant with the necessary information to contact you directly to resolve the complaint. You shall identify a representative for the purposes of receiving such communications.

Privacy

Because the Internet is an inherently open and insecure means of communication, any data or information a User transmits over the Internet may be susceptible to interception and alteration. Subject to our Online Privacy Policy, we make no guarantee regarding, and assume no liability for, the security and integrity of any data or information a User transmits via the Service or over the Internet, including any data or information transmitted via any server designated as "secure."

Claims of Copyright Infringement

Please click here ______ for information relating to the Interim Designation of Agent to Receive Notification of Claimed Infringement pursuant to the Digital Millennium Copyright Act, 17 U.S.C. Section 512(c).

Additional Terms and Conditions

The use of the Seller Network by a Customer of Seller is subject to the terms and conditions of any agreements entered into by such Customer and Seller. This Policy is incorporated into such agreements by this reference.

Complaints and Contact Information

Any complaints regarding prohibited use or other abuse of the Seller Network, including violations of this Policy, should be sent to Seller. Please include all applicable information that will assist Seller in investigating the complaint, including all applicable headers of forwarded messages.

Sites experiencing live attacks from Seller Customers should call into our Customer Care Center (telephone ____________) to submit a complaint as quickly as possible. Describe the urgency of the situation should you need immediate attention. If you are unsure whether any contemplated use or action is permitted, please submit questions or comments to Seller at __________.

Example Clause #10: In this clause there is more than a general statement that the AUP applies to Customer's users. Seller requires that Customer obtain from each of its users an agreement in writing that they will comply with the AUP. Further, the Customer must actively enforce the AUP on its customers.

ACCEPTABLE USE POLICY

All Seller customers are responsible for reviewing and complying with this Acceptable Use Policy. Seller customers who provide services to their own users ***must affirmatively and contractually pass on the restrictions of this Acceptable Use Policy to their users, and take steps to ensure compliance by their users with this Acceptable Use Policy*** including, without limitation, termination of users who violate this policy. For the purposes of this Policy, "Customer" shall be defined to include Seller's customers and also a customer's users and account holders. This Policy is subject to change with notice by publication on the web at ________________. Customers are responsible for monitoring this web site for changes. This Policy was last changed on _________.

The actions described below are defined by Seller as "system abuse" and are strictly prohibited under this Policy. The examples named below are not exhaustive and are provided solely for guidance to Customers. If any Customer is unsure of whether a contemplated use or action is permitted, it is Customer's responsibility to determine whether the use is permitted by contacting Seller via electronic mail. The following activities are expressly prohibited, and Seller expressly reserves the right, at its discretion, to pursue any remedies that it believes are warranted which may include, but is not limited to, filtering, ***suspending, or terminating accounts***, end-users or Customers that engage in system abuse or who refuse to address user abuse.

In general, Seller Customers may not use Seller's network, machines, or services in any manner which:

- violates any applicable law, regulation, treaty, or tariff;
- violates the acceptable use policies of any networks, machines, or services which are accessed through Seller's network;
- infringes on the intellectual property rights of Seller or others;
- violates the privacy of others;
- involves the resale of Seller's products or services, unless specifically documented in a separate written agreement or in the initial Customer contract with Seller;

- involves deceptive online marketing practices including, without limitation practices that violate the United States Federal Trade Commission's guidelines for proper online marketing schemes; or
- otherwise violates this Acceptable Use Policy.

Prohibited activities also include but are not limited to, the following:

- unauthorized use (or attempted unauthorized use) of any machines or networks,
- attempting to interfere with or denying service to any user or host (e.g. denial of service attacks);
- falsifying header information or user identification information;
- introduction of malicious programs into the network or Server (e.g. viruses, worms, Trojan horses, etc.);
- monitoring or scanning the networks of others without permission;
- attempted or successful security breaches or disruption of Internet communication including, but not limited to, accessing data of which Customer is not an intended recipient or logging into a Server or account that Customer is not expressly authorized to access;
- executing any form of network monitoring (e.g. packet sniffer) which will intercept data not intended for the Customer;
- attempting to circumvent Customer authentication or security of any host, network, or account ("cracking");
- Using any program/script/command, or sending messages of any kind, designed to interfere with a third party customer terminal session, via any means, locally or via the Internet;
- sending unsolicited bulk email;
- maintaining an open mail relay;
- collecting email addresses from the Internet for the purpose of sending unsolicited bulk email or to provide collected addresses to others for that purpose;
- transmitting or receiving copyright-infringing or obscene material;
- furnishing false or incorrect data on the signup form; or
- attempting to circumvent or alter the process or procedures to measure time, bandwidth utilization, or other methods to document "use" of Seller's products and services.

Dedicated and Webhosting Customers

Seller dedicated Internet and web hosting customers who provide services to their own users ***must affirmatively and contractually pass on the restrictions of this Acceptable Use Policy to its users, and take steps to ensure compliance by their users*** with this Acceptable Use Policy including, without limitation, termination of the user for violations of this policy.

Seller dedicated Internet and web hosting customers who provide services to their own users also must maintain valid postmaster and abuse addresses for their domains, comply with all applicable Internet RFCs, maintain appropriate reverse DNS information for all hosts receiving connectivity through Seller's network for which DNS responsibility has been delegated to the customer, maintain accurate contact information with the InterNIC and any other appropriate domain and IP address registries, take reasonable steps to prevent IP spoofing by their users and downstream customers, provide a 24/7 contact address to Seller for dealing with security and abuse issues, and act promptly to ensure that users are in compliance with Seller's Acceptable Use Policy. Reasonable steps include, but are not limited to, using ip verify unicast reverse path wherever appropriate and using IP address filtering wherever appropriate.

Email

Sending unsolicited ("opt-out") bulk email is prohibited. Sending unsolicited bulk email from another provider advertising or implicating, directly or indirectly, the use of any service hosted or provided by Seller, including without limitation email, web, FTP, and DNS services, is prohibited and is grounds for termination of those services to Customers or users who engage in the practice. Customers or users who send unsolicited bulk email from Seller accounts will be charged the cost of labor to respond to complaints, with a minimum charge of $200. Customers or users who send bulk email to "opt-in" lists must have a method of confirmation or verification of subscriptions and be able to show evidence of subscription for users who complain about receiving unsolicited email. The following actions are likewise prohibited:

- Using e-mail to engage in harassment, whether through language, frequency, or size of messages. Continuing to send someone email after being asked to stop is considered harassment.
- Using email to disrupt (e.g., mail bombing, "flashing," etc.) is prohibited. Sending email with falsified header information.
- Creating or forwarding chain letters, pyramid schemes, and hoaxes
- Using the Seller or customer account to collect replies to messages sent from another provider which violate these rules or those of the other provider.

Usenet newsgroups

Customers and users should be familiar with the workings of Usenet by reading FAQs regarding Usenet at:
http://www.faqs.org/faqs/by-newsgroup/news/news.announce.newusers.html before becoming active participants. Seller places the following restrictions on newsgroup postings by its users:

- no illegal content, including pyramid/Ponzi schemes, infringing materials, or child pornography, is permitted;
- all postings should conform to the various conventions, guidelines and local culture found in each respective newsgroup and Usenet as a whole.
- Commercial advertising is typically off-topic and/or a violation of charter in most Usenet newsgroups.

- Posting 20 or more copies of the same article in a 45-day period ("spamming") or continued posting of off-topic articles after being warned is prohibited. Users who engage in spamming using Seller accounts will be charged the cost of labor to issue cancellations and respond to complaints, with a minimum charge of $200. Users who engage in spamming from another provider advertising or implicating, directly or indirectly, the use of any service hosted or provided by Seller, including without limitation email, web, FTP, and DNS services, is prohibited and is grounds for termination of those services to those users.
- Excessive cross-posting (Breidbart Index of 20 or greater in a 45-day period) is prohibited. The Breidbart Index (BI) is calculated by taking the sum of the square roots of the number of newsgroups each copy of an article is cross-posted to. If two articles are posted, one cross-posted to 9 newsgroups and the other cross-posted to 16 newsgroups, the BI = sqrt(9)+sqrt(16)=3+4=7. Cross-posting articles to newsgroups where they are off-topic is prohibited; a good rule of thumb is that if you are cross-posting to more than five newsgroups, it's likely to be off-topic on at least one of them.
- Posting articles with falsified header information is prohibited. "Munging" header information to foil email address harvesting by "spammers" is acceptable provided that a reasonable means of replying to the message originator is given. Use of anonymous remailers is acceptable, so long as the use is not otherwise a violation of this policy.
- Users may not issue cancellations for postings except those which they have posted themselves, those which have headers falsified so as to appear to come from them, or in newsgroups where they are the official moderator.

The World Wide Web and FTP

The web space and public FTP space included with a dialup account may not be resold or used for adult-oriented material.

Seller reserves the right to require that sites using web or FTP space which receive high amounts of traffic be moved to other servers.

Web pages and FTP files may not contain any material, text, or images, whether hosted on Seller servers or "transclusioned" (images from another site displayed on the page) which violate or infringe any copyright, trademark, patent, statutory, common law, or proprietary rights of others. Web pages and FTP files may not contain links that initiate downloads of copyright-infringing or other illegal material.

Digital Millennium Copyright Act Policy

It is the policy of Seller to respond expeditiously to claims of intellectual property infringement. Seller will promptly process and investigate notices of alleged infringement and will take appropriate actions under the Digital Millennium Copyright Act ("DMCA") and other applicable intellectual property laws. Upon receipt of notices complying or substantially complying with the DMCA, when it is under its control, Seller will act expeditiously to remove or disable access to any material claimed to be infringing or claimed to be the subject of infringing activity, and will act expeditiously to remove or disable access to any reference or link to material or activity that is claimed to be infringing. Seller will terminate access for Customers who are repeat infringers. For our Webhosting customers, you must adopt and implement a Digital Millennium Copyright Act policy that reserves the necessary rights to remove or disable infringing material.

If you believe that a copyrighted work has been copied and is accessible on our site in a way that constitutes copyright infringement, you may notify us by providing our registered copyright agent with the following information:

- an electronic or physical signature of the person authorized to act on behalf of the owner of the copyright interest;
- a description of the copyrighted work that you claim has been infringed;
- a description of where the material that you claim is infringing is located on the site;
- your address, telephone number, and e-mail address;
- a statement by you that you have a good faith belief that the disputed use is not authorized by the copyright owner, its agent, or the law;
- a statement by you, made under penalty of perjury, that the above information in your Notice is accurate and that you are the copyright owner or authorized to act on the copyright owner's behalf.

Notices of claimed infringement should be directed to ________________.

When Seller removes or disables access to any material claimed to be infringing, Seller may attempt to contact the user who has posted such material in order to give that user an opportunity to respond to the notification. Any and all counter notifications submitted by the user will be furnished to the complaining party. Seller will give the complaining party an opportunity to seek judicial relief in accordance with the DMCA before Seller replaces or restores access to any material as a result of any counter notification.

Internet Relay Chat

Using IRC bots is prohibited. Flooding, cloning, spoofing, harassment, or otherwise hindering the ability of others to properly use IRC is prohibited. Impersonating other users, advertising, and spamming via IRC is prohibited.

Servers and Proxies

Users may not run on Seller servers any program which makes a service or resource available to others, including but not limited to port redirectors, proxy servers, chat servers, MUDs, file servers, and IRC bots. Users may not run such programs on their own machines to make such services or resources available to others through a Seller dialup account; a dedicated access account is required for such purposes.

Customers are responsible for the security of their own networks and machines. Seller will assume neither responsibility nor accountability for failures or breach of customer-imposed protective measures, whether implied or actual. Abuse that occurs as a result of a compromised customer's system or account may result in suspension of services or account access by Seller, for example, if a system is abused after becoming infected with Back Orifice or the NetBus Trojan horse programs as a result of an Internet download or executing an email attachment.

Any programs, scripts, or processes which generate excessive server load on Seller servers are prohibited and Seller reserves the right to terminate or suspend any such program, script, or process.

Unsolicited Bulk Email Support Services and Email Address Harvesting

Customers and users may not advertise, distribute, or use software intended to facilitate sending unsolicited bulk email or harvest email addresses from the Internet for that purpose. Customers and users may not sell or distribute lists of harvested email addresses for that purpose.

Referral ID Services

Customers who provide or make use of a service employing referral IDs will be considered responsible for unsolicited bulk email sent by members of the referral ID service that makes reference to services hosted by Seller. Customers must be able to provide visible evidence from the UBE-advertised link that a member of their referral ID service has had their membership terminated.

Dialup Connections

Customers may not run programs or configure machines in such a way as to keep a dialup connection active when not in use or otherwise bypass automatic disconnection for inactivity, unless they have a dedicated access account. Seller users may not have multiple simultaneous connections with a single dialup account. Seller reserves the right to impose restrictions on or terminate accounts deemed to be in violation of these conditions. Seller's dialup access servers will disconnect after 30 minutes of inactivity and after 12 hours of continuous access.

Storing files

The storage of any program, utility or file on Seller's servers the use of which would constitute a violation of this policy is prohibited. For example, it is a violation to store hacker scripts, IRC bots, or spamming software on Seller's servers.

How to Contact Us

To contact us with questions or comments regarding this Acceptable Use Policy or claimed violations of this Policy, please email ___________.

Example Clause #11: The following AUP encompasses not only use of lit fiber services, but also use of Seller's physical plant.

In the event of a violation of the Acceptable Use Policy, this clause sets forth a procedure for escalation of notices and Seller's remedies against Customer.

EXHIBIT AUP
ACCEPTABLE USE POLICY

This acceptable use policy ("AUP") covers (a) Seller's customers ("Customer") (and the Customer's customers, agents and users) ***use of and access to Seller's facilities*** (e.g. Internet Data Centers); (b) Customer's (and its customers, agents and users) ***use of Seller's online services***; and (c) Seller's maintenance of the services it provides to its Customers.

Access to Internet Data Centers

Only those individuals identified in writing by Customer on the Customer Registration Form ("Representatives") may access the Internet Data Centers. Customer shall deliver prior written notice to Seller of any changes to the Customer Registration Form and the list of Representatives. Customer and its Representatives shall not allow any unauthorized persons to have access to or enter any Internet Data Centers. Customer and its Representatives may only access the Customer Area, unless otherwise approved and accompanied by an authorized Seller representative.

Use of Internet Data Center Facility

Conduct at Internet Data Centers. Customer and its Representatives agree to adhere to and abide by all security and safety measures established by Seller. Customer and its Representatives shall also not do or participate in any of the following: (a) misuse or abuse any Seller property or equipment or third party equipment; (b) make any unauthorized use of or interfere with any property or equipment of any other Seller Customer; (c) harass any individual, including Seller personnel and representatives of other Seller customers; or (d) engage in any activity that is in violation of the law or aids or assists any criminal activity while on Seller property or in connection with the Internet Data Center Services.

Prohibited Items. Customer and its Representatives shall keep each Customer Area clean at all times. It is each Customer's responsibility to keep its area clean and free and clear of debris and refuse. Customer shall not, except as otherwise agreed to in writing by Seller, (a) place any computer hardware or other equipment in the Customer Area that has not been identified in writing to Seller; (b) store any paper products or other combustible materials of any kind in the Customer Area (other than equipment manuals); and (c) bring any Prohibited Materials (as defined below) into any Internet Data Center. "Prohibited Materials" shall include, but not limited to, the following and any similar items:

- food and drink;
- tobacco products;
- explosives and weapons;
- hazardous materials;
- alcohol, illegal drugs and other intoxicants;
- electro-magnetic devices which could unreasonably interfere with computer and telecommunications equipment;
- radioactive materials; and
- photographic or recording equipment of any kind (other than tape back-up equipment).

Equipment and Connections

Customer Equipment. Each connection to and from a piece of Customer Equipment shall be clearly labeled with Customer's name (or code name provided in writing to Seller) and the starting and ending point of the connection. Customer Equipment must be configured and run at all times in compliance with the manufacturer's specifications, including power outlet, power consumption and clearance requirements. Customer must use its best efforts to provide Seller with at least 48 hours prior notice any time it intends to connect or disconnect any Customer Equipment or other equipment.

Maintenance

Seller will conduct routine scheduled maintenance of its Internet Data Centers only according to its regularly scheduled maintenance schedule. In the event a mission critical maintenance situation arises, Seller may be required to perform emergency maintenance at any time. During these scheduled and emergency maintenance periods, Customer's Equipment may be unable to transmit and receive data, and Customer may be unable to access the Customer Equipment. Customer agrees to cooperate with Seller during the scheduled and emergency maintenance periods.

Online Conduct

Customer Content. Customer acknowledges that Seller exercises no control whatsoever over the content of the information passing through Customer's site(s) and that it is the sole responsibility of Customer to ensure that the information it and its users transmit and receive complies with all applicable laws and regulations and this AUP.

Prohibited Activities. Customer will not, and will not permit any persons ("Users") using Customer's online facilities and/or services, including, but not limited to, Customer's Web site(s) and transmission capabilities, to do any of the following ("Prohibited Activities"):

- send unsolicited commercial messages or communications in any form ("SPAM");

- engage in any activities or actions that infringe or misappropriate the intellectual property rights of others, including, but not limited to, using third party copyrighted materials without appropriate permission, using third party trademarks without appropriate permission or attribution, and using or distributing third party information protected as a trade secret information in violation of a duty of confidentiality;
- engage in any activities or actions that would violate the personal privacy rights of others, including, but not limited to, collecting and distributing information about Internet users without their permission, except as permitted by applicable law;
- send, post or host harassing, abusive, libelous or obscene materials or assist in any similar activities related thereto;
- intentionally omit, delete, forge or misrepresent transmission information, including headers, return mailing and Internet protocol addresses;
- engage in any activities or actions intended to withhold or cloak Customer's or its Users' identity or contact information;
- use the Seller connectivity services for any illegal purposes, in violation of any applicable laws or regulations or in violation of the rules of any other service providers, web sites, chat rooms or the like; and
- assist or permit any persons in engaging in any of the activities described above.

If Customer becomes aware of any Prohibited Activities, Customer will use best efforts to remedy such Prohibited Activities immediately, including, if necessary, limiting or terminating User's access to Customer's online facilities.

Third Party Complaint Process. Seller routinely receives written complaints ("Complaints") from third parties regarding Prohibited Activities allegedly being conducted by a Customer or its Users. Due to the nature of Seller's business, in Seller's experience, most legitimate complaints and actual Prohibited Activity is conducted by customers and users of Seller's customers, not by Seller's customers themselves. Seller requires its Customers to have in effect and enforce policies substantially similar to this AUP and will work with its customers to resolve violations. Seller will take the following actions to document and resolve each Complaint received by Seller related to a Customer or its Users.

<u>First Complaint</u>. Upon receipt of the initial complaint from a third party regarding Prohibited Activity by a Customer or its User, ***Seller will send a letter*** (the "First Letter") to the complaining third party that describes Seller's policies relating to the Prohibited Activity and lists the contact information for the Customer and encloses a copy of the original Complaint received by Seller. Seller also will deliver notice of the Complaint to the Customer by sending a copy of the same letter to the Customer via e-mail to its abuse address so that Customer can identify and remedy the Prohibited Activity. Seller's goal is to put the complainant directly in touch with the party in the best position to remedy the problem, Seller's Customer who has the relationship with the alleged violator.

<u>Second Complaint</u>. Upon receipt of a second complaint after the date of the First Letter related to the same or similar Prohibited Activity of Customer described in the First Letter that indicates that the Prohibited Activity has continued after the date of the First

Letter, ***Seller will send a second letter*** (the “Second Letter”) with a copy of the second complaint to the Customer and request that Customer respond in writing to Seller with an explanation and timeline of the actions to be taken by Customer to remedy Prohibited Activity. If Customer does not respond to the Seller’s Second Letter and remedy the Prohibited Activity within ten (10) business days, ***Customer will pay Seller $500*** (pursuant to invoice) in the following month to cover Seller’ administrative costs associated with the Prohibited Activities of Customer.

Third Complaint. Upon receipt of a third complaint after the date of the Second Letter relating to the same or similar Prohibited Activity of Customer described in the Second Letter that indicates that the Prohibited Activity has continued after the date of the First Letter, ***Seller will send a third and final letter*** (the “Third Letter”) with a copy of the third complaint to the Customer and request again that the Prohibited Activity cease immediately. If the Prohibited Activity does not cease within five (5) business days, Seller will terminate or suspend its connectivity service to its Customer, and will only resume providing service when it receives adequate assurances that such activity will not continue. ***Customer will pay Seller $5,000*** (pursuant to invoice) to cover Seller’s administrative costs associated with the Prohibited Activities.

Suspension and Termination of Service. Seller reserves the right to suspend and/or terminate a Customer’s Service at any time for any material failure of Customer, its Representatives or its Users to comply with this AUP.

Part 4, Chapter 7 - Termination Agreement

For many reasons a Seller and a Customer may wish to terminate a service order. Typically a Master Services Agreement will require a customer to pay a termination charge if a service order is terminated before the normal expiration of its term. However, if the Seller and the Customer agree to terminate a service order before the normal expiration of its term, they would enter into a termination agreement.

A termination agreement might call for a termination payment to be made by Customer, as described in the forms below, or there might be no termination payment, in which case Section 2 may be omitted.

A termination agreement will set forth the effective date of termination and will describe the obligations, if any, that are to survive the termination of the order. If it is intended that no obligations will survive the termination, that should also be stated explicitly. (See Example Agreement #1, Section 1 below.)

Example Agreement #1:

Service Order Form
Termination Agreement

This Termination Agreement (this "Termination Agreement") is entered into as of ____________ (the "Effective Date") by and between _______________, a ________ ("Customer") and _______________, a _______________ ("Seller").

RECITALS:

A. The parties entered into that certain Master Services Agreement dated as of __________ (the "MSA").

B. Pursuant to the MSA, Customer and Seller entered into that certain Service Order Form dated as of _____, ____ (the "Service Order Form"), by the terms of which Seller agreed to provide ________ service for Customer between Customer's facilities in __________, __________ and at ________, __________.

C. Customer and Seller wish to terminate the Service Order Form on the terms hereinafter set forth.

NOW, THEREFORE, for valuable consideration, the parties agree as follows:

1. Termination. Customer and Seller agree that the Service Order Form shall be terminated effective as of the close of business on ________________ (the "Effective Date"). From and after the Effective Date, the Service Order Form shall be of no further force and effect, and Customer and Seller shall be ***released*** of and from any ***further*** obligations under the Service Order Form, ***except for those obligations which have accrued prior to the Effective***

Date, including the obligation of Customer to pay for services made available to Customer prior to the Effective Date.

2. Termination Payment. Customer ***has paid*** to Seller the sum of $_______ as payment in full for monthly recurring charge for the month of __________. In addition, Customer agrees to pay to Seller the sum of $____________ as a termination fee and as full and final payment under the Service Order Form. Customer agrees to make such payment promptly following the execution and delivery of this Termination Agreement by both parties.

3. Applicable Laws. This Termination Agreement and the rights hereunder shall be interpreted in accordance with the laws of the State of ___________, and all rights and remedies shall be governed by such laws without regard to principles of conflict of laws.

4. Entire Agreement. This Termination Agreement and the MSA constitute the entire agreement between the parties pertaining to the subject matter hereof and supersede all prior agreements and understandings between the parties pertaining thereto. In deciding to become a party to this Termination Agreement, no party has relied upon any representations or warranties other than those expressly set forth in this Termination Agreement.

5. Construction. The words "hereof," "herein" and "hereunder" and words of similar import when used in this Termination Agreement shall refer to this Termination Agreement as a whole and not to any particular provision of this Termination Agreement. The parties have participated jointly in the negotiation and drafting of this Termination Agreement, and in the event of an ambiguity or question of intent or interpretation arises, this Termination Agreement shall be construed as if drafted jointly by the parties, and no presumption or burden of proof shall arise favoring or disfavoring any party by virtue of the authorship of any provision of this Termination Agreement.

6. Enforceability. The invalidity or unenforceability of any particular provision of this Termination Agreement shall not affect the other provisions hereof, and this Termination Agreement shall be construed in all respects as if such invalid or unenforceable provision were omitted.

7. Counterparts. This Termination Agreement may be executed in any number of counterparts, each of which when executed and delivered shall be deemed to be an original and all of which together shall be deemed to be one and the same instrument binding upon all of the parties notwithstanding the fact that all parties are not signatory to the original or the same counterpart. For purposes of this Termination Agreement, facsimile and .PDF signatures shall be deemed originals, and the parties agree to exchange original signatures as promptly as possible.

8. Representations and Warranties.

(a) Seller represents and warrants to Customer that, as of the date hereof: (1) subject to and except as otherwise provided in this Termination Agreement, it has full corporate power and authority to execute this Termination Agreement; (2) it has been duly organized and is validly existing as a corporation in good standing under the laws of the state of its organization and is duly qualified to do business in each state in which it is required to be so qualified to perform its obligations hereunder; (3) no litigation or governmental proceeding is pending or, to Seller's knowledge, threatened in writing, against Seller which might have a material adverse effect on this Termination Agreement, the transactions contemplated hereby or the rights of the parties hereunder; and (4) the execution and delivery of this Termination Agreement, and the performance of its obligations hereunder, have been duly authorized by all necessary corporate actions and do not violate any provision of law or its articles of incorporation or other governing documents.

(b) Customer represents and warrants to Seller that, as of the date hereof: (1) it has full limited liability company power and authority to execute this Termination Agreement; (2) it has been duly organized and is validly existing as a limited liability company in good standing under the laws of the state of its organization and is duly qualified to do business in each state in which it is required to be so qualified to perform its obligations hereunder; (3) no litigation or governmental proceeding is pending or, to Customer's knowledge, threatened in writing, against Customer which might have a material adverse effect on this Termination Agreement, the transactions contemplated hereby or the rights of the parties hereunder; and (4) the execution and delivery of this Termination Agreement, and the performance of its obligations hereunder, have been duly authorized by all necessary limited liability company actions and do not violate any provision of law or its articles of organization (or equivalent charter document) or its limited liability company operating agreement.

9. The Effective Date. This Termination Agreement shall be effective as of the Effective Date.

10. Confirmation of MSA. Except as modified hereby, the MSA remains in full force and effect, and Customer and Seller ratify and affirm the MSA in all respects.

Customer: ________________	Seller: ________________
By: ________________	By: ________________
Name: ________________	Name: ________________
Title: ________________	Title: ________________
Date: ________________	Date: ________________

Example Agreement #2:

Service Order
Termination Agreement

This Termination Agreement (this "Agreement") is entered into as of ____________ (the "Effective Date") by and between _________________, a ______ ______________ ("Customer") and ____________________________, a _______________________ ("Seller").

RECITALS:

A. Customer and Seller entered into that certain Master Services Agreement dated ______________________ (the "MSA"). Pursuant to the MSA, Customer and Seller entered into certain service order forms for telecommunications service orders described as follows by which Seller agreed to provide service for Customer between Customer's facilities described therein.

(1) ________________, dated _______________ for _____________ Service ("Order A"), with an MRC of $_____;
(2) ___________________, dated _____________________ for _________________ Service ("Order B"), with an MRC of $_______;
(3) ___________________, dated _______________________ for ___________________ Service ("Order C"), with an MRC of $______________;
(4) ___________________, dated ______________________, for ____________________ Service ("Order D"), with an MRC of $_____.

B. All of the service orders described in Recital A are collectively referred to in this Agreement as the "Service Orders."

C. Customer is in arrears in payment of amounts due under the Service Orders.

D. Customer and Seller have reached an agreement with respect to amounts past due.

NOW, THEREFORE, for valuable consideration, the parties agree as follows:

1. Settlement Payments. The parties agree that the amount of $__________ is presently due and owing under the Service Orders for services provided through ____________. Customer agrees to pay to Seller the entire amount past due in installment payments (the "Settlement Payment"). The Settlement Payment shall be payable in four (4) monthly installments as follows:

(a) _________ on or before ________________; and
(b) ___________ on or before ________________, and
(c) ___________ on or before ________________, and
(d) ___________ on or before ______________.

2. Termination. Provided that Customer pays in full the Settlement Payment according to the installment schedule set forth above, and pays all other amounts due under the MSA and the Service Orders as and when they become due, Seller agrees to waive its rights under the MSA for Customer's past breaches of the Service Orders.

   If Customer fails to pay any installment of the Settlement Payment as and when due, or fails to pay any other amount due under the MSA or a Service Order as and when due, such default shall be considered a termination by Seller for Customer's Default pursuant to Section ___ of the MSA, and Seller may, in addition to all other remedies, terminate for Default the Service Orders and all other Service Orders entered into between the parties without notice, and Customer shall be obligated to pay the early termination charges with respect to all of the terminated Service Orders, and Seller may apply any and all amounts previously paid by Customer hereunder toward the payment of any other amounts then or thereafter payable by Customer hereunder.

3. Confirmation of MSA. Except as modified hereby, the MSA remains in full force and effect, and Customer and Seller ratify and affirm the MSA in all respects.

4. Definitions. For purposes of this Agreement, all capitalized terms used in this Agreement that are not defined in this Agreement have the meanings assigned to such terms in the MSA.

5. General.

   (a) All actions, activities, consents, approvals and other undertakings of the parties in this Agreement shall be performed in a reasonable and timely manner, it being expressly acknowledged and understood that time is of the essence in the performance of obligations required to be performed by a date expressly specified herein.

   (b) No delay or omission by either party to exercise any right or power occurring upon non-compliance or failure of performance by the other party, and no custom or practice which may arise between the parties in the administration of any part of this Agreement, shall be construed to waive or lessen the right of a party to insist on the performance by the other party in strict accordance with the provisions hereof, nor shall it impair that right or power or be construed to be a waiver.

   (c) This Agreement shall be governed by and construed in accordance with the domestic laws of the State of ____________ without reference to its choice of law principles.

   (d) This Agreement constitutes the entire and final agreement and understanding between the parties with respect to the subject matter hereof and supersedes all prior agreements relating to the subject matter hereof which are of no further force or effect.

(e) This Agreement may be executed in one or more counterparts, all of which taken together shall constitute one and the same instrument. Electronic transmission of any signed original document and/or retransmission of any signed electronic transmission will be deemed the same as delivery of an original. At the request of either party, the parties will confirm electronic transmission by signing a duplicate original document.

Customer: ______________________

By: ______________________
Name: ______________________
Title: ______________________
Date: ______________________

Seller: ______________________

By: ______________________
Name: ______________________
Title: ______________________
Date: ______________________

Part 4, Chapter 8 - Third Party Vendor

Often the provision of services requested by a customer requires the use of facilities of another telecommunications company. In such cases a seller may not wish to accept responsibility for the performance of another carrier's network to the same extent that it would for services provided entirely on its own network. In such instances the seller and the customer may agree to adopt the acceptance testing specifications, credits for interruptions of service, and other terms and conditions rather than those of the seller with whom the customer has contracted.

Example Clause #1: In the first example the Seller makes reference to two sets of acceptance testing specifications. One set of acceptance testing specifications applies to services that are delivered over Seller's owned or leased network. A second set of specifications are those of the carrier over whose network the services are delivered.

The customer (Buyer) is not entitled to outage credits from Seller for interruptions of service that occur on network that is not owned by Seller. Outage credits, if any, are allowed only to the extent that they are available from the third party vendor.

> Testing and Acceptance of Services.
>
> Testing of Service. Upon completion of installation or connection of facilities and/or equipment necessary for the provision of each Service to be provided to Buyer, Seller shall conduct appropriate ***tests to demonstrate that the Service meets the applicable specifications*** set forth in Exhibit ___ ***for any Services provided entirely on Seller's Network Facilities or*** meets the applicable ***specifications set forth in the applicable third party's tariff or contract***, whichever is applicable, ***for any Services provided on Third Party Facilities***. Buyer shall cooperate and provide reasonable assistance to Seller in the initial testing of the Service in accordance with industry standard testing procedures and protocols, or such other procedures and protocols as the Parties may agree. Upon successful completion of such tests, Seller shall notify Buyer in writing that such Service is available for use (the "Circuit Installation Notification").
>
> Point of Demarcation. The Point of Demarcation for all Services originating and terminating on Seller's Network Facilities shall be set forth in the FOC. The Point of Demarcation for any Services originating or terminating on Third Party Facilities shall be ***determined by the third party which owns and/or operates such facilities***, for which Seller shall provide Buyer notification thereof as soon as practicable following the receipt of such determination by the third party. Seller shall provide Buyer with any other information required (including operational data) to interconnect and test the Services provided on Seller's Network Facilities and Seller shall use all commercially reasonable efforts to cause third party suppliers to provide Buyer with similar information for Services provided on Third Party Facilities. Seller shall be responsible for performing monitoring up to the Point of Demarcation and Buyer shall be responsible for performing monitoring beyond the Point of Demarcation.

Adjustments to Payments. If any overcharge or undercharge in any form whatsoever shall at any time be found and the bill therefor has been paid, Seller shall refund the amount of any overcharge received by Seller and Buyer shall pay the amount of any undercharge, within thirty (30) days after final determination thereof; provided, there shall be no retroactive adjustment of any overcharge or undercharge if the matter is not brought to the attention of the other Party in writing within one hundred twenty (120) days following the date the Services were provided under this Agreement regarding which the overcharge or undercharge applies; provided that such adjustment period shall be adjusted to coincide with the period in which adjustments are permitted under a third party's contract or tariff, whichever is applicable, with respect to Services provided on Third Party Facilities to the extent that such contract or tariff provides for different adjustment periods than those set forth above.

Operating and Performance Standards. The operating and performance standards for each Service that are provided on the Network Facilities are set forth in Exhibit __. ***The operating and performance standards for each Service that are provided on Third Party Facilities shall be in accordance with the terms and conditions set forth in the applicable carrier's tariff or contract, whichever is applicable.***

Operation and Maintenance of Facilities. Seller will operate and maintain, or cause the operation and maintenance of its Network Facilities, in compliance with all Applicable Laws, Prudent Industry Practices and otherwise in accordance with this Agreement. Seller will, during the Term, only employ or contract with appropriately qualified (as determined in Seller's reasonable opinion) personnel for the purposes of operating and maintaining the Network Facilities and coordinating operations with its network operating center. ***Seller will use commercially reasonable efforts to cause any Third Party Facilities to be operated and maintained in accordance with all Applicable Laws, Prudent Industry Practices and otherwise in accordance with this Agreement.***

If any portion of a Service is provided on Third Party Facilities, then in no event shall any Force Majeure Event occurring on such Third Party Facilities or the unavailability, incompatibility, delay in installation, or other impairment of Third Party Facilities excuse Buyer's obligation to pay Seller all rates and charges applicable to the Service, whether or not such Services are useable by Buyer.

"Network," "Network Facilities" or "Seller's Network Facilities" shall mean the telecommunications network facilities that are (a) ***owned or leased by Seller*** and (b) on which it owns and operates its optronic and electronic equipment. For avoidance of doubt, the Network Facilities ***shall not include any facilities through which a third party***, including without limitation an incumbent local exchange carrier, ***provides lit services to Seller.***

"Point of Demarcation" shall mean the interface between the Network Facilities and the facilities of a third party, including Buyer, the customers of Buyer or another carrier. Such point will be identified on a Service Order, which will designate the point at which

Seller's responsibility to provide Service ends and the Buyer's responsibilities commence.

"Third Party Facilities" shall mean the telecommunication network facilities owned and/or operated by third parties over which all or a portion of the Services may be provided under this Agreement.

Notwithstanding the above, a Service Outage shall not be deemed to have occurred and no Service Outage Credits will apply:

(a) During periods (A) of less than ten (10) minutes, (B) in which Seller is not given access to its facilities or equipment that are required to provide the Services or to remedy any Service Outage, (C) in which planned or scheduled maintenance and repair activities are occurring, (D) in which Buyer continues to use the Services on an impaired basis, or (E) that are not reported to Seller within thirty (30) days of the date the Service was affected; or

(b) ***For interruptions that are caused by or due to (A) acts or omissions of Buyer or a third party, including without limitation an interruption on the Third Party Facilities that may provide a portion of the Services***, (B) the failure or malfunction of facilities or equipment not owned or operated by Seller, including without limitation the failure of the power supply, or (C) a Force Majeure Event or (D) disconnections by Seller for non-payment or other contract default or breaches by Buyer.

Example Clause #2: In the following clause, the use of third party facilities to provide services affects a service in two ways. First, if delivery of the service is delayed, the Customer may not terminate the service contract. Second, the third party' vendor's service level agreement and outage credits apply to the service, rather than Seller's SLA.

Discontinuance of Customer Order by Customer.

(a) If Seller's installation of Service is delayed for more than thirty (30) business days beyond the Customer Commit Date for reasons other than an Excused Outage, Customer may terminate and discontinue the affected Service upon written notice to Seller and without payment of any applicable termination charge; provided such written notice is delivered prior to Seller delivering to Customer the Connection Notice for the affected Service. ***This Section _____ shall not apply to any Off-Net*** Local Loop Service, including, without limitation, Seller Metropolitan Private Line (Off-Net Service, provisioned by Seller through a third party carrier for the benefit of Customer.

(b) The Availability Service Levels and associated credits set forth in this Section ________ shall not apply to Off-Net Local Loop Service, including, without limitation, Seller Metropolitan Private Line (Off-Net) Service, provisioned by Seller through a third party carrier for the benefit of Customer. ***Seller will pass through to Customer any availability service level and associated credit*** (if applicable) provided to Seller by the third party carrier for such Off-Net Local Loop Service.

Example Clause #3: In the following clause the Seller disclaims responsible for matters relating to third party facilities, including facilities that are owned by any third party. Also, if services are provided using the network of a third party, installation intervals and credits for service interruptions are governed by the terms of the agreement with that third party vendor.

However, this clause is specific in that for such other terms to govern instead of the Seller's, Customer must have expressly requested that services be provided using the facilities of the third party vendor.

> Disclaimers. Seller will not be liable for claims or damages resulting from or caused by:
>
> (a) Customer's fault, negligence or failure to perform Customer's responsibilities;
> (b) Claims against Customer by any other party (except for claims indemnified under Section ___);
> (c) Any act or omission of any other party, including End Users; or
> (d) Equipment or ***services furnished by a third party***, including End Users.
>
> OUTAGES
>
> GENERAL
>
> As described below, Seller offers Service Credits related to ***installation intervals and Service Availability***. These credits are the Customer's sole and exclusive remedy for Service-related claims. To qualify for a service credit, Customer must not have any invoices that are past due, and must notify Seller that a trouble ticket should be opened to document the event. In no event shall the total amount of service credits per month exceed the total MRC (or NRC if applicable) for the affected Service for such month. ***If at Customer's request Seller provides through a third-party vendor a service*** that is not a standard Seller Service, and if there is a delay in installation or interruption of such service, ***Customer shall be entitled to a Service credit only if and to the extent of the service credit to which Seller is entitled under its agreement with such third-party vendor***.
>
> Off-Net" means any telecommunications capacity and related ancillary services provided by a third party vendor ordered by Seller on Customer's behalf.

Example Clause #4: The use of third party services will constitute an exception to the standard terms of service. If third party services are used, the Customer will not have the right to terminate the agreement in the event that the Seller fails to meet the agreed Service Activation Date. Also, credits for interruption of service are not applicable to services provided with the use of third party facilities.

> If the Service Activation Date for a Service has not occurred within thirty (30) days following the FOC date, except in the case of a delay caused by Customer or a Force Majeure Event, and ***except in the case of an Off-Net Service***, Customer may cancel the

Service without penalty, provided that Customer notifies Seller in writing of cancellation before Seller has completed the acceptance testing.

"Off-Net" means a service that originates from or terminates to a location that is not on the Seller network;

Availability Service Level for Unprotected Fast 10G Optical Wavelength Service.

If a Fast 10G Optical Wavelength Service becomes Unavailable (as defined above) for reasons other than an Excused Outage, Customer will be entitled to a service credit. The amount of the credit shall be based on the cumulative Unavailability for the affected Service in a given calendar month, as set forth in the following table:

Cumulative Monthly Unavailability	Service Level Credit for Unprotected Fast 10G Optical Wavelength Service (% of MRC)
Less than __ hours	No Credit
__ hours or longer but less than __ hours	__% of the MRC
__ hours or longer but less than __ hours	__% of the MRC
__ hours or longer but less than __ hours	__% of the MRC
__ hours or longer but less than __ hours	__% of the MRC
__ hours or longer but less than __ hours	__% of the MRC
__ hours or longer but less than __ hours	__% of the MRC
__ hours or longer	100% of the MRC

The Availability Service Levels and associated credits set forth in this Section shall not apply to Off-Net Service provisioned by Seller through a third party carrier for the benefit of Customer. Seller will pass through to Customer any availability service level and associated credit (if applicable) provided to Seller by the third party carrier for such Off-Net Service.

Part 4, Chapter 9 - Intellectual Property

For the provision of lit services, a seller will purchase electronic equipment manufactured by another company. The warranties and indemnities typically extended by equipment vendors are very limited and highly qualified. Because a seller does not control the design or manufacture of equipment, a seller will be reluctant to accept responsibility for claims that might arise that the design of equipment infringes on the intellectual property rights of another party, especially since recourse, if any, against the equipment vendor would be very limited.

There is a broad range of clauses that may address a seller's liability for claims of infringement on intellectual property rights of third parties. In some examples, a seller will disclaim any responsibility at all, or will avoid mention of such potential claims. In other examples, a seller will accept limited liability subject to any number of conditions or qualifications.

When an indemnity or other responsibility for intellectual property claims is applicable, it will often be part of a broader array of liability provisions that cover many subjects relating to liability and limitation of liability. For instance, those liability provisions may limit the total value of all claims permitted under the agreement, or may be subject to a waiver of consequential damages. The question then arises, do those limitations apply as well to obligations relating to indemnity for intellectual property claims, or is the IP indemnity an exception to the limitation of liability?

Example Clause #1: In this example discussion of intellectual property has been omitted. The contract makes no mention of responsibility for intellectual property rights. Further, the Seller has disclaimed any representation or warranty except those that are expressly set forth in the agreement. The result is that the Seller would have disclaimed any responsibility for claims relating to intellectual property.

In contrast, the indemnity by Customer, which is found at the beginning of this clause, was drafted very broadly, and could be interpreted as imposing on Customer responsibility for claims relating to intellectual property rights.

> Indemnification and Limitations on Liability. ***Customer will indemnify***, defend and hold Seller, its affiliates and each of its respective owners, directors, officers, employees and agents, harmless from and against any and all claims, suites expenses, losses, demands, actions, causes of action, judgments, fees and costs, of any kind or nature whatsoever (Claims), arising from or related to any ***use, attempt to use or resale of Service or otherwise arising in connection with any Service or this Agreement***. IN NO EVENT WILL EITHER PARTY BE LIABLE FOR ANY DAMAGES WHATSOEVER FOR LOST PROFITS, LOST REVENEUS, LOSS OF GOODWILL, LOSS OF ANTICIPATED SAVINGS, LOSS OF DATA, THE COST OF PURCHASEING REPLACEMENT SERVICES, OR ANY INDIRECT, INDCIDENTAL, SPECIAL, CONSEQUENTIAL, EXEMPLARY OR PUNITIVE DAMAGES ARISING OUT OF THE PERFORMANCE OR FAILURE TO PERFORM UNDER THIS AGREEMENT OR ANY ORDER. SELLER WILL HAVE NO LIABILITY FOR ANY CLAIM

AGAINST CUSTOMER BY A THIRD PARTY IN CONNECTION WITH OR FOR RESPONDING TO EMERGENCY 911 OR OTHER EMERGFENCY REFERRAL CALLS.

Warranties. ***SELLER MAKES NO WARRANTIES OR REPRESENTATIONS, EXPRESS OR IMPLIED***, IN FACT OR BY OPERATION OF LAW, STATUORY OR OTHERWISE, WITH RESPECT TO THE SERVICE, INCLUDING BUT NOT LIMITED TO ANY WARRANTIES OF MERCHANTABLILITY OR FITNESS FOR A PARTICULAR PURPOSE.

Example Clause #2: In this form as well, the Seller has disclaimed any representation regarding the Service, except for those expressly stated in this agreement, which omits any mention of responsibility for intellectual property rights. Furthermore, Seller's liability to Customer for all claims under the agreement is limited to one month of monthly recurring charges.

Seller is not liable for any indirect, incidental, consequential, special or punitive damages (including without limitation, lost profits or revenue) arising out of or related to the provision of Services hereunder, including any claims made by or through third parties. ***Seller's liability to Customer may not exceed one month's calculation of monthly charges for the applicable Services***. Seller has no liability whatsoever for the content of information passing through its Network.

Representations and Warranties: Each party represents and warrants that it is fully authorized to enter into this Agreement. Seller represents and warrants to Customer that any Services provided hereunder will be performed in a professional manner by qualified and trained personnel. UNLESS SPECIFICALLY STATED HEREIN OR IN ANY SERVICE ORDER, ***SELLER MAKES NO WARRANTIES, REPRESENTATIONS OR AGREEMENTS, EXPRESS OR IMPLIED***, EITHER IN FACT OR BY OPERATION OF LAW, STATUTORY OR OTHERWISE, INCLUDING WARRANTIES OF MERCHANTABILITY OR FITNESS FOR A PARTICULAR PURPOSE.

Example Clause #3: The following clause generally limits Seller's liability for various claims relating to the agreement. Seller has incorporated into this general limitation of liability a specific disclaimer of any liability for claims of infringement of intellectual property rights. Further, this paragraph includes a limit on all claims that may be asserted by Customer under the contract.

Limitation of Liability and Disclaimer of Warranties. Except as set forth in Section __, neither party shall be liable to the other party for any special, incidental, indirect, punitive or consequential damages arising out of or in connection with such party's failure to perform its respective obligations hereunder, including, but not limited to, damage or loss of property or equipment, loss of profits, goodwill or revenue (whether arising out of Outages, transmission interruptions or problems, any interruption or degradation of the functioning of the granted IRU(s) or otherwise), cost of capital, cost of replacement services, or claims of customers, whether occasioned by any construction, reconstruction, relocation, repair or maintenance performed by, or failed to be performed by, the other party UNDER ANY THEORY OF TORT, CONTRACT, WARRANTY, STRICT LIABILITY OR NEGLIGENCE, all claims for which damages are hereby specifically waived, EVEN IF THE PARTY HAS BEEN ADVISED, KNEW OR SHOULD HAVE KNOWN OF THE POSSIBILITY OF SUCH DAMAGES. NOTWITHSTANDING THE FOREGOING, ***SELLER'S TOTAL LIABILITY HEREUNDER SHALL IN NO EVENT EXCEED THE AGGREGATE AMOUNT OF ANY APPLICABLE OUTAGE CREDITS*** DUE HEREUNDER. ***SELLER MAKES NO WARRANTIES***, EXPRESS OR IMPLIED, AS TO ANY SERVICE PROVISIONED HEREUNDER. SELLER SPECIFICALLY DISCLAIMS ANY AND ALL WARRANTIES, EXPRESS OR IMPLIED, WITH RESPECT TO THE CUSTOMER CAPACITY AND THE GRANTED IRU(s), INCLUDING BUT NOT LIMITED TO ANY WARRANTIES OF MERCHANTABILITY, FITNESS FOR A PARTICULAR PURPOSE, ***OR TITLE OR NON-INFRINGEMENT OF THIRD PARTY RIGHTS***. Nothing contained herein shall operate as a limitation on either party to bring an action for damages against any third party, including indirect, special or consequential damages, based on any acts or omissions of such third party as such acts or omissions may affect the construction, operation or use of the granted IRU(s) or the Seller Network; provided, however, that each party shall reasonably cooperate to the extent necessary to enable the other party to pursue such action against such third party.

Example Clause #4: In the following clause the Seller does agree to indemnify the Customer for claims of infringement of intellectual property rights, but the indemnity was drafted very narrowly. The Seller's indemnity does not extend to all aspects of the Service; it is limited to "network configuration or design," elements for which the Seller has control.

Customer has accepted the risk of claims based on other elements of the Service.

Seller will indemnify, hold harmless, and defend Customer, in accordance with the procedures described in Section ____, Indemnification Procedures, hereof, against all Losses arising out of, in connection with, resulting from or based on allegations of, any of the following:

(a) A claim that Seller's ***network configuration or design infringes*** on a patent or copyright of a third party. ***Seller's indemnification shall not extend to a claim that network equipment, hardware or software infringes*** on another party's patent or copyright. ***This represents Seller's entire obligation to Customer regarding any claim of infringement*** of any intellectual property right. Furthermore, Seller has no obligation regarding any claim based on any of the following: (1) anything Customer provides which is incorporated into the network equipment, hardware or software; (2) functionality provided by Seller at the instruction of Customer; (3) Customer's modification of hardware or software; (4) the combination, operation, or use of hardware or software with other products not provided by Seller as a system, or the combination, operation, or use of hardware or software with any product, data, or apparatus that Seller did not provide; or (5) infringement arising from an item provided by a third party vendor, as distinguished from its combination with Services that Seller provides to Customer as a system.

Example Clause #5: In the example that follows, Seller has agreed to make reasonable efforts to enforce a vendor warranty for the benefit of Customer. However, Seller has made no other representation or warranty regarding intellectual property rights. Of course, that is not an indemnity, but it would afford some limited protection to the Customer.

> With respect to components and materials that are incorporated into Seller's network, Seller agrees to use commercially reasonable efforts to enforce for the benefit of Customer any indemnity with respect to intellectual property rights which Seller may have received from the manufacturers or sellers.

Example Clause #6: In the following example, Seller has made a very narrow representation that at present there are no outstanding claims for infringement of intellectual property rights. However, Seller has made no other representation. Seller has not represented that there is no basis on which such a claim could be asserted.

> Representations of Seller. As of the Effective Date of this Agreement, Seller represents that there are no actions, suits or proceedings pending, or to the knowledge of Seller threatened, against Seller, Seller's Representatives and subcontractors alleging infringement, misappropriation or other violation of any intellectual property rights related to any Service contemplated by this Agreement.

Example Clause #7: Although the following clause is short, the representation and warranty is broad enough to encompass a broad range of potential liabilities in Seller. Not only has the Seller represented that the service will not infringe on the intellectual property rights of any other party, but Seller has also represented that there is no current actual or threatened lawsuit relating to such a claim.

> Seller represents and warrants that Seller's design and implementation of the Network and Customer's use of the Network as contemplated herein ***does not and shall not infringe*** on any patent, trademark, copyright, trade secret or other intellectual property or proprietary right of any third party, and ***there is currently no actual or threatened suit*** against Seller by any said third party based on an alleged violation of said right. This warranty shall survive the expiration or termination of this Agreement.

Example Clause #8: The following clause addresses liability for claims of infringement, and other matters that might arise under the service contract. In addition to the right of indemnification, if the Vendor is unable to resolve a claim of infringement, the Customer would be entitled to obtain substitute service at Vendor's expense.

> Intellectual Property Infringement. Vendor shall ***indemnify***, defend and hold harmless Customer, its parents, subsidiaries and affiliates, and its and their respective directors, officers, partners, employees, agents, successors and assigns from all third party claims, suits, demands, damages, liabilities, expenses (including, but not limited to, reasonable fees and disbursements of counsel and court costs), judgments, settlements and penalties of every kind ("IP Claim") arising from or relating to any actual or alleged ***infringement***

or misappropriation of any patent, trademark, copyright, trade secret or any actual or alleged violation of any other intellectual property or proprietary rights ***arising from or in connection with the Service*** provided (including related products furnished hereunder) under this Agreement. The indemnification procedures set forth below shall apply in the case of any claims brought under this paragraph. Without limiting any of the foregoing, if the sale or use of any Service becomes subject to an IP Claim, Vendor shall, at Vendor's option and Vendor's expense:

(a) Procure for Customer the right to use the Service (including related products furnished hereunder);

(b) Replace the Services (including related products furnished hereunder) with equivalent, non-infringing Service;

(c) Modify the Service (including related products furnished hereunder) so it becomes non-infringing; or

(d) If Vendor is unable to resolve the IP claim or provide non-infringing Services then ***Customer may suspend or terminate the Service*** (including use of related products furnished hereunder) and the ***Vendor shall pay all cost incurred by Customer to replace the infringing Services***.

Example Clause #9: There are three elements to the following clause. First, the Service Provider has made a representation and warranty that the Services do not infringe on the intellectual property rights of any third party. Second, the Service Provider has agreed to indemnify the Customer from any such claims. Finally, the Service Provider has agreed that, if in the reasonable opinion of Customer a claim of infringement is likely, the Service Provider is required to address Customer's concern or terminate the Service.

The last provision would allow the Customer to anticipate a possible claim of infringement and require the Service Provider to seek a solution to Customer's concern before any claim actually arises.

Service Provider ***represents and warrants*** that the Services and their component parts do not infringe on the rights of others, including, but not limited to, trademarks, copyrights, patents and any other intellectual property rights.

Service Provider shall ***indemnify, defend and hold harmless Customer***, its Affiliates, the employees, directors, officers, agents and contractors of Customer and Customer Affiliates from and against all claims, demands, actions, causes of actions, damages, liabilities, losses, and expenses (including reasonable attorney's fees and the costs of in-house counsel) incurred as a result of:

(a) ***claims for intellectual property infringement relating to the Services*** or their component parts;

In the event the Services or any one or more of a Service's component parts becomes, ***or in the reasonable opinion of Customer or Service Provider is likely to become, the subject of a claim of intellectual property infringement***, or should an injunction restricting Customer's full use and enjoyment of the Services be issued, Service Provider shall, at its expense, (a) immediately procure for Customer the right to continue using the infringing item, or, if that cure is not made available to Service Provider following exertion of its best efforts, (b) replace or modify the infringing item so as to make it non-infringing; provided, however, that the specifications set forth in this Agreement shall not be compromised. If neither (a) nor (b) is attainable, then Service Provider shall (1) discontinue providing such Service to Customer, and (2) refund to Customer any charges for the affected Services paid in advance of the rendering thereof.

Example Clause #10: In the following clause the Provider's has disclaimed liability for claims of infringement that arise from the Customer's use of the Services in combination with equipment furnished by Customer.

Further, the Customer has given a reciprocal representation and warranty that it will not violate an intellectual property right in its use of the Service.

INFRINGEMENT

(a) Provider represents and warrants that the equipment and facilities that will be used in furnishing Service to Customer pursuant to this Agreement shall not infringe or violate any copyright, patent, trade secret or any other intellectual property rights or similar property rights, ***excluding, however, any such infringement which may arise due to combining such equipment and facilities with (1) equipment or facilities furnished by Customer***, or (2) equipment or facilities located off of Provider's Network and furnished by a third party unrelated to Provider. Provider shall indemnify, defend (by counsel reasonably acceptable to Customer) and hold Customer and its customers harmless from and against any claims made and/or loss suffered (including reasonable attorneys' fees which shall include the allocable costs of in-house counsel) as a consequence of any such infringement or violation of any copyright, patent, trade secret or any other intellectual property rights or similar property rights. Moreover, should the equipment or facilities furnished by Provider hereunder become, or in Provider's opinion is likely to become, the subject of a claim of infringement, or should Customer's use of the equipment and facilities be finally enjoined, Provider shall, at its expense:

(1) Procure for Customer the right to continue using the equipment or facilities; or,

(2) Replace or modify the equipment or facilities to make it non-infringing.

(b) ***Customer represents and warrants*** that the services it will provide to its Customers in connection with Service obtained pursuant to this Agreement shall not infringe or violate any copyright, patent, trade secret or any other intellectual property rights or similar property rights. Customer shall indemnify, defend (by counsel reasonably acceptable to Provider) and hold Provider harmless from and against any claims made and/or loss suffered (including reasonable attorneys' fees which shall include allocable costs of in-house counsel) as a consequence of any such infringement or violation of any copyright, patent, trade secret or any other intellectual property rights or similar property rights.

Example Clause #11: In this clause the Customer has accepted broad responsibility for claims that the Service infringes on the intellectual property rights of another party. The phrase "any claim that Customer's use of the Service is illegal and/or infringes" is not entirely clear. This could be interpreted as encompassing all network components and other service elements that are used by the Seller to provide the Service, or it could be interpreted only as the discrete elements of the service that are within Customer's control and separate and apart from the Seller's service elements.

Compliance with Laws: Customer is solely responsible for obtaining all licenses, approvals, and regulatory authority for its operations. ***Customer shall indemnify***, defend and hold harmless Seller from and against any third party claims, actions, damages, liabilities, costs, judgments or expenses (including attorney fees) arising out of or relating to any ***claim that Customer's use of the Service is illegal and/or infringes*** upon a third party's intellectual property rights. In the event Customer resells (and/or uses the Services in connection with its own service offerings) any of the Services, it shall further comply with sections (a) and (b) below.

(a) In connection with any ***resale of Services to any End Users***, Customer understands and agrees that it is solely responsible for all billing, billing adjustments/credits, customer service, first level (i.e., tier 1 user support and initial troubleshooting of any network difficulty to determine whether it is an Seller caused network problem, creditworthiness and other service-related requirements of its End Users, and Seller shall have no liability to Customer's End Users under this Service Form or otherwise. Customer shall comply with all terms and conditions of this Service Form, including, but not limited to, its payment obligations, regardless of Customer's ability to collect payments or charges from its End Users, Affiliates, agents, brokers or re-sellers.

(b) Seller reserves the right to suspend any or all of the Services immediately and/or terminate this Service Form for Cause if: (1) Customer fails to comply in all material respects with any foreign, federal, state or local law or regulation applicable to any resale of the Services to its End Users; or (2) its End Users commit any illegal acts relating to the subject matter of this Service Form, including but not limited to, use of the Services for illegal purposes, dissemination of obscene material, or infringement of a third party's intellectual property rights). Customer shall be liable to Seller for: (3) any damages caused by its own acts or omissions, illegal or otherwise, (e.g., slamming) in connection with its resale of the Services to its End Users; and (4) any damages caused by Customer's refusal to assist Seller in complying with any obligations or requirements imposed upon Seller by any regulatory or governmental agency applicable to the use of the Services by Customer's End Users. ***Customer shall indemnify***, defend and hold harmless Seller, and its Affiliates, agents, officers, directors, shareholders and employees from and against any third party/End User claims, actions, damages, liabilities, costs, judgments or expenses (including attorney fees) arising out of or ***relating to its End User's use of the Services*** (including but not limited to any claims by third parties that its

End Users' use of the Service included dissemination of obscene material and/or ***infringement of a third party's intellectual property rights***).

Example Clause #12: The following clause gives the Customer a very broad range of protections against claims of various kinds, including infringement of intellectual property rights. An indemnity for infringement claims is combined with several other indemnification provisions relating to Provider's performance under the agreement.

In addition to the indemnity obligation, if Provider is unable to resolve an infringement claim successfully, Provider must refund to Customer "an appropriate amount of the compensation and expense paid hereunder…" The phrase "appropriate amount" is not entirely clear. In addition, Provider would be liable to pay Customer's costs of replacing the Service that is subject to the infringement claim.

INDEMNITY

13.1 To the extent of the negligence, gross negligence or willfulness of Provider or any party under the direction or control of Provider, or to the extent any deliverables or materials used in or arising from provision of the Services are defective or unreasonably dangerous, Provider will indemnify and hold harmless Customer, its owners, parents, affiliates, subsidiaries, agents, directors and employees from and against all judgments, orders, awards, claims, damages, losses, liabilities, costs and expenses, including, but not limited to, court costs and reasonable attorneys' fees ("***Liabilities***") arising from the performance of the Services hereunder or the acts or omissions of Provider, its agents and employees and others under its direction or control. Such Liabilities will include, but not be limited to, those which are attributable to personal injury, sickness, disease or death; and/or result from injury to or destruction of real or personal property including loss of use thereof, theft, misuse or misappropriation.

13.2 To the extent of the negligence, gross negligence or willfulness of Customer or any party under the direction or control of Customer, Customer will indemnify and hold harmless Provider, its parent, affiliates, subsidiaries, agents, directors and employees from and against all Liabilities arising from the acts or omissions of Customer, its agents and employees and others under its direction or control. Such Liabilities will include, but are not limited to, those which are attributable to personal injury, sickness, disease or death; and/or result from injury to or destruction of real or personal property including loss of use thereof, theft, misuse or misappropriation.

13.3 In addition, Provider agrees to indemnify and hold harmless Customer, its owners, parents, affiliates, subsidiaries, agents, directors and employees from and against Liabilities that arise out of Provider's breach of any of the terms and conditions of this Agreement.

PATENT, TRADEMARK, COPYRIGHT OR TRADE SECRET INDEMNIFICATION

14.1 ***Provider will indemnify*** and hold harmless Customer, its owners, parents, affiliates, subsidiaries, agents, directors and employees from and against all ***Liabilities that may result by reason of any infringement*** or claim of infringement of any patent, trademark, copyright, trade secret or other proprietary right relating to deliverables or materials used in or arising from provision of Services and/or the use thereof. Provider will defend and/or settle at its own expense any action brought against Customer to the extent that it is based on a claim that Services and/or the use thereof, infringe any patent, trademark, copyright, trade secret or other proprietary right. Customer may have its own counsel participate in the defense of any such claim or action, provided that the cost of such counsel will be borne exclusively by Customer.

14.2 If a preliminary or final judgment will be obtained against Customer's use of any Services or any part thereof by reason of alleged infringement or if in Provider's opinion, such Services are likely to become subject to a claim for infringement, Provider will, at its expense and option and without any effect or waiver of any right Customer may possess at either law or equity, either:

(a) Procure for Customer the right to continue using such Services; or

(b) Replace or modify the Services so that they become non-infringing but only if the modification or replacement does not adversely affect Customer's rights or ability to use same as specified herein.

If neither of those options is reasonably possible, Provider will ***refund to Customer an appropriate amount*** of the compensation and expenses paid hereunder, based on considering the amount of time Customer actually received the benefit of Services compared to the amount of time Customer expected to be able to receive the benefit of said Services. ***Provider will also pay all expenses of removing the Service(s) and any expenses incurred by Customer to install alternatives to the Service(s).***

Example Clause #13: The following clause contains mutual indemnifications by both Customer and Supplier for infringement claims. The mutual infringement provisions are combined with other indemnifications extended by the parties to each other. The Customer has agreed to indemnify the Supplier for claims of infringement of intellectual property rights arising from Customer's use of Supplier's network, and Supplier has agreed to indemnify the Customer for claims of infringement arising from equipment or software used by Supplier.

A. Customer will indemnify, defend and hold harmless the Supplier and its parent company, affiliates, employees, directors, officers, and agents from and against all claims, demands, actions, causes of actions, damages, liabilities, losses, and expenses (including reasonable attorney's fees) incurred as a result of:

1. Claims for libel, slander, infringement of copyright or unauthorized use of trademark, trade name or service mark arising out of Customer's or its customers' use of the Services;

2. ***Claims for patent infringement*** arising from ***combining or connection of facilities to use Supplier's network***; and

3. Claims for damage to property and/or personal injuries (including death) arising out of the gross negligence or willful act or omission of Customer and/or End User.

B. Supplier will indemnify, defend and hold harmless Customer, its affiliates, employees, directors, officers, and agents from and against all claims, demands, actions, causes of actions, damages, liabilities, losses, and expenses (including reasonable attorney's fees) incurred as a result of:

1. ***Claims for patent or copyright infringement*** relating to the ***equipment or software used by Supplier*** to provide the Services hereunder; and

2. Claims for damages to property and/or personal injuries (including death) arising out of the gross negligence or willful act or omission of the Supplier.

Example Clause #14: In the following clause the Seller has disclaimed any warranty or responsibility for claims of infringement of intellectual property rights. The disclaimer is combined with waivers of other types of claims as well, and is reinforced by a reference to Seller's "Liability Limitations," as if there could be a question about the disclaimer.

The Seller has also reserved the right to suspend a service for any of several reasons, including Customer's use of the Service in a manner which infringers on the intellectual property rights of another party.

> In addition to all rights contained in the MSA or Sales Order, ***Seller may temporarily suspend or terminate this Exhibit or a Service*** with ten (10) days prior written notice if: (a) Customer fails to materially comply with any foreign, federal, state or local law or regulation applicable to use of the Service by Customer; or (b) Customer commits any illegal acts relating to the subject matter of this Exhibit, including but not limited to, use of the Services for illegal purposes, dissemination of obscene material, ***or infringement of a third party's intellectual property rights***; or (c) a regulatory body, governmental authority, or a court of competent jurisdiction changes any material rate, charge or term of the Services, or restricts or prohibits Seller, or Customer, from providing a Service; or (d) with respect to any non-Seller facility where a circuit terminates, Seller or Customer is unable to obtain or to maintain upon acceptable terms and pricing, building access rights, permits and rights of way necessary to provision the Services; or (e) Seller or Customer becomes aware of exigent circumstances resulting from Customer's use of the Services, including but not limited to circumstances posing a danger to the Customer Network or Seller's network as applicable. Initial and continued provisioning of Services are subject to Customer's compliance at all times with Seller's standard credit terms, as they may be modified from time to time as well as those special credit terms, if any, set forth in this Exhibit. All Services are provisioned by Seller, Customer, its Affiliates, subcontractors and vendors, pursuant to the terms set forth herein.

Warranty and Liability Limitation. Subject to the limitations contained herein, Seller warrants that its Services will conform to the requirements of the Service Level Agreement. Customer's sole and exclusive remedy for breach of the above warranty or any claim related to Services shall be in accordance with the Service Level Agreement pursuant to Section ___ above or payment of any Service Level Credits, if any, provided by Customer. WITH RESPECT TO THE SERVICES, ANY EQUIPMENT AND/OR THE SUBJECT MATTER OF THIS AGREEMENT, THE FOREGOING WARRANTY IS IN LIEU OF ALL OTHER WARRANTIES, EXPRESS, IMPLIED OR STATUTORY, INCLUDING, BUT NOT LIMITED TO, THE IMPLIED WARRANTIES OF MERCHANTABILITY, FITNESS FOR A PARTICULAR PURPOSE, TITLE, AND NON-INFRINGEMENT. ALL SUCH WARRANTIES ARE HEREBY EXPRESSLY DISCLAIMED TO THE MAXIMUM EXTENT ALLOWED BY LAW. Notwithstanding anything to the contrary contained herein, the Liability Limitations in the MSA shall govern this agreement and shall apply with equal force to any Customer.

Example Clause #15: In contrast to many agreements in which a Seller's warranties are very limited and narrowly defined, in the following clause the Seller has extended to Customer a long litany of representations and warranties. However, the Seller has expressly disclaimed responsibility for claims of infringement of intellectual property, except for "Material created by or for Seller."

Seller represents, warrants and covenants to Customer that:

(a) The Fiber Connectivity Solution will conform to the description in this Agreement (including Exhibit __) throughout the Term, as such description may be modified from time to time based on changes to the Fiber Connectivity Solution made in accordance with this Agreement. This warranty shall not be construed as a warranty of the actual performance of any Optical Wavelength Service, it being understood and agreed that the performance of each Optical Wavelength Service shall be measured against the applicable Service Levels in Exhibit ___.

(b) The Services will be performed with promptness and diligence and executed in a workmanlike manner in accordance with the practices and professional standards used in a well-managed operation performing services similar to the Services. Seller will use adequate numbers of qualified individuals with suitable training, education, experience and skill to perform the Services.

(c) Seller will provide the Services using proven, current technology.

(d) ***No Material created by or for Seller that is used in connection this Agreement will infringe the Intellectual Property Rights of any third party***, nor will Seller fail to secure any licenses or other rights that are required for Seller to use Material of a third party in providing the Services.

The foregoing shall not extend to any infringement that arises from the manufacturer, licensor or distributor of any commercially available product used by Seller in connection with this Agreement having violated the Intellectual Property Rights of a third party.

EXCEPT AS EXPRESSLY PROVIDED IN THIS AGREEMENT, NO EXPRESS OR IMPLIED WARRANTY IS MADE BY EITHER PARTY IN CONNECTION WITH THIS AGREEMENT, INCLUDING ANY IMPLIED WARRANTY OF MERCHANTABILITY OR FITNESS FOR A PARTICULAR PURPOSE.

Indemnities by Seller:

Seller shall indemnify, defend and hold harmless Customer and its Affiliates, and their respective officers, directors, employees, agents, successors and assigns, from and against any and all Losses and threatened Losses arising out of or in connection with any of the following:

(a) The death or bodily injury of any agent, employee, customer, business invitee, or business visitor or other person caused by the tortious conduct of Seller or its subcontractors;

(b) The damage, loss or destruction of any real or tangible personal property caused by the tortious conduct of Seller or its subcontractors;

(c) Any claim asserted against Customer but resulting from an act or omission of Seller (or its subcontractors) in its capacity as an employer of a person. For clarification, the indemnity under Paragraph (c) shall not apply to any claim asserted against Customer that is based solely on acts or omissions of Customer and does not include as an element of the damages or other relief sought by the claimant any loss or injury arising from any acts or omissions of Seller (or its subcontractors) in its capacity as the employer of the claimant;

(d) ***Any Material created by or for Seller infringing the Intellectual Property Rights of a third party or the failure by Seller to secure necessary licenses or other rights to use Material of a third party in providing the Services***. The indemnity under this Paragraph (d) ***shall not apply to any claim of infringement that arises from the manufacturer, licensor or distributor of any commercially available product*** used by Seller in connection with this Agreement having violated the Intellectual Property Rights of a third party. To the extent permitted under the applicable agreement, Seller shall pass through to Customer all warranties, indemnities and other protections against infringement provided by such manufacturer, licensor or distributor;

(e) Breach of confidentiality obligations in Section _______ by Seller or its subcontractors;

(f) Violations of applicable laws, rules or regulations by Seller or its subcontractors;

(g) Any claims between Seller and any of its subcontractors;

(h) Any claims arising out of or related to occurrences Seller is required to insure against pursuant to Section _____ (to the extent of the required coverage for such insurance); and

(i) Any third party claim seeking compensation or damages from Customer based on Customer's use of the Services to the extent such third party claim arises out of any failure by Seller to obtain or maintain any Required Right.

"Material" shall mean systems, software, hardware, documentation, reports, methodologies, processes, workflows and other information and material used by Seller in connection with this Agreement.

Example Clause #16: In the following clause the Service Provider has agreed to a very broad series of indemnifications, which include an indemnification for claims of infringement of intellectual property rights. However, there are some limitations to the indemnification for infringement. The Service Provider has disclaimed liability for claims that arise from a modification of the Service by Customer or claims resulting from a combination of the Service with other network or service elements by anyone other than Supplier.

Service Provider shall indemnify defend and hold harmless the Customer and its parent company, affiliates, employees, directors, officers, and agents from and against all claims, demands, actions, causes of actions, damages, liabilities, losses, and expenses (including reasonable attorney's fees) incurred as a result of third party:

(a) Claims for ***patent or copyright infringement*** relating to the Service Provider's equipment or Service Provider's software used by Service Provider to provide the Services hereunder; and

(b) Claims for damages to property and/or personal injuries (including death) arising out of the gross negligence or willful act or omission of the Service Provider.

If as the result of a claim or the threat of a claim related to intellectual property infringement, (1) the Services or any essential component thereof are held by a court of competent jurisdiction, or in Service Provider's reasonable judgment may be held to infringe, or (2) either party receives a valid court order enjoining that party from using the Services or any essential component thereof, or in Service Provider's reasonable judgment such order may be received, Service Provider shall in its reasonable judgment, and at its expense, (a) replace or modify the effected Service or essential component to be non-infringing; (b) obtain for Customer a license to continue using the effected Service or essential component; or (c) if Service Provider cannot reasonably obtain the remedies in (a) or (b), terminate the effected Service without penalty to Customer. This Section states Service Provider's entire liability and Customer's exclusive remedy for any claim of infringement. Notwithstanding the foregoing, ***Service Provider shall have no liability***

for any claim of infringement based on (x) the use of the Services in a manner not contemplated or otherwise not in accordance with this Agreement and documentation related to the Services; (y) ***combinations of the Service or essential component thereof with other elements or the modification of a Service by anyone other than Service Provider where, but for such combination or modification, no claim of infringement would exist*** or (z) the use of the Services. Any and all claims for indemnification shall be conditioned upon (i) prompt written notification by the indemnified party to the indemnifying party; (ii) assumption of sole control of the defense of such claim and all related settlement negotiations by the indemnifying party; and (iii) provision by the indemnified party of the assistance, information and authority necessary for the indemnifying party to perform its obligations under this Section.

Part 4, Chapter 10 - Bilateral Carrier Services Agreement

This form of Bilateral Carrier Services Agreement may be used by two telecommunications carriers that are selling services to each other. Under this form either party could be a seller and a buyer.

In creating this form, the parties have settled on approaches to common questions that arise in a Master Services Agreement in a manner that they believe is balanced. Other clauses that might be found in a Master Services Agreement, such as preferred provider, minimum purchase obligations, or chronic outage termination clauses, can be easily substituted for the following clauses or tailored and incorporated into this form to apply to either or both of the parties.

Bilateral Master Carrier Services Agreement
Agreement No.______________

This Bilateral Master Carrier Services Agreement is entered into as of ___________________, 2____ (the "Effective Date") by and between ________________________, a ____________ ("__________") whose principal place of business is located at __________________________ and ___________________, a ___________ ("________") whose principal place of business is located at ________________________. ________ and ________ are individually referred to as "Party" and collectively referred to herein as "Parties."

ARTICLE 1. DELIVERY OF SERVICE

1.1 General. This Agreement states the terms and conditions by which one Party will deliver ("Seller") interstate telecommunications services (the "Services") and the other Party has requested and will purchase ("Buyer") the Services. The specific Services to be provided hereunder are identified in the Service Order Form(s) submitted by Buyer and accepted by Seller. Each Service Order Form submitted, accepted and executed by both Parties is hereby incorporated by reference into this Agreement. This Agreement is intended to cover any and all Services ordered by Buyer and provided by Seller. The Parties may prepare and execute additional Service Order Forms from time to time, which shall also be made a part of this Agreement.

1.2 Buyer Premises. Buyer shall allow Seller access to the Buyer premises to the extent reasonably determined by Seller for the installation, inspection and scheduled or emergency maintenance of Facilities relating to the Service. Seller shall notify Buyer two (2) business days in advance of any regularly scheduled maintenance that will require access to the Buyer premises. Buyer will be responsible for providing and maintaining, at its own expense, the level of power, heating and air conditioning necessary to maintain the proper environment for the Facilities on the Buyer premises. In the event Buyer fails to do so, Buyer shall reimburse Seller for the actual and reasonable cost of repairing or replacing any Facilities damaged or destroyed as a result of Buyer's failure. Buyer will provide a safe place to work and comply with all laws and regulations regarding the working conditions on the Buyer Premises.

1.3 Seller Facilities. Except as otherwise agreed, title to all property owned or leased by Seller ("Facilities") and used to deliver Service, including wires, lines, ports, routers, switches, channel service units, data service units, cabinets, racks, private rooms, other equipment, and collocation shall remain with Seller. Buyer shall not, and shall not permit others to, rearrange, disconnect, remove, attempt to repair, or otherwise tamper with any Seller Facilities, without the prior written consent of Seller. The Facilities shall not be used for any purpose other than that for which Seller provides them. Buyer shall not take any action that causes the imposition of any lien or encumbrance on the Facilities. In no event will Seller be liable to Buyer or any other person for interruption of Service or for any other loss, cost or damage caused or related to improper use or maintenance of the Seller Facilities by Buyer or third parties provided access to the Facilities by Buyer in violation of this Agreement, and Buyer shall reimburse Seller for any damages incurred as a result thereof. Buyer agrees (which agreement shall survive the expiration, termination or cancellation of any Service Addendum) to allow Seller to remove the Facilities from the Buyer Premises: (a) after termination, expiration or cancellation of this Agreement in connection with which the Facilities were used; or (B) for repair, replacement or otherwise as Seller may determine is necessary or desirable, but Seller will use reasonable efforts to minimize disruptions to the Service caused thereby.

1.4 Buyer-Provided Equipment. Seller may install certain Buyer-provided communications equipment upon installation of Service, but Seller shall not be responsible for the operation or maintenance of any Buyer-provided communication equipment. Seller undertakes no obligations and accepts no liability for the configuration, management, performance or any other issue relating to Buyer's routers or other Buyer-provided equipment used for access to or the exchange of traffic in connection with the Service.

ARTICLE 2. BILLING AND PAYMENT

2.1 Commencement of Billing. Upon installation and testing of a segment of Service ("Segment") specified in the Service Addendum, Seller will deliver to Buyer a connection notice ("Connection Notice") confirming that the Service or Segment (as the case may be) has been installed and is properly functioning. The commencement of billing is the first to occur of (i) the date of the Connection Notice, unless Buyer delivers written notice to Seller within two (2) business day for Services in the United States or Europe and five (5) business for circuits outside the United States and Europe after the date of the Connection Notice that the Service or Segment (as the case may be) is not installed in accordance with the Service Addendum and functioning properly, (ii) the date upon which Buyer acknowledges in writing that the Service or Segment (as the case may be) has been installed and is functioning properly, or (iii) the date Buyer begins using the Service or Segment (other than for testing purposes); regardless of whether Buyer has procured services from other carriers needed to operate the Service, and regardless of whether Buyer is otherwise prepared to accept delivery of the Service or Segment (as the case may be).

2.2 Payment of Invoices. Seller shall bill monthly in advance on the first of the month for Service to be provided during the upcoming month, except for charges that are dependent upon usage of Service, which are billed in arrears. Billings for partial months are prorated. All invoices are due thirty (30) days after invoice date. Past due amounts shall be subject to an interest charge at the rate of 1.0% of the late payment amount per month or the maximum lawful rate allowable under applicable law, whichever is lower. All payments shall be made in United States dollars.

2.3 Taxes and Fees. Any applicable federal, state or local use, excise, sales or privilege taxes, Universal Service Fee, duties, franchise fees, right of way fees and right of entry fees or similar liabilities charged to or against Seller or Buyer because of Service furnished by Seller (collectively, "Taxes and Additional Charges"), shall be paid by Buyer in addition to the regular charges under each Service Addendum, excluding charges for the use of public rights of way and taxes based upon Seller's net income or property. This shall include any new Taxes and Additional Charges that shall be imposed after the effective date of this Agreement. In the event Buyer believes it is exempt from any Taxes and Additional Charges, Buyer will provide Seller with an exemption certificate evidencing such claimed exemption. Notwithstanding anything to the contrary in this Article, Buyer shall be entitled to protest and/or contest by appropriate proceedings any such tax for which it may be liable hereunder; provided that Buyer shall indemnify, defend (by counsel reasonably acceptable to Seller) and hold harmless Seller against any damages, losses, claims or judgments arising out of such protest and/or contest, including, without limitation, any liens or attachments.

2.4 Disputed Invoices. In the event that Buyer reasonably disputes any portion of a Seller invoice, Buyer must pay the undisputed portion of the invoice by the due date and submit a written claim for the disputed amount. All claims must be submitted to Seller within sixty (60) days of the date of the invoice. Buyer waives the right to dispute any charges not disputed within the time frame set forth above. In the event that the dispute is resolved against Buyer, Buyer shall pay the disputed amount plus interest at the rate set forth in Section 2.2 above. If the dispute is not resolved during this period, either Party may seek resolution of the dispute in accordance with Article 14.10.

2.5 Demand for Reasonable Assurances. If at any time there is a material adverse change in Buyer's creditworthiness, as defined in this Article 2.5, then, in addition to any other remedies available to Seller under this Agreement or applicable law, Seller may elect, in its sole discretion, to demand reasonable assurance of payment from Buyer. A "material adverse change in Buyer's creditworthiness" shall include, but not be limited to: (a) Buyer's default of its obligations to Seller under this Agreement; (b) acquisition of Buyer (whether in whole or by majority or controlling interest) by an entity which is insolvent, which is subject to bankruptcy or insolvency proceedings, which owes past due amounts to Seller or any entity affiliated with Seller or which is a materially greater credit risk than Buyer; (c) Buyer's being subject to or having filed for bankruptcy or insolvency proceedings or the legal insolvency of Buyer; or (d) negative Buyer net worth or working capital. If Buyer's financial statements are not public information, upon Seller's demand for reasonable assurance of payment, Buyer shall provide Seller with copies of its most recent financial statements. After receipt and review of Buyer's financial information, Seller may require a deposit or other similar means to establish reasonable assurance of payment. If Buyer has not provided Seller with its financial information and with reassurance satisfactory to Seller within thirty (30) days of Seller's notice of demand for reassurance, then, in addition to any other remedies available to Seller, Seller shall have the option, in its sole discretion, to exercise one or more of the following remedies: (i) cause the start of any Services described hereunder or in a previously executed Service Addendum to be delayed pending satisfactory reassurance; (ii) decline to accept any Service Addendum or other requests from Buyer to provide Services hereunder; or (iii) terminate this Agreement with forty-eight (48) hours written notice.

2.5.1 If Buyer is delinquent making payments three (3) times in any calendar year, Seller may require Buyer to post a deposit, at Seller's option, to secure Buyer's payment for the term of this Agreement in the amount of two (2) months recurring charges for the Services. Should Seller determine that a deposit is necessary, Buyer and Seller will execute an appropriate deposit agreement with terms satisfactory to Seller. Should Buyer fail to post such deposit or fail to execute the appropriate agreement within fifteen (15) days after Seller's notice of such requirement, or should Buyer fail to abide by the terms of the deposit agreement, Buyer shall be deemed to be in default of this Agreement and Seller may terminate this Agreement in accordance with Article 11 (Termination).

ARTICLE 3. TERM; RENEWAL OPTIONS

3.1 Term of Agreement. The term of this Agreement shall begin on the Effective Date and shall continue for an initial term of three (3) years unless terminated earlier by either Party in accordance with Article 11 (Termination) of this Agreement. Provided that either Party is not in default or has not been in default in the twelve (12) months previous, the term of the Agreement will automatically be renewed for successive one (1) year periods thereafter unless one of the Parties gives written notice to the other at least sixty (60) days prior to the end of the then current term that the Agreement will not be automatically renewed. In the event a particular Service(s) ordered by Buyer will extend beyond the effective termination date of this Agreement, and provided that Buyer is not in breach or default of this Agreement, such Services(s) shall remain in effect for the agreed-upon time of Service, subject to all of the terms and conditions of this Agreement as if it were still in effect with respect to such Service(s).

3.2 Circuit Term. With respect to each Circuit provided to Buyer under this Agreement, the term shall begin on the Start of Service Date and shall continue as specified in the Service Addendum for each Circuit but in no event less than one (1) year unless otherwise agreed. If no term is specified in a Service Addendum, it shall be one (1) year. The term of the Circuit shall automatically renew for successive one (1) month periods thereafter unless (i) one of the Parties gives written notice to the other at least thirty (30) days prior to the end of the then current term that the Circuit order will not be automatically renewed or (ii) the Agreement has been terminated by either Party.

ARTICLE 4. SERVICE OUTAGE CREDITS

See Appendix No. 1-A.

ARTICLE 5. PROVISIONING REQUIREMENTS

See Appendix No. 3.

ARTICLE 6. WARRANTIES

6.1 DISCLAIMER OF WARRANTIES. EXCEPT AS SPECIFICALLY SET FORTH IN THIS AGREEMENT AND ANYAPPENDICES HERETO, NEITHER PARTY MAKES AND EACH PARTY HEREBY SPECIFICALLY DISCLAIMS ANY REPRESENTATIONS OR WARRANTIES, EXPRESS OR IMPLIED, ORAL OR WRITTEN, REGARDING THE PRODUCTS AND SERVICES CONTEMPLATED BY THIS AGREEMENT, INCLUDING BUT NOT LIMITED TO ANY IMPLIED WARRANTY OF MERCHANTABILITY OR FITNESS FOR A PARTICULAR PURPOSE, EXCEPT TO THE EXTENT THAT SUCH DISCLAIMERS ARE HELD TO BE LEGALLY INVALID.

6.2 Seller reserves the right to substitute, change or rearrange any equipment or circuit used in delivering Service, provided that the quality, cost or type of Service is not adversely affected; provided, however, Buyer will receive at least thirty (30) days prior written notice of any such change to any Service.

6.3 Buyer represents and warrants that to its knowledge none of Buyer's equipment, services, its customer's equipment or services, or their respective networks shall interfere with, impair the use of, damage any Service, damage any of Seller's equipment, or create a hazard to Seller's personnel or to the public.

ARTICLE 7. LIMITATION OF LIABILITY; CONSEQUENTIAL DAMAGES WAIVER.

7.1 CONSEQUENTIAL DAMAGES WAIVER. EXCEPT AS EXPRESSLY SET FORTH IN THIS SECTION 7, IN NO EVENT SHALL EITHER PARTY BE LIABLE TO THE OTHER PARTY OR SUCH PARTY'S CUSTOMER OR CUSTOMER'S CUSTOMERS OR CLIENTS OR TO ANY OTHER PERSON, FIRM OR ENTITY IN ANY RESPECT, INCLUDING, WITHOUT LIMITATION, FOR ANY DAMAGES, EITHER DIRECT, INDIRECT, CONSEQUENTIAL, SPECIAL, INCIDENTAL, ACTUAL, PUNITIVE, OR ANY OTHER DAMAGES, OR FOR ANY LOST PROFITS OF ANY KIND OR NATURE WHATSOEVER, ARISING OUT OF MISTAKES, ACCIDENTS, ERRORS, OMISSIONS, INTERRUPTIONS, DELAYS, OR DEFECTS IN TRANSMISSION INCLUDING THOSE WHICH MAY BE CAUSED BY REGULATORY OR JUDICIAL AUTHORITIES, ARISING OUT OF OR RELATING TO THIS AGREEMENT OR THE OBLIGATIONS OF SUCH PARTY PURSUANT TO THIS AGREEMENT. The foregoing limitation shall not be construed to limit a party's indemnification obligations hereunder with respect to third party claims against an Indemnified Party and covered by indemnification hereunder. Additionally, liability for damages will be limited and excluded, even if any exclusive remedy provided for in this Agreement fails of its essential purpose.

SECTION 8. INDEMNIFICATION.

8.1 Each Party (the "Indemnifying Party") hereby agrees to indemnify, defend and hold harmless the other Party and its affiliates, and their respective employees, officers, directors and agents (collectively, " Indemnified Persons") from and against and in respect of all demands, claims, actions or causes of action, liabilities and expenses, including, without limitation, interest, penalties and attorneys' fees and disbursements (collectively, "Claims"), to the extent any such Claim is asserted against, resulting to, imposed upon or incurred by the Indemnified Person, directly or indirectly, by reason of or resulting from any (i) personal injury, death or property damage caused by the negligence or willful misconduct of the Indemnifying Party, its employees, agents or subcontractors, (ii) misrepresentation or noncompliance with any covenant, condition or other agreement, given or made pursuant to this Agreement, or (iii) Claims by a Party's customer or a customer's customer.

8.2 Notice. The Indemnified Party shall notify the Indemnifying Party to describe the claim or action to the Indemnifying Party within fourteen (14) days of becoming aware of the claim or action itself. The Indemnifying Party may undertake the defense of any such claim or action and permit the Indemnified Party to participate therein at the Indemnified Party's own expense. The settlement of any such claim or action by an Indemnified Party without the Indemnifying Party's prior written consent, which consent shall not be unreasonably withheld or delayed, shall release the Indemnifying Party from its obligations hereunder with respect to such claim or action so settled. The Indemnifying Party shall not enter into a settlement of any claim which is prejudicial to the Indemnified Party.

8.3 Buyer shall make all arrangements with copyright holders, music licensing organizations, performers' representatives or other parties for necessary authorizations, clearances or consents with respect to transmission contents ("Consents"). Buyer shall indemnify and hold harmless Providers (as defined below) against and from any court, administrative or agency action, suit or similar proceeding, whether civil or criminal, private or public, brought against Providers arising out of or related to the contents transmitted hereunder (over Seller's network or otherwise) including, but not limited to, claims, actual or alleged, relating to any violation of copyright law, export control laws, failure to procure Consents, failure to meet governmental or other technical broadcast standards, or that such transmission contents are libelous, slanderous, an invasion of privacy, pornographic, or otherwise unauthorized or illegal. "Providers" shall be defined to include a Party and its parent, subsidiary and affiliated companies, shareholders, directors, officers, employees and agents, or any third party or affiliated provider, operator or maintenance/repair contractor of facilities employed in connection with the provision of Services under this Agreement. Seller may terminate or restrict any transmissions over its network if, in its judgment, (a) such actions are reasonably appropriate to avoid violation of applicable law; or (b) there is a reasonable risk that criminal, civil or administrative proceedings or investigations based upon the transmission contents shall be instituted against Providers. Buyer agrees not to use any Services provided hereunder for any unlawful purpose, including without limitation any use that constitutes or may constitute a violation of any local, state or federal obscenity law.

8.4 With respect to third parties that use the Services provided hereunder through Buyer, Buyer shall defend, indemnify and hold harmless Providers against any claims by such third parties for damages arising or resulting from any defect in, failure to provide or interruption or Outage of any of the Services.

SECTION 9. INSURANCE

The Parties shall, at their own expense, obtain and keep in full force and effect at all times for the duration of this Agreement, insurance policies of the following kinds and in the following amounts:

(a) Workers' Compensation Insurance in accordance with all applicable laws;
(b) Employer's liability insurance with limits for employer's liability of $500,000 per accident;
(c) Comprehensive bodily injury and property damage liability insurance, including automobile insurance and contractual liability insurance, in at least the following amounts:

Bodily injury to any one person: $5,000,000

Bodily injury aggregate per occurrence:	$5,000,000
Property damage in any one accident:	$1,000,000
Property damage aggregate per occurrence:	$5,000,000

Upon request of either Party, the Parties shall furnish certificates of such insurance and/or copies of the applicable policies and/or certificates of self-insurance. Each policy shall provide that no material change or cancellation shall become effective except on thirty (30) days' prior notice to the other Party of such change or cancellation. Failure by a Party to obtain replacement coverage in the event of cancellation of a policy within thirty (30) days after such demand shall constitute an event of default and entitle the non-defaulting to exercise its rights under Article 13 (Default).

SECTION 10. CONFIDENTIALITY AND PUBLICITY

10.1 Confidential Information. If either Party desires that information provided to the other Party under this Agreement be held in confidence, that Party will, prior to or at the time of disclosure, identify the information in writing as confidential or proprietary ("Information"). The recipient may not disclose Information, may use it only for purposes specifically contemplated in this Agreement, and must treat it with the same degree of care as it does its own similar information, but with no less than reasonable care. These obligations do not apply to Information which: (a) is or becomes known by the recipient without breach of any obligation to maintain its confidentiality; (b) is or becomes known to the public through no act or omission of the recipient; (c) is independently developed by the recipient without the use of Information; or (d) is disclosed in response to a valid order by a court or governmental body, if prior to such disclosure, the recipient gives written notice to the discloser, so as to afford it the opportunity to object. This Section 10 will not affect any other nondisclosure agreement between the Parties.

10.2 Publicity. Neither Party shall have the right to use the other Party's or its affiliates' trademarks, service marks or trade names or to otherwise refer to the other Party in any marketing, promotional or advertising materials or activities without the prior written consent of the other Party, such consent not to be unreasonably withheld. Neither Party shall issue any publication or press release relating to the terms and conditions of this Agreement or any contractual relationship between Seller and Buyer, except as may be required by law without the prior written consent of the other Party, such consent not be unreasonably withheld.

ARTICLE 11. TERMINATION

11.1 This Agreement may be terminated by:

(a) By a written amendment to this Agreement executed by the Parties;
(b) By the Party as set forth in Articles 2 (Billing and Payment and Article 3 (Term; Renewal Options).
(c) Upon default by either party, in accordance with Section 13.1 of this Agreement.
(d) In addition to the termination provisions set forth in (a) through (c) above, either party may terminate this Agreement upon written notice if the other party fails to cure a material breach within thirty (30) days of written notice of such breach.

11.2 In the event that Buyer and Seller elect to terminate any particular Service Addendum pursuant to this Agreement, such termination shall not affect any other Service Addendum and Buyer shall continue to pay the Service charges for any Service Addendum(s) not terminated.

11.3 In addition to any other rights of termination specified in this Article 11, this Agreement, or any Services provided pursuant to a Service Addendum, may be terminated upon thirty (30) days prior written notice, as follows:

(a) By Seller, with respect to particular Service Addendum, in the event that a final order or judgment is entered in any lawsuit or regulatory proceeding restraining performance under this Agreement, declaring or otherwise rendering performance unlawful or compelling removal, or discontinuation or divestiture of all or part of the network, or requiring Seller to pay an exorbitant or grossly disproportional amount, in Seller's reasonable judgment, for the acquisition of any easement or right-of-way and such order or judgment has not been vacated, reversed, or stayed within thirty (30) days from the date of entry thereof.

(b) By either Party, if the transaction contemplated by the Parties hereunder become subject to regulation of any kind whatsoever under any law to a greater or different extent than that existing on the effective date of this Agreement, and such regulation either (i) renders this Agreement illegal or unenforceable or (ii) materially adversely affects the business of either Party, with respect to its financial position or otherwise, then, either Party shall at such time have the right to seek to renegotiate the pricing for the Services and if a mutually agreeable rate cannot be reached, shall have a right to terminate this Agreement without penalty or liability.

11.4 Portability. With respect only to Services provided by Seller using solely Seller-owned facilities (on-net Services), Buyer shall have the option to disconnect a Service and order a new Service of like type for the remaining term of the initial service as a replacement without incurring a termination charge; provided that Seller has available capacity to provide the Service. Any payments related to the replacement Service shall be credited against any termination charges; except for payments for nonrecurring charges related to connection of the replacement Service. The replacement Service must be ordered by Buyer within ten (10) calendar days after the date on which Buyer provides Seller with written notice of disconnection. If Buyer exercises this on-net portability option, Buyer will be liable for any and all applicable one-time charges and any new monthly recurring charges associated with the disconnection of the original service and the connection of the new Service, including any third-party charges. In the event disconnection of any on-net Service should result in an off-net Service being rendered unusable or "stranded," Buyer will continue to be liable for ongoing costs associated with the continued availability of said off-net Services. Seller, at Buyer's request, will work with the provider of said off-net Service to determine terms for disconnection and termination of said Service; and, upon Buyer's written agreement to be liable for any costs associated with such off-net termination, Seller will facilitate the termination of said Service.

ARTICLE 12. EFFECT OF TERMINATION

Upon termination of this Agreement and payment of any amount required pursuant to Article 2 (Billing and Payment), except as provided below, this Agreement, or the Service Addendum(s) if only a portion of the Service(s) are affected, shall be terminated and neither Party shall have any continuing performance obligations under the terminated Agreement or Service Addendum; provided, however, that the obligations of the Parties under Article 7 (Limitation of Liability) and Article 10 (Confidentiality) of this Agreement, and the obligations of Buyer to pay Service charges through the effective date of such termination or the date of the outage giving rise to the termination (whichever first occurs), shall remain in full force and effect, and no termination pursuant to Article 11 (Termination) shall entitle Buyer to the return of any Service charges theretofore paid or afford to Buyer any defense to the payment of Service charges then due and payable, except to the extent any such Service fee(s) have been unearned by Seller, as, for example, when termination occurs in the middle of the month.

ARTICLE 13. DEFAULT

13.1 Events of Default. A Party shall be deemed in default of this Agreement upon the occurrence of any one or more of the following events:

(a) A Party makes a general assignment or arrangement for the benefit of creditors for the benefit of creditors; become bankrupt, becomes a debtor in a bankruptcy proceeding, becomes insolvent, however evidenced, or becomes unable to pay its debts as they fall due; files a petition or otherwise commences a proceeding under any bankruptcy, insolvency, reorganization or similar law, or has any such petition filed or commenced against it; or has a liquidator, administrator, receiver, trustee, conservator or similar official appointed with respect to it or any portion of its property or assets;

(b) A Party violates any applicable laws, statutes, ordinances, codes or other legal requirements with respect to the Service and such violation(s) are not remedied or a Party has not diligently pursued a remedy within thirty (30) days after written notice thereof provided, however, that each Party reserves the right to contest and/or appeal any such claim of violation in which event the existence of any default shall be stayed_pending resolution of the contest and/or appeal; or

(c) A Party fails to perform its obligations under this Agreement and such nonperformance is not remedied within thirty (30) days after notice thereof, except for payment defaults, for which the period to remedy is ten (10) days.

13.2 Rights upon Default

(a) Upon the occurrence of an Event of Default by Buyer, Seller may terminate this Agreement and/or all or any portion of the Services provided and/or the affected Service in accordance with Article 11 (Termination). Seller may also declare immediately due and payable all the remaining charges for the remainder of the term of the terminated Service Addendum.

(b) Upon the occurrence of an Event of Default by Seller, Buyer shall be entitled to terminate the affected Service(s) in accordance with Article 11, and recover any amounts paid in advance of service. This shall constitute Buyer's sole remedy for a Seller default. Seller shall have the right to remove all of its equipment from Buyer's premises within five (5) business days of Buyer's termination of a Service.

(c) Seller's right to terminate a specific Service or this Agreement shall be in addition to, and not in substitution for, any other rights that Seller may have as a result of an Event of Default by Buyer. In the exercise of its right of termination as herein provided, Seller may, at its option, elect to terminate this Agreement in its entirety or only with respect to the particular Service to which the Event of Default pertains.

SECTION 14. ADDITIONAL PROVISIONS

14.1 Relationship. Nothing in this Agreement shall be deemed to create any relationship between Seller and Buyer other than that of independent parties contracting with each other solely for the purpose of carrying out the provisions of this Agreement. Neither of the Parties hereto shall be deemed or construed, by virtue of this Agreement, to be the agent, employee, representative, partner, or joint venturer of the other. Neither Party is authorized, by virtue of this Agreement, to represent the other Party for any purpose whatsoever without the prior written consent of the other Party. Nothing in this Agreement shall prevent either Party from entering into similar arrangements with, or otherwise providing services to or buying services from any other person or entity.

14.2 Compliance with Laws. In connection with the matters provided for in this Agreement, each

Party hereto, where applicable, shall comply with all applicable laws and regulations, including, but not limited to, the Communications Act of 1934, as amended, and the policies, rules and regulations of the FCC. Additionally, the Parties agree that in the event of a decision by a telecommunications regulatory authority at the federal, state or local level necessitates material modifications in this Agreement, the Parties will negotiate in good faith to modify this Agreement in light of such decision. If a mutually acceptable resolution cannot be achieved, the Parties may elect to exercise the termination provisions contained in Section 11.3 of this Agreement. In the event of any change in applicable law, regulation, decision, rule or order that materially increases the costs or other terms of delivery of Service, Seller and Buyer will negotiate regarding the rates to be charged to Buyer to reflect such increase in cost and, in the event that the Parties are unable to reach agreement respecting new rates within thirty (30) days after Seller's delivery of written notice requesting renegotiation, then (a) Seller may pass such increased costs through to Buyer, and (b) Buyer may terminate the affected Service Addendum without termination liability by delivering written notice of termination no later than thirty (30) days after the effective date of the rate increase

14.3 Force Majeure. The performance of the obligations set forth in this Agreement shall be suspended or excused in the event such performance is materially and adversely affected due to causes beyond its reasonable control, including, but not limited to, acts of God, suppliers, fire, explosion, vandalism, cable cut, storm or other similar occurrences; any law, order, regulation, direction, action or request of the United States government, or of any other government, including state and local governments having jurisdiction over either of the Parties, or of any department, agency, commission, court, bureau, corporation or other instrumentality of any one or more of said governments, or of any civil or military authority; national emergencies; insurrections; riots; wars; or strikes, lock-outs, work stoppages or other labor difficulties (each, a "Force Majeure").

14.4 Assignment. This Agreement shall be binding on and inure to the benefit of the Parties hereto and their respective successors and permitted assigns. Neither Party shall assign or transfer its rights or obligations under this Agreement without the prior written consent of the other Party, which consent shall not be unreasonably withheld, conditioned or delayed; provided, however, either Party may assign this Agreement without the other Party's consent (i) to any parent company, subsidiary or affiliate of such Party, (ii) to any successor to all or substantially all of such Party's business, whether by merger, consolidation or otherwise, or (iii) in connection with obtaining financing. In addition, either Party may without the other Party's consent assign some or all of its rights and obligations hereunder to lenders in connection with a financing by the assigning Party of construction of its network, and such lenders, may further assign this Agreement as collateral for such financing.

14.5 Notices. All notices, demands, requests, or other communications to be given by either Party shall be in writing and mailed by first-class, registered or certified mail, return receipt requested, postage prepaid, or transmitted by hand delivery or by facsimile. Each Party may designate by notice in writing a new person and/or address to which any notice, demand, request or communication may thereafter be so given, served or sent. Each notice, demand, request, or communication which shall be mailed, delivered or transmitted in the manner described above shall be deemed sufficiently given at such time as it is delivered to the addressee or at such time as delivery is refused by the addressee. Notices shall be sent to the following address:

If to _____:

If to ___________:

______________ ______________ ______________ Tel. No.:__________ Fax No. __________ Attn.: __________	______________ ______________ ______________ Tel. No.:__________ Fax No. __________ Attn.: __________
With a Copy To:	With a copy to:

14.6 Recovery of Costs. In the event that a Party brings suit or engages an attorney to enforce the terms of this Agreement or to collect money due hereunder, the prevailing Party may be entitled to recover, in addition to any other remedy, reimbursement for reasonable attorneys' fees, court costs, and other related expenses incurred in connection therewith.

14.7 Right to Set-Off. Seller hereby reserves the right at all times to set off any amount owed to Buyer or its affiliated companies in connection with any transaction or occurrence against any amount owed by Buyer to Seller under this Agreement.

14.8 Governing Law. This Agreement, the rights and obligations of the Parties hereto, and any claims or disputes relating thereto, shall be governed by and construed in accordance with the laws of New York (but not including the choice of law rules thereof).

14.9 Arbitration. Any dispute relating to the Agreement shall be resolved by binding arbitration governed by the then-current rules of the American Arbitration Association (the "Rules"). The Parties expressly waive any right of appeal to any court. Only damages allowed pursuant to the Agreement may be awarded. The arbitration shall be conducted in a location mutually acceptable to the Parties and such arbitration, and any related award shall be confidential and not disclosed, except to a Party's affiliates, accountants and counsel. Each Party shall bear its own expenses, but the Parties will share equally the expenses of Arbitration. This Agreement will be enforceable, and any arbitration award will be final, and judgment thereon may be entered in any court of competent jurisdiction.

14.10 Severability. If any part of any provision of this Agreement or any other agreement, document or writing given pursuant to or in connection with this Agreement shall be invalid or unenforceable under applicable law, said part shall be ineffective to the extent of such invalidity only, without in any way affecting the remaining parts of said provision or the remaining provisions of this Agreement and the Parties hereby agree to negotiate with respect to any such invalid or unenforceable part to the extent necessary to render such part valid and enforceable. Additionally, the Service provided by Seller as set forth in each separate Service Addendum attached hereto is severable, and upon termination of the Service with respect to any Service Addendum whether pursuant to the provisions of Section 1 or Section 3 hereof, the Service with respect to other Service Order Forms shall continue unaffected.

14.11 Waiver. The terms of this Agreement may only be amended or modified by an instrument in writing executed by the Parties hereto. Neither the waiver by either of the Parties hereto of a breach under any of the provisions of this Agreement, nor the failure of either of the Parties, on one or more occasions, to enforce any of the provisions of this Agreement or to exercise any right hereunder shall thereafter be construed as a waiver of any subsequent breach of a similar nature, or as a waiver of any such provisions or rights hereunder.

14.12 Counterparts; Originals. This Agreement may be executed in two or more counterparts, each of which will be deemed an original, but all of which together shall constitute one and the same instrument. Once signed, any reproduction of this Agreement made by reliable means (e.g., photocopy, facsimile) is considered an original. For purposes of this Agreement, the term "written" means anything reduced to a tangible form by a Party, including a printed or hand written document, e-mail or other electronic format.

14.13 Entire Agreement. This Agreement (including all Service Order Forms) constitutes the entire agreement between the Parties hereto with respect to the subject matter hereof, and it supersedes all prior oral or written agreements, commitments or understandings with respect to the matters provided for herein. The terms of this Agreement may only be amended or modified by an instrument in writing executed by the Parties hereto. Neither the waiver by either of the Parties hereto of a breach under any of the provisions of this Agreement, nor the failure of either of the Parties, on one or more occasions, to enforce any of the provisions of this Agreement or to exercise any right hereunder shall thereafter be construed as a waiver of any subsequent breach of a similar nature, or as a waiver of any such provisions or rights hereunder. In the event of any ambiguity and/or inconsistency, the following descending order of precedence shall control: (1) Service Order Forms; (2) this Agreement and any amendments hereunder; (3) Schedules, if any.

14.14 Appendices. The following Appendices shall be attached to and incorporated within this Agreement as necessary. In the event of any inconsistency between the terms contained in the Appendices and this Agreement, the Appendices shall control only with respect to the specific inconsistency.

Appendix No. 1	Testing
Appendix No. 1-A	SLA & Network Performance
Appendix No. 2	Maintenance and Repair
Appendix No. 3	Ordering Procedures for Services
Appendix No. 4	Service Descriptions
Appendix No. 5	Pricing for On-Net Service

IN WITNESS WHEREOF, the Parties have executed this Agreement on the date set forth above

______________________________	______________________________
(Corporate Name)	(Corporate Name)
By:___________________________	By:___________________________
(Authorized Representative)	(Authorized Representative)
______________________________	______________________________
(Print Name and Title)	(Print Name and Title)

APPENDIX NO. 1

Testing

1.0 Testing - Interface Requirements.

The Point of Demarcation means the point at which Seller's responsibility to provide equipment and Service ends and where Buyer's responsibilities begin. The Point of Demarcation will be one of the following: the Seller DSX or OSX panel located in the Seller rack; the facility's meet-me room; Buyer's cage or rack.

Seller Interface hand-offs are BNC in the case of DS-3 Services; Single Mode Fiber, SC Connector in the case of OC-N Services; Single or Multi-Mode Fiber, SC Connector in the case of Gigabit Ethernet Services; Single Mode Fiber, SC Connector in the case of Wavelength Services.

2.0 Transmission Performance Specification.

2.1 Performance for Service connections - see Appendix 1-A.

3.0 System Acceptance Criteria

3.1 End-to-end system performance.

The following acceptance tests will be conducted:

DS-3 and OC-N: Testing shall be performed by using accepted clear channel standards. Seller will use a PRBS pattern that will be of a time length between 24 to 72 hours to ensure that circuits are operating within Seller standards and to the satisfaction of Buyer. Buyer will receive test results upon completion, and is encouraged to perform their own testing from demarc to demarc to ensure approval. The Buyer will be provided 48 hours to perform this test and notify Seller of any issues with the circuit, before billing commences.

4.0 Maintenance

Maintenance action is required when the Out-of-Service Limit (as defined below) or the Maintenance Limit (as defined below) is exceeded. Buyer shall advise Seller when it becomes aware that maintenance action is required. Seller shall respond as soon as practicable when it becomes aware of such events.

Out of Service Limit (major) indicates that Service is seriously affected, and is no longer available. This can be due to an outright failure of transmission medium (equipment failure, cable cuts, etc.) or due to degradation of performance limit below the specifications set forth in this Agreement.

Maintenance Limit (minor) indicates that performance has degraded to a point that performance objectives are not met. If possible, the line should remain in Service while maintenance is taking place.

APPENDIX NO. 1-A
Service Level Agreement ("SLA") & NETWORK PERFORMANCE

1.0 Related Sections

1.1 Service Outages: Refer to Article 2

1.2 Provisioning Intervals: Refer to Article 2

1.3 Repair Times: Refer to Appendix No. 2

2.0 Service Level Agreement ("SLA")

2.1 Service Credits

Buyer acknowledges the possibility of unscheduled Service Outages. In the event that Seller is unable to restore the Service as required hereunder, Buyer shall be entitled to a credit for the affected Circuits for all unplanned outages in excess of the following schedule:

Service	**Service Outages in Excess of**	**Credit***
Standard Circuit Services (DS3, OCn with 1+0 handoff)	2 Hours	1 day per circuit
Carrier Class Circuit Services (OCn with 1+1 handoff)	15 Minutes	1 day per circuit
Ethernet Services	2 hours	1 day per circuit
UWave Unprotected Wavelength Services	3 Hours	1 day per circuit
PROWave Protected Wavelength Services	2 Hours	1 day per circuit
RADWave Redundant and Diverse Wavelength Services	30 Minutes	1 day per circuit

*For the service outages outlined, credits may reach up to a maximum of 5 days per circuit per month

2.2 Provisioning Intervals

Seller will bring Service Order location "on net" on or before the schedule outlined in Appendix No. 3, or the Nonrecurring Installation charge for that month will be credited to Buyer's account.

3.0 Performance Objectives

3.1 Network (On-Net) Availability Service Objective

Network Availability is the ratio of the total amount of time during which an individual Circuit is fully operational, between such Circuit's point of demarcation, over a given period. The availability objective for circuits is to provide performance levels over a twelve-month period as follows:

Network Availability Objective by Service
Standard Circuit Services (DS3, OCn with 1+0 handoff): 99.950%
Carrier Class Circuit Services (OCn with 1+1 handoff): 99.999%
Ethernet Services: 99.950%
UWave Unprotected Wavelength Services: 99.50%
PROWave Protected Wavelength Services: 99.95%

RADWave Redundant and Diverse Wavelength Services: 99.997%

3.2 TDM Services

Errored Seconds (ES) and Severely Errored Seconds (SES) are the primary measure of error performance. SELLER applies the ANSI Standard (ANSI T1.231 – 1997) for definition of Errored Seconds and Severely Errored Seconds. Seller objectives are:

Not more than 212 Errored Seconds/Day
Not more than 10 Severely Errored Seconds/Day

3.3 Packet Services

Network Latency and Bit Error Rate are the measures Seller utilized for Packet Services performance.

Networks Latency (Delay): The average network latency is measured as the average of 15-minute samples taken throughout the month. The latency objective does not include the Buyer's LAN (Buyer equipment included) scheduled maintenance events, or Buyer caused outages. Seller's objective is to have:

Maximum average over a calendar month: One-way latency of 2 milliseconds (Max).

Bit Error Rate (BER): BER is the number of bits with errors divided by the total number of bits that have been transmitted, received or processed over a 24-hour period.

Seller's objective is to have a BER of 1 in 10^12 or less.

APPENDIX NO. 2
MAINTENANCE AND REPAIR

1.0 Performance Monitoring and Reporting.

1.1 Seller will be responsible for performing surveillance of its network infrastructure and network elements.

1.2 Seller will sectionalize faults occurring within the system localized to the Buyer system elements as follows: Seller Transmission equipment on the End-User Premises; equipment between Seller and Seller's Point of Demarcation.

2.0 Maintenance and Repair

2.1 Any maintenance required on the Seller's system, on Seller or Buyer Premises, shall be performed by Seller or its designated contractors at no additional cost to Buyer if the failure is due to the failure of facilities or employees of Seller.

2.2 Scheduled maintenance which may place the Buyer circuit in jeopardy or require system down time will normally be performed during the "Maintenance Window" of 12:00 midnight and 6:00 a.m. local time at the affected site, on Wednesdays or Sundays. Seller will work with the Buyer to ensure the activity occurs during the window that is most advantageous to the Buyer. The Buyer will receive notification of scheduled maintenance no less than fourteen (14) calendar days in advance. Additionally, the Buyer will receive two (2) courtesy notifications prior to the maintenance, and one (1) notification at the completion of the scheduled maintenance window.

2.2.1 In the event that Immediate Maintenance is required, Seller will make every effort to notify Buyer as soon as possible before the maintenance begins. Seller will continue to update Buyer during this maintenance period, and notify Buyer when the immediate maintenance work has been completed.

2.3 Specifications: Maintenance of the Seller network will be performed so as to meet the equipment manufacturers' specifications and Appendix 1. Buyer shall have the right to review Seller's maintenance procedures and policies and edited maintenance records.

2.4 Any maintenance or service function performed by Seller on the Seller network which will or could affect service provided by Buyer to End Users will be coordinated and scheduled through Buyer surveillance system operations center whenever possible for Seller. Buyer shall provide and update a list of Buyer contacts for maintenance and escalation purposes.

2.5 Response & Repair Times. In the event of a Service failure, Seller shall target to have repair personnel on site within two (2) hours after failure occurring. Seller shall restore the Service on the failed system as follows:

(i) Electronic Restoration. In the event of an electronic failure, Seller shall use commercially reasonable efforts to restore service to the affected electronics within two (2) hours.

(ii) Cable Restoration. Seller shall use commercially reasonable efforts to restore the cable within four (4) hours of failure.

2.6 Seller shall maintain a twenty-four (24) hours a day, seven (7) days a week point-of-contact for Buyer to report to Seller service interruption, system faults or updates of service or maintenance activities.

2.7 Equipment Spares. Seller will provide all maintenance spares required to support its network elements. In general, Buyer need not provide equipment storage space in Buyer facilities over and above storage space available in Seller's equipment racks.

2.8 Access to Equipment and Facilities.

2.8.1 Employees or agents of Seller shall have unrestricted twenty-four (24) hour, seven (7) day a week escorted access to any Seller equipment or facilities at a Buyer End-User Premises or Buyer Premises, subject to End-User's or Buyer's access and security regulations at all times. These shall include, but not be limited to:

Proper Identification
Seller Authorized Personnel List
Restricted Area Access Provisions
Accompaniment by End-Users/Buyer personnel

Seller employees or agents while on Buyer End User Premises or Buyer Premises shall comply with applicable state, federal, End User and/or Buyer Facility rules and regulations.

2.8.2 Upon request, employees or agents of Buyer shall be given escorted access, for viewing only, to areas at Seller locations containing facilities and/or equipment associated with Buyer's Service, subject to Seller's access and security regulations. These shall include, but not be limited to:

Proper Identification
Buyer Authorized Personnel List
Restricted Area Access Provisions
Accompaniment by Seller personnel

Buyer employees or agents, while on Seller premises, shall comply with Seller's Facility rules and regulations. This access shall be coordinated through the Seller sales management team assigned to Buyer.

APPENDIX NO. 3
ORDERING PROCEDURES FOR SERVICES

1.0 Building Lists

Seller shall provide its On-Net building list, if requested to do so. Seller will provide information to Buyer in a mutually agreeable format. Buyer will treat this information as Seller confidential information in accordance with Article 10.

2.0 Ordering Vehicle

If a standard electronic format becomes available to transmit Service Orders or ASRs from Buyer to Seller, this vehicle will be used. If an electronic format is utilized, Seller will follow any OBF (Ordering and Billing Forum) standards for use thereof. If an electronic format cannot be utilized, Buyer will transmit ASR to Seller via facsimile or email. Facsimile and email information will be provided to Buyer and updated as needed.

3.0 Contacts and Escalation

Seller will provide a complete escalation list of contacts for the Services provided to Buyer, including the Buyer & Network Support Center (CNSC).

4.0 Service Order Intervals

As used in this paragraph 4 "shall" or "will" with respect the performance of Seller shall mean, "use its commercially reasonable efforts to."

4.1 On-Net Service requests:

4.11 Service Order and/or ASR Issuance - Refer to Article 1.1

4.12 Firm Order Commitment ("FOC") - Refer to Article 1.1

4.13 Design Layout Record ("DLR") – Upon request, Seller will provide DLR information within five (5) business days of acceptance of the Service Order or ASR.

5.0 Installation Intervals

The standard installation interval for all On-Net Services will be in accordance with Section 4. If a shorter installation interval is required that is less than the standard fifteen (15) calendar days, Seller will use commercially reasonable efforts to meet the expected Service date, and if Seller cannot meet the expected Service date, then Seller will make commercially reasonable efforts to negotiate in good faith the earliest Start of Service date possible with Buyer. Should the Buyer request demarc extensions, cross-connects or other services that are not controlled by Seller, the fifteen (15) calendar day installation intervals may no longer apply.

6.0 Cancellation Charges for On-Net Services

There will be no order cancellation charge if the On-Net Service is canceled 48 hours or more prior to the FOC date. If the On-Net Service is canceled less than 48 hours prior to the FOC date but prior to the Start of Service Date, Buyer shall pay one month's recurring changes, plus any applicable service ordering and installation charges. If the On-Net Service has been activated, Buyer may terminate prior to the expiration of an agreed term by providing Seller sixty (60) days prior written notice of its intent to cancel such Circuit. No later than sixty (60) days after Seller's receipt of such notice of termination, Buyer shall pay to Seller, as liquidated damages and not as a penalty, a termination charge equal to:

(a) 100% of the monthly recurring charge that would have been incurred for the Circuit for months 1-12 of the agreed term, plus

(b) 50% of the monthly recurring charge that would have been incurred for the Circuit for months 13-24 of the agreed term, plus

(c) 25% of the monthly recurring charge that would have been incurred for the Circuit for months 25 through the end of the agreed term.

The parties agree that Seller's damages in the event of Service cancellation shall be difficult or impossible to ascertain. These provisions are intended, therefore, to establish liquidated damages in the event of cancellation and are not intended as a penalty.

APPENDIX 4
SERVICE DESCRIPTIONS

1.0 Seller Services

Seller offers Buyers a range of interconnectivity options — SONET services, packet-based service, and wavelength services.

Services are available between Buyer-designated locations on a point-to-point or point-to-multipoint basis.

2.0 On-Net

On-Net Services are those which connect two (2) or more Points of Presence and are provided entirely by Seller. The backbone connecting two (2) locations shall normally have a physically diverse service/protect path with a single fiber pair handoff unless otherwise specified by Seller. Dual fiber pair handoffs are available from Seller for an additional cost. Unless otherwise specified, all prices provided to Buyer will pertain to single fiber pair handoffs.

A list of On-Net buildings will be provided to Buyer.

3.0 Off-Net

Off-Net Services are services that connect one or more Premises which are not currently served by the Seller network but that are Target Buildings. Off-Net Services are also referred to as Type II Services and are typically provided by Seller in partnership with a third party. Under no circumstances may Type II Services be provided to Buyer without Buyer's explicit acknowledgement and consent.

3.1 Off-Net Services for Target Buildings: Seller reserves the right to provision ordered Services to Buyers for Target Buildings Type II Services as opposed to using services entirely provided by Seller until such time as Seller's own facilities are operation. If Seller offers Type II Services to Buyer, Buyer will be notified of applicable provisioning intervals which are different from those described in this Agreement for On-Net Services. Network standards and maintenance and repair provisions associated with Type II Services are the same as those described in this Agreement as applicable for On-Net Services.

3.3 Off-Net Services for Non-Targeted Buildings: Seller reserves the right to provision ordered Services to Buyers with Type II Services as opposed to using services entirely provided by Seller. If Seller offers Type II Services for non-targeted buildings to Buyer, Buyer will be notified of applicable provisioning intervals, network standards, and maintenance and repair provisions associated with Type II Services which are different from those described in this Agreement as applicable for On-Net Services.

4.0 SONET Services

SONET circuits can be provisioned as DS-3, OC-3, OC-12, OC-48 or OC-192 services with either Standard (1+0) or Carrier Class (1+1) handoff configurations.

A fractional bandwidth option, available with OC-12 or OC-48 circuits, allows provisioning as little as 25% capacity between ports, allowing Buyer to purchase increased network capacity in STS-1 (51.8 Mbps) increments as demand dictates.

4.1 OC-3/3C Service

This service is a full duplex channel that operates over a SONET OC-3 interface. Each interface operates at 155.52 Mbps and supports concatenated and non-concatenated payloads on an STS-1 basis.

4.2 OC-12/12C Service

This service is a full duplex channel that operates over a SONET OC-12 interface. The interface operates at 622.08 Mbps and supports concatenated and non-concatenated payloads on an STS-1 basis.

This circuit can be provisioned initially at OC-3 with bandwidth incrementally increased on an STS-1 basis.

4.3 OC-48/48C Service

This service is a full duplex channel that operates over a SONET OC-48 interface. The interface operates at 2.488 Gbps and supports concatenated and non-concatenated payloads on an STS-1 basis.

This circuit can be provisioned initially at OC-12 with bandwidth incrementally increased on an STS-1 basis.

4.4 OC-192 Service

This service is a full duplex channel that operates over a SONET OC-192 interface. The interface operates at 9.952 Gbps.

This circuit is available only as a point-to-point circuit. Provisioning less than fully capacity is not currently available.

4.5 SONET Services Specifications

SONET services comply with Bellcore GR-253-CORE and ANSI T1.105 standards.

The OC-n signals will terminate using SC connectors on a Seller's provided Fiber Distribution Panel (FDP).

4.6 DS-3 Service

Full duplex channel at 44.74 Mbps terminating at a DSX with a BNC connector.

5.0 Packet Services

Ethernet services will be delivered over a fully protected optical network.

5.1 Gigabit Ethernet Service

Gigabit Ethernet (1000 Base-SX) can be purchased point-to-point at wirespeed (1000 Mbps) or as a flexible service that delivers bandwidth starting at 100 Mbps and bursting on demand in 100 Mbps increments.

6.0 Wavelength Services

Wavelength services are provisioned in either protected (PROWave) or unprotected (UWave) 1.25 Gbps or 2.5 Gbps optical path. Additionally, the RADWave product provides dual unprotected 2.5 Gbps wavelength services to be delivered over diverse paths and over separate electronics. A 10 Gbps wavelength service is available only in an unprotected configuration.

7.0 Custom Services

Dedicated Services or non-standard configurations not described above will be evaluated on an individual case basis.

8.0 Service Options

Beyond these basic products, certain additional Services may be made available to Buyer on an individual case basis and mutual agreement.

9.0 Change of Service Offerings

Seller reserves the right to offer new services in addition to those listed above as well as deleting specific offerings. Should an existing Service offering be deleted as an option, any pre-existing orders or installed services will continue to be supported by Seller; however, new orders for that terminated Service will not be accepted.

Seller also retains the ability to add features and capabilities to its existing catalogue of Service offerings.

Part 4, Chapter 11 - Reseller Agreement

A reseller is a vendor, a provider of telecommunications services to its customers, but is not a provider of services by means of its own networks or facilities. A reseller must purchase telecom services from another telecommunications carrier. Therefore, a reseller must enter into a purchase agreement, a "Reseller Agreement," with a telecommunications carrier to purchase services to be resold to its customers.

In this form, the telecommunications company that owns the network and from which the services originate is referred to as the "Carrier." The reseller is the "Reseller." And the reseller's customers that ultimately use the services are referred to as the "Customers."

In many ways a Reseller Agreement is similar to a master services agreement in that it states the general terms and conditions under which services will be provided to the reseller's customers using the Carrier's network. And many clauses that are typically found in a master services agreement may be found in a Reseller Agreement as well, such as order acceptance and installation procedures, security deposit, billing and payment procedures, and portability.

For example, although the Carrier has authorized Reseller to sell the Services, the Carrier retains the right to accept or reject orders submitted by the Reseller. (See Section 1.1(a)) This is a provision that is typical of master services agreements.

On the other hand, although the services to the ultimate customers will be provided over the Carrier's network, the Reseller would be responsible for managing the contractual relationship with Customers. (See Section 3) And a Reseller Agreement would require the reseller to protect the Carrier from all claims of Customers. (See Section 16.2)

RESELLER AGREEMENT

This Reseller Agreement (this "**Agreement**") is made and entered into as of __________ (the "**Effective Date**") by and between ____________________, a __________________________ ("**Carrier**"), and _________________________, a ________________________________ ("**Reseller**"). The terms and conditions of this Agreement, and any exhibits and attachments will govern the provision of data and communications services (the "**Services**") from Carrier to Reseller.

Recitals

A. Carrier is engaged in the business of providing telecommunication and network services;

B. Reseller is engaged in the business of providing telecommunication services;

C. Carrier, on its own, or through one or more of its Affiliates, has the capability to provide or procure the services that are described in ***Exhibit A*** (the "**Services**") and agrees to appoint Reseller to sell and distribute such Services; and

D. Reseller desires to purchase for resale the Services and Carrier desires to provide said Services to Reseller.

E. Each defined term shall have the meaning set forth in this Agreement where such term is first used or, if no meaning is so set forth, the meaning assigned to such term in the Glossary of Terms which is attached hereto and incorporated herein by this reference.

Now, therefore, in consideration of the mutual promises set forth below, and other good and valuable consideration, the receipt and sufficiency of which are hereby acknowledged, the Parties hereby agree as follows:

1. TERM, DEFINITIONS, STATUTORY PROVISIONS AND CONSTRUCTION

1.1 Term. The term of this Agreement shall commence on the Effective Date, and shall terminate twelve (12) months following the ***Service Activation Date*** (the "**Term**"), unless terminated earlier as provided in this Agreement. At the expiration of the Term this Agreement shall be automatically extended on a month-to-month basis at the same rates, terms and conditions, until terminated by either Party on not less than thirty (30) days' notice to the other Party.

1.2 Order Acceptance Procedures. Reseller may resell Carrier's Services to Reseller's Customers based on and subject to the terms of this Agreement.

(a) Reseller shall place orders ("**Service Orders**") substantially in the form of Exhibit A hereto, for its Customers by facsimile transmission to a Carrier representative to be identified within ten (10) days following the Effective Date. ***All Service Orders are subject to acceptance by Carrier, in its sole discretion***.

(b) Reseller shall be entitled to receive Services at the Carrier Pricing set forth in Exhibit A and with a minimum Service Term of 12 months for Service Orders placed for Customers.

(c) Estimated service activation dates will depend on the particular circumstances and be stated in each Service Order. Such dates shall not be a commitment on Carrier's part to activate Service by such date. Carrier agrees to provide Firm Order Commitment Dates ("**FOC Dates**") for On-Net buildings no more than thirty (30) days following receipt of Service Orders, subject to construction issues which Carrier will communicate to Reseller on a timely basis; provided however, that in the event either: (1) inside wire/cable of more than 15 floors; or (2) construction of any riser facilities is required, Carrier will provide, on an individual case basis, FOC Dates and non-recurring charges ("**ICB NRC**") within ten (10) days following receipt of the Service Order.

(d) Reseller has provided Carrier with a quarterly forecast of the number of circuits and bandwidth of expected service for expected sales through the next full calendar quarter. Thereafter, Reseller shall provide a quarterly forecast within thirty (30) days from the beginning of each quarter. Such forecasts shall not be binding on Reseller provided Reseller prepares such forecasts in good faith and in a commercially reasonable manner.

(e) Reseller shall be responsible for Customer billing and all issues associated therewith.

1.3 Order Installation. After installing a Service, Carrier will email an order completion notification to Reseller. If Reseller does not notify Carrier in writing within five (5) business days following receipt of the order completion notification that the Services do not conform to Carrier's specifications (with evidence of such non-conformance included in the notice), or if Carrier is unable to perform the acceptance testing due to Reseller's failure to satisfy any of its obligations under this Agreement, or if Reseller or a Reseller begins using the Service for any purpose other than testing, the Service shall be deemed accepted, and such date shall constitute the "**Service Activation Date**." If Reseller notifies Carrier of a failure to conform to specifications, Carrier will take such action as reasonably necessary, and as expeditiously as practicable, to correct or cure the non-conformity and then repeat the process set forth in this Section.

1.4 Service Terms. Notwithstanding any other provision of this Agreement to the contrary, on thirty (30) days' notice to Reseller, Carrier may change, abandon or add Services, in Carrier's sole discretion, for orders entered into by the Parties following the effective date of the notice.

2. APPOINTMENT OF RESELLER

2.1 Grant. Carrier hereby grants to Reseller, and Reseller hereby accepts, subject to the terms and conditions of this Agreement, the right to sell the Services directly to Customers. Carrier, shall provide to Reseller the Services that Reseller is ordering in accordance with this Agreement.

2.2 No Exclusivity. The Reseller acknowledges and agrees that during the Term, Carrier will conduct sales and marketing activities (including through or in cooperation with other resellers) with respect to services that are the same as or similar to the Services, and that no agreement has been reached between the Parties to make any division of area in which Carrier, the Reseller or another reseller will be conducting sales, whether by customer, industry, product or service line or geographical location. Reseller further acknowledges and agrees that Carrier has granted Reseller the right to resell the Services only on a non-exclusive basis and furthermore that Carrier retains (on behalf of itself and its Affiliates) the right, at any time, to, among other things, market, sell and provision any services similar or identical to the Services, directly or indirectly, to third parties or customers on terms and conditions acceptable to Carrier in its sole and absolute discretion.

3. RESELLER'S OBLIGATIONS

3.1 Reseller shall market, sell and distribute the Services in accordance with the terms and conditions of this Agreement and shall:

(a) Have the sole responsibility for interacting with its Customers in all matters pertaining to the Services, including Service installation, operation, maintenance and termination, dispute handling and resolution, billing and collection matters. Carrier shall incur no obligation, nor shall it be deemed to have any obligation, to interact with the Customers for any reason or purpose;

(b) Obtain and maintain, throughout the Term, at its own expense, and comply with all necessary Authorizations required to enable Reseller to purchase, operate, use, market, maintain and sell the Services and to ensure the full and legal operation of this Agreement. Should Reseller fail to maintain its necessary Authorizations throughout the Term, such failure shall be deemed a breach of a material term and shall invoke the termination rights set forth under ***Section 6*** herein;

(c) Allow Carrier access to and use of Reseller's premises to the extent reasonably necessary for the installation, connection, inspection, scheduled or emergency maintenance, removal of Facilities and systems relating to the Services, and for repair, replacement or otherwise as Carrier may determine is necessary; provided however, that Carrier shall use reasonable efforts to minimize disruptions to the Services caused thereby and provide reasonable notice prior to accessing Reseller's premises;

(d) Allow Carrier to remove the Facilities from Reseller's premises upon termination, expiration or cancellation of this Agreement, on notice to Reseller;

(e) Be responsible for providing and maintaining at its own expense the level of power, heating and air conditioning necessary to maintain the proper environment for all Facilities incident to providing the Services on Reseller's premises;

(f) Comply with all applicable laws and regulations regarding the working conditions on Reseller's premises;

(g) If any Customer Contract is terminated for any reason whatsoever and the same results in the termination of the Services provided under this Agreement, continue to be liable to Carrier for the Services provided by Carrier to Reseller, until such Service is terminated in accordance with the terms and conditions of this Agreement. The Reseller shall also be liable for any pass-through costs or expenses incurred by Carrier, including but not limited to costs and expenses relating to local access charges, in fulfillment, by Carrier, of the Reseller's orders; and

(h) At all times comply with international standards of acceptable use and comply with its own acceptable use and privacy policies and warrant that the Services sold pursuant to this Agreement will not be used, by the Reseller or its Customers, in willful contravention of any national or international telecommunications regulations, laws or tariffs.

3.2 Without limiting the foregoing, Reseller shall not and shall not permit others to:

(a) Rearrange, disconnect, remove, attempt to repair, or otherwise tamper with any Facilities of Carrier, without the prior consent of Carrier which shall be in writing, if feasible; or

(b) Take any action that causes the imposition of any lien or encumbrance on the Facilities.

3.3 In no event will Carrier be liable to Reseller, its Customers or any other persons for interruption of Service or for any other loss, cost or damage caused or related to improper use or maintenance of the Facilities by Reseller or third parties that are provided access to the Facilities by Reseller in violation of this Agreement.

3.4 Reseller shall be solely responsible for solicitation, service requests, creditworthiness, customer service, billing and collection of its Customers. Reseller remains responsible for compliance with all terms and conditions of this Agreement, including without limitation, payment obligations, without regard to Reseller's ability to charge for Services used by or purchased from it by Customers or its ability to collect payment from its Customers.

4. CARRIER'S OBLIGATIONS

Carrier agrees to:

4.1 Obtain and maintain and comply with, at its own expense, throughout the Term all necessary Authorizations required to provide the Services to the Reseller and to ensure the full and legal operation of this Agreement;

4.2 Provide and maintain the Facilities in good working order;

4.3 If Carrier determines it necessary to interrupt Services for the performance of routine systems maintenance, Carrier will use good faith efforts to provide Reseller with notice prior to conducting such maintenance and to conduct such maintenance during non-peak hours. In no event shall interruption for system maintenance constitute a failure of performance by Carrier. Carrier shall use reasonable endeavors to keep such interruption to a minimum; and

4.4 Provide the Services in accordance with the applicable Service Schedule as agreed upon by the Parties.

5. SECURITY DEPOSIT

Carrier's obligation to provide Services to Reseller pursuant to this Agreement is subject to approval by Carrier of Reseller's credit status. At any time during the Term, if Reseller fails three (3) times during any twelve-month period to make payment when due in accordance with Section 7, then Carrier may require a security deposit or, if Reseller has previously provided security, Carrier may require additional security (the "**Security Deposit**"). Reseller's failure to provide the requested Security Deposit within five (5) days following Carrier's request shall constitute a default; provided, however, that in no event shall the amount of the Security Deposit exceed the greater of two (2) months' estimated or actual usage charges, the MRC, and other amounts payable by Reseller to Carrier under this Agreement. Any such Security Deposit shall be maintained as security for Reseller's performance of its obligations under this Agreement. In its sole discretion, Carrier may offset any amounts past due from Reseller to Carrier against the Security Deposit without waiving any additional rights or remedies or making an election of remedies. At expiration of the Term the amount of the Security Deposit will be credited to Reseller's account, and any remaining credit balance will be refunded to Reseller within thirty (30) days thereafter.

6. TERMINATION

6.1 Either Party may terminate a Service for Default by the other Party. In the event of a Default by Reseller, Carrier shall have the right to (a) suspend Service(s) to Reseller; or (b) terminate this Agreement or any applicable Service. If Carrier terminates this Agreement due to a Default by Reseller, Reseller shall pay Carrier the early termination charge as described below within thirty (30) days of receipt of invoice for such early termination charge.

6.2 Except as otherwise provided in Section 6.3, if Reseller terminates a Service before the expiration of the Term for any reason other than a Default by Carrier or as expressly permitted by this Agreement (including its exhibits, schedules or order forms), in addition to any charges incurred for Service provided herein, Reseller shall incur an early termination charge. The Parties acknowledge and agree that the early termination charge is a reasonable estimate of the likely loss and damage suffered by Carrier and is not a penalty. Notice of early termination must be delivered to Carrier in writing, and will be effective thirty (30) days after receipt. The early termination charge shall be an amount equal to: (a) 100% of the MRC for each terminated Service multiplied by the number of months remaining in the first year of the Term, if any; plus (b) 50% of the MRC for each terminated Service multiplied by the number of months remaining in the second and succeeding years of the Term; plus (c) all documented third party charges incurred by Carrier which are directly related to the installation or termination of the Service; plus (d) all supplemental charges and NRC charges (including all nonrecurring charges that were waived by Carrier at the commencement of the Service), if not already paid; plus (e) in the case of collocation space, the costs incurred by Carrier in returning the collocation space to a condition suitable for use by other parties.

6.3 Either Party may terminate this Agreement without penalty if the other Party: (a) becomes or is declared insolvent or bankrupt; (b) is the subject of any proceedings related to its liquidation, insolvency or for the appointment of a receiver or similar officer for it; (c) makes an assignment for the benefit of its creditors; (d) enters into an agreement for the composition, extension, or readjustment of all or substantially all of its obligations.

6.4 Carrier may suspend or terminate this Agreement or a Service without penalty if: (a) Reseller or a Customer fails to comply with any foreign, federal, state or local law or regulation related to the Service, or Carrier has a reasonable belief that Reseller or a Customer has committed any illegal act relating to the Service, including but not limited to, use of the Services for illegal purposes; or (b) a regulatory body, governmental authority, or a court of competent jurisdiction, restricts or prohibits Carrier from providing a Service on the same terms and conditions as agreed herein; or (c) Carrier is unable to obtain or maintain, on acceptable terms and pricing, any access right, permit or right of way necessary to provision the Services; or (d) if such suspension or termination is necessary to protect the technical integrity of the Carrier network due to actions by Reseller or a Customer. Any termination pursuant to subsection (a) or (d) shall constitute a Default by Reseller without notice to Reseller.

7. BILLING AND PAYMENT; RATES AND CHARGES

Beginning on the Service Activation Date, Carrier will invoice Reseller monthly in advance, and all undisputed amounts shall be due thirty (30) days following receipt of the invoice. Past due balances are subject to an interest charge calculated from the date thirty-one (31) calendar days from the invoice date through the date of receipt of payment at the lesser rate of one and one-half percent (1.5%) per month or the maximum lawful rate allowable under applicable law. All stated charges herein do not include, and Reseller agrees to pay, any and all applicable foreign, federal, state and local taxes (other than taxes on Carrier's net income or property), including without limitation, all sales, use, value-added, surcharges, excise, franchise, commercial, gross receipts, license, privilege and other taxes, levies, surcharges, duties, fees, or other tax-related surcharges (including the Universal Service Fund surcharge) or those charges resulting from Regulatory Activity, whether charged to or against Carrier or Reseller with respect to the sale or use of the Services or the facilities provided by Carrier. Reseller shall provide Carrier with appropriate tax exemption certificates demonstrating that it maintains tax-exempt status from collection of all or part of these types of charges. Reseller shall keep its billing address and contact information current, and shall be responsible for paying all reasonable collection costs incurred by Carrier (including without limitation, reasonable attorneys' fees) related to unpaid undisputed invoices. Pricing of each Service includes all applicable discounts.

8. BILLING DISPUTES

8.1 Reseller must pay according to the terms of this Agreement all invoiced charges that are not properly disputed pursuant to this section.

8.2 An invoiced charge will be deemed properly disputed by Reseller only if: (a) Reseller believes in good faith that the charge was invoiced in error; (b) Reseller provides Carrier written notice of the disputed charge no later than 120 days from the date the charge first appeared on an invoice; and (c) Reseller's notice of the disputed charge includes the amount of the disputed charge, the reason the charge is disputed, and documentation supporting the dispute.

8.3 Reseller may withhold payment of a disputed charge only if notice of the dispute is received by Carrier on or before the due date of the invoice on which the charge first appears.

8.4 Carrier will investigate all billing disputes and notify Reseller in writing that: (a) a credit will be issued to reverse any amount that Carrier determines was incorrectly billed, or (b) Carrier has determined that the disputed charge was invoiced correctly. Reseller will notify Carrier in writing if it disagrees with Carrier's determination and the parties will make a good faith effort to expeditiously resolve the dispute. If the dispute cannot be resolved expeditiously following Reseller's notice, then either party may initiate the dispute resolution procedures set forth in Section 17.11 of this Agreement.

8.5 After a billing dispute is resolved:

(a) If the dispute is resolved in Carrier's favor, Reseller will, within five (5) business days following such resolution, remit to Carrier any required payment.

(b) If the dispute is resolved in Reseller's favor, and Reseller properly withheld payment of the disputed amount under Section 8.3, then Carrier will issue a credit to reverse the amount incorrectly billed.

(c) If the dispute is resolved in Reseller's favor, and Reseller previously paid the disputed amount, then Carrier will issue a credit to reverse the amount incorrectly billed, and refund the amount previously paid by Reseller.

8.6 Carrier will bill Reseller for Services rendered within one hundred and twenty (120) days following the date of Service, unless otherwise agreed to by Carrier and Reseller.

8.7 Collection from Customers. Reseller is responsible for amounts it cannot collect from Customers, including, but not limited to, fraudulent charges, billing adjustments or credits it grants Customers, including adjustments for fraudulent charges, and Reseller's inability to charge or collect for Services used by Customers. Reseller is solely responsible for Customer solicitation, service requests, creditworthiness, customer service, billing and collection.

9. TAXES AND ASSESSMENTS

Reseller is responsible for the collection and remittance of all governmental assessments, surcharges and fees pertaining to its purchase and resale of the Services, including, but not limited to, all value added taxes or similar taxes, or any other taxes or surcharge as may be may be legally imposed or required on the amount payable to Carrier hereunder (other than taxes on Carrier's net income) (collectively, "**Taxes**"). All such Taxes shall be borne by Reseller and not deducted from such charges. If any such Taxes are required to be paid, Reseller shall pay such Taxes as are necessary to ensure that Carrier receives a net charged amount equal to the charge which Carrier would have received had the payment not been subject to such Taxes.

10. UNIVERSAL SERVICE FUND

Reseller expressly acknowledges and agrees that Reseller is solely responsible for payment of any Universal Service Fund assessments imposed on, related to, or otherwise arising out of the Services, and Reseller agrees to give Carrier a Universal Service Fund exemption certificate each year, or upon request from the Carrier, that fully satisfies applicable standards and requirements pertaining to such certificates such that Reseller is subject to no Universal Service Fund assessments or other Federal regulatory fees and assessments pertaining to the Services. Reseller agrees to indemnify and hold Carrier harmless from and against all claims or demands made upon Carrier as a result of Reseller's failure to pay any such taxes, assessments or Universal Service Fund assessments for any reason, regardless of whether such failure was the result of negligence, gross negligence, willful misconduct, or fraud.

11. LEGAL AND REGULATORY COMPLIANCE

11.1 Regulations. This Agreement is made expressly subject to all present and future valid orders, rules, regulations and laws of any Government Authority having jurisdiction over the subject matter hereof and any Regulatory Requirements. In the event this Agreement, or any of its provisions, shall be found contrary to or in conflict with any such order, rule, regulation, law or Regulatory Requirement, this Agreement shall be deemed modified to the extent necessary to comply with any such order, rule, regulation, law, or Regulatory Requirement, and shall be modified in such a way as is consistent with form, intent, and purpose of the Agreement.

11.2 Reseller Certificates. Reseller shall not commence providing any Services until such time as Reseller has delivered to Carrier a certificate acceptable to Carrier indicating that Reseller is purchasing Carrier's Services for resale. Carrier may from time to time request delivery of updated certificates to confirm such status. If Reseller does not promptly provide such updated certificates, Carrier reserves the right to suspend the Services provided under this Agreement until it receives such updated certificate.

11.3 Legal Compliance. Reseller warrants that (a) it shall be the carrier of record for all Customers; (b) it will maintain all necessary tariffs, permits, certifications, authorizations, licenses or similar documentation as may be required by any Government Authority having jurisdiction over its business; and (c) comply with all Regulatory

Requirements and all applicable local, federal, state and international laws and regulations applicable to the performance of its duties and obligations under this Agreement.

12. WARRANTIES; DISCLAIMER AND LIMITATION OF LIABILITY

12.1 Disclaimers and Limitation of Liability: (a) CARRIER MAKES NO EXPRESS OR IMPLIED WARRANTY AS TO ANY SERVICE PROVISIONED HEREUNDER. CARRIER SPECIFICALLY DISCLAIMS ALL IMPLIED WARRANTIES, INCLUDING WITHOUT LIMITATION, IMPLIED WARRANTIES OF MERCHANTABILITY, FITNESS FOR A PARTICULAR PURPOSE, TITLE, NON-INFRINGEMENT OF THIRD PARTY RIGHTS, AND PERFORMANCE OR INTEROPERABILITY OF THE SERVICE WITH ANY RESELLER OR RESELLER PROVIDED EQUIPMENT.

12.2 NEITHER PARTY SHALL BE LIABLE TO THE OTHER FOR ANY INDIRECT, CONSEQUENTIAL, EXEMPLARY, SPECIAL, INCIDENTAL, COVER-TYPE OR PUNITIVE DAMAGES, INCLUDING, WITHOUT LIMITATION, LOSS OF USE OR LOST BUSINESS, OR GOODWILL, ARISING IN CONNECTION WITH THIS AGREEMENT OR CARRIER'S PROVISIONING OF THE SERVICES (INCLUDING BUT NOT LIMITED TO: (A) ANY SERVICE IMPLEMENTATION DELAYS OR FAILURES; (B) LOST, DELAYED OR ALTERED MESSAGES OR TRANSMISSIONS; OR (C) UNAUTHORIZED ACCESS TO OR THEFT OF RESELLER'S TRANSMITTED DATA), UNDER ANY CLAIM OR CAUSE OF ACTION, INCLUDING ANY THEORY, CAUSE OF ACTION OR CLAIM, INCLUDING TORT, CONTRACT, WARRANTY, STRICT LIABILITY OR NEGLIGENCE, EVEN IF THE PARTY HAS BEEN ADVISED, KNEW OR SHOULD HAVE KNOWN OF THE POSSIBILITY OF SUCH DAMAGES.

12.1 EXCEPT FOR RESELLER'S PAYMENT OBLIGATIONS UNDER SECTION 6.2 AND SECTION 7, THE TOTAL LIABILITY OF EITHER PARTY TO THE OTHER IN CONNECTION WITH THIS AGREEMENT SHALL BE LIMITED TO THE LESSER OF DIRECT DAMAGES OR THE PREVIOUS TWELVE (12) MONTHS' MRCs ASSOCIATED WITH THE AFFECTED SERVICE. THE FOREGOING LIMITATION APPLIES TO ALL CAUSES OF ACTIONS AND CLAIMS, INCLUDING WITHOUT LIMITATION, BREACH OF CONTRACT, BREACH OF WARRANTY, NEGLIGENCE, STRICT LIABILITY, MISREPRESENTATION AND OTHER TORTS. RESELLER ACKNOWLEDGES AND ACCEPTS THE REASONABLENESS OF THE FOREGOING DISCLAIMER AND LIMITATIONS OF LIABILITY. NEITHER PARTY MAY ASSERT ANY CAUSE OF ACTION AGAINST THE OTHER PARTY UNDER ANY THEORY WHICH ACCRUES MORE THAN ONE (1) YEAR PRIOR TO THE INSTITUTION OF A LEGAL PROCEEDING ALLEGING SUCH CAUSE OF ACTION. For purposes of this Section, all references to Carrier and Reseller include their respective officers, directors, shareholders, members, managers, and employees, Affiliates, customers, agents, lessors and providers of service to Carrier.

12.2 The foregoing notwithstanding, the waiver of claims and limitation of liability described above shall not apply to termination charges described in Section 3, and shall not apply to a claim arising from a breach of the restrictions on transfer or assignment that are described in Section 16, or a claim for property damage, or for bodily injury or death.

13. CONNECTIONS

Carrier will provide Service between Carrier's fiber distribution panels at the specified locations. Reseller is responsible for all costs incurred on Reseller's side of the demarcation points, including but not limited to, costs for customer equipment, interconnections, cross connects, hand-offs, installation charges, and any costs incurred at Reseller's request.

14. SERVICE CREDITS

Carrier may offer service credits related to installation intervals, Service availability, latency, and time to restore Service, which shall be set forth in the applicable exhibits. These credits and any applicable termination rights are the Reseller's sole and exclusive remedy for Service related claims. To qualify for a service credit, Reseller must not have any invoices that are past due, and must notify Carrier that a trouble ticket should be opened to document the event. In no event shall the total amount of all service credits in a month exceed the total MRC (or NRC if applicable) for the affected Service for such month. If Carrier specifies that a Service will be provided through the use of a third-party vendor, and if there is a delay in installation or interruption of such service obtained from such third party vendor, Reseller shall be entitled to remedies for such delay or interruption of service only if and to the extent of the service credit to which Carrier is entitled under its agreement with such third-party vendor.

15. CONFIDENTIALITY

15.1 Duties. As used herein, "**Confidential Information**" shall mean the pricing and other terms of this Agreement, and any information relating to the disclosing Party's technology, business affairs, and marketing or sales plans, provided that such Confidential Information is marked as confidential or, given the nature of the information or the circumstances surrounding its disclosure, such information reasonably should be considered as confidential (collectively the "**Confidential Information**").

15.2 Exceptions. The foregoing notwithstanding, Confidential Information shall not include, and the receiving Party shall not have an obligation to preserve the proprietary nature of information that: (a) was previously known to the receiving Party free of any obligation to keep it confidential; (b) is or becomes publicly available by means other than unauthorized disclosure; (c) is developed by or on behalf of the receiving Party independently of any Confidential Information furnished under this Agreement; or (d) is received from a third party whose disclosure does not violate any confidentiality obligation.

15.3 Confidentiality. Each Party agrees to hold the Confidential Information of the other Party in confidence. Neither Party shall divulge or otherwise disclose the Confidential Information of the other Party to any third party without the prior written consent of the other Party, except that either Party may make disclosure to those required for the implementation of this Agreement, and to purchasers and prospective purchasers, auditors, attorneys, financial advisors, lenders and prospective lenders, investors and prospective investors, provided that in each case the recipient has agreed in writing to be bound by the confidentiality provisions set forth in this Article. In addition, either Party may make disclosure as required by a court order or as otherwise required by law or in any legal or arbitration proceeding relating to this Agreement.

15.4 Required Disclosure. If either Party is required by law or by interrogatories, requests for information or documents, subpoena, civil investigative demand or other legal process to disclose the provisions of this Agreement or the Confidential Information of the other Party, it will provide the other Party with prompt prior written notice of such request or requirement so that such Party may seek an appropriate protective order or waive compliance with this Article. The Party whose consent to disclose information is requested shall respond to such request, in writing, within five (5) business days following the request by either authorizing the disclosure or advising of its election to seek a protective order, or if such Party fails to respond within the prescribed period the disclosure shall be deemed approved.

15.5 Return of Confidential Information. Upon termination of this Agreement for any reason or upon request of either Party, the Parties shall return all Confidential Information, together with all copies of same, to the disclosing Party. The requirements of confidentiality set forth herein shall survive return of such Confidential Information and the expiration of the Term.

15.6 No Implied Rights. Nothing herein shall be construed as granting any right or license under any copyrights, inventions, or patents, or enhancements thereto, now or hereafter owned or controlled by Carrier.

15.7 Promotions. Neither Party shall, without first obtaining written consent of the other Party, use any trademark or trade name of the other Party or refer to the subject matter of this Agreement or the other Party in any promotional activity or otherwise, nor disclose to others any specific information about the content of this Agreement.

15.8 Survival. The provisions of this Article shall survive expiration or other termination of this Agreement.

16. ASSIGNMENTS

16.1 Reseller may not assign its interest in this Agreement without the prior written consent of Carrier, which consent may not be unreasonably withheld, conditioned or delayed, except that Reseller may assign all, but not less than all, of its interest in this Agreement without the consent of Carrier (a) to any parent, Affiliate or subsidiary of Reseller; (b) pursuant to a merger, acquisition, reorganization, sale or transfer of all or substantially all of the assets of Reseller; or (c) as a collateral assignment for purposes of financing of Reseller.

16.2 Reseller may resell the Services, but may resell to its end user Customers only, and may not resell to other telecommunications carriers or resellers, and may not divide or multiplex the Services. Any use or resale of the Services shall be subject to the terms and conditions of this Agreement. Reseller agrees to indemnify, defend and hold harmless Carrier, its Affiliates and their employees, agents, officer and directors, from and against, and assumes all liability for, all suits, actions, damages and claims of any nature arising out of or resulting from a contractual or other relationship between Reseller and any such third parties as it relates to this Agreement or the use of a Service.

(a) Reseller may incorporate the Services into integrated, or bundled, packages that include one or more products or services not developed by Carrier or Reseller. However, Reseller shall not, and shall use its best efforts not to allow any other party to, connect, interconnect or merge the Services with the network or network services of any other telecommunications company, including Reseller, if such connection, interconnection or merging would result in a Customer receiving a service that terminates at a location other than the termination points of the Services.

(b) Reseller shall require any Customer to be bound by restrictions on further resale that are substantially similar to Section.

16.3 Subject to the foregoing, this Agreement shall be binding on Reseller and its respective Affiliates, successors, and permitted assigns.

17. GENERAL

17.1 Force Majeure. Neither Party shall be liable to the other for any delay or failure in performance of any part of this Agreement to the extent that a Force Majeure Event causes such delay or failure. Further, Carrier shall not be liable for any delay or failure in performance to the extent caused by Reseller's failure to perform any of its obligations under this Agreement.

17.2 Governing Law. This Agreement is governed by and shall be construed in accordance with the laws of the State of New York without regard to its choice of law principles, except and to the extent that the Communications Act of 1934, as amended by the Telecommunications Act of 1996, and as interpreted by the FCC, applies to this Agreement.

17.3 Waiver. The terms, covenants, representations and warranties of this Agreement may be waived only by a written instrument executed by an authorized representative of a Party waiving compliance. Except as otherwise provided for herein, neither Party's failure, at any time, to enforce any right or remedy available to it hereunder shall be construed as a continuing waiver of such right or a waiver of any other provision hereunder.

17.4 Authority. Reseller represents and warrants that the full legal name of the legal entity intended to receive the benefits and Services hereunder is accurately set forth herein. Each Party represents and warrants that: (a) the person signing this Agreement has been duly authorized to execute on its behalf; and (b) the execution hereof is not in conflict with law, the terms of any charter or bylaw, or any agreement to which such Party is bound or affected.

17.5 Headings. The headings used in this Agreement are for convenience only and do not in any way limit or otherwise affect the meaning of any terms herein.

17.6 Third Party Beneficiaries. Except as expressly set forth in Section 12 and subsection 17.16, the representations, warranties, covenants and agreements of the Parties set forth herein are not intended for, nor shall they be for the benefit of or enforceable by, any third party or person not a party hereto, including without limitation, Reseller's Customers.

17.7 Relationship. Neither Party shall have the authority to bind the other by contract or otherwise or make any representations or guarantees on behalf of the other. Both Parties acknowledge and agree that the relationship arising from this Agreement is one of independent contractor, and does not constitute an agency, joint venture, partnership, employee relationship or franchise.

17.8 Severability. If any provision of this Agreement is held to be invalid or unenforceable, the remainder of the Agreement terms and conditions will remain in full force and effect, unless such survival would be inconsistent with any express termination right provided for herein. If any such provision may be made enforceable by a limitation of its scope or time period, such provision will be deemed to be amended to the minimum extent necessary to render it enforceable.

17.9 Integration. This Agreement and the exhibits hereto set forth the entire agreement of the Parties with respect to the subject matter hereof, and supersede and merge all prior agreements and understandings whether written or oral. No amendment, modification, or waiver of any provision of this Agreement shall be effective unless it is in writing and signed by the Party granting such waiver or consent. The attachments and exhibits applicable to this Agreement are hereby incorporated by reference as though fully set forth herein. If there is a conflict or inconsistency between this Agreement and an exhibit, the order of precedence shall be as follows: (a) the applicable Service description; (b) this Agreement; (c) the applicable exhibit attached hereto (other than the Service description); and (d) the applicable tariff or schedule of terms and conditions published by Carrier, if any.

17.10 Survival. The expiration or termination of this Agreement shall not relieve either Party of those obligations that by their nature are intended to survive.

17.11 Remedies; Arbitration; Jurisdiction. Any dispute arising between the Parties in connection with this Agreement shall be resolved by binding arbitration in New York City, New York, in accordance with the Commercial Arbitration Rules of the American Arbitration Association. In addition to such Rules, the arbitration shall be conducted in accordance with the Federal Rules of Civil Procedure, including, without limitation, the applicable rules therein with respect to discovery and the introduction of evidence. The arbitration shall be conducted by a panel of three arbitrators. Each Party shall select one arbitrator. The two chosen arbitrators shall then select the third arbitrator. The arbitrators shall have experience in telecommunications matters. Such award shall be final when rendered. The Parties shall not file any lawsuit or seek judicial review unless in accordance with this Section. Judgment on any award rendered by the arbitrators under this Section may be entered in any court having jurisdiction thereof. Any court having jurisdiction shall enforce as a binding and final arbitral award any interim measures ordered by the arbitral tribunal. In rendering an award under this Section, no indirect, consequential, special or punitive damages will be payable. Where a dispute involves a monetary claim, each Party acknowledges and agrees that it shall be required to place all disputed sums in an arbitrator approved escrow account during the pendency of the arbitration proceeding.

17.12 Planned System Maintenance. Carrier usually conducts Planned System Maintenance outside of normal working hours, on weekdays between 8:00 p.m. and 3:00 a.m. Eastern time, and on weekends after 5:00 p.m. Eastern time on Friday and before 5:00 p.m.

Eastern time on Sunday. Carrier will use reasonable efforts to minimize any Service interruptions that might occur as a result of Planned System Maintenance and will give Reseller at least seven (7) days' advance notice of any Planned System Maintenance, if practical under the circumstances.

17.13 Notice Information. Except as otherwise specifically provided herein, all notices required to be given by either Party hereunder shall be in writing and (except invoices) delivered by hand, courier, overnight delivery service or registered or certified mail return receipt requested, or sent by facsimile transmission to the facsimile telephone number listed below. Any notice or other communication shall be deemed given when received or refused and shall be sent to the addresses below:

To Carrier: ____________________________
____
____
Attn: ______________________
Email Address: ______________
Facsimile: __________________

With a copy to: ____________________________
____
____
Attn: ______________________
Email: _____________________
Facsimile: __________________

To Reseller: ____________________________
____
____
____
____

The foregoing notwithstanding, each Party agrees that delivery of this Agreement by electronic mail or by facsimile shall have the same force and effect as delivery of original signatures and that each Party may use such facsimile signature as evidence of the execution and delivery of the Agreement by all Parties to the same extent that an original signature could be used. Reseller shall send Carrier the original executed version as soon as is reasonably practicable.

17.1 Public Disclosures. All media releases, public announcements, and public disclosures relating to this Agreement or the subject matter of this Agreement, including promotional or marketing material, but not including announcements intended solely for internal distribution or disclosures to the extent required to meet legal or regulatory requirements beyond the reasonable control of the disclosing Party, shall be coordinated with and shall be subject to approval by each Party prior to release.

17.2 No Personal Liability. Every action or claim against any Party arising under or relating to this Agreement or a Service shall be made only against such Party as a corporation or company, and any liability relating thereto shall be enforceable only against the corporate or company assets of such Party. No Party shall seek to pierce the corporate veil or otherwise seek to impose any liability relating to, or arising out of, this Agreement or a Service against any shareholder, employee, officer, director, member, agent or representative of the other Party. Each of such persons is an intended beneficiary of the mutual promises set forth in this Section and shall be entitled to enforce the obligations of this Section.

In witness whereof, the Parties hereto have caused this Agreement to be executed as of the Effective Date.

Carrier: ____________________	Reseller: ____________________
By: ____________________	By: ____________________
Name: ____________________	Name: ____________________
Title: ____________________	Title: ____________________
Date: ____________________	Date: ____________________

Glossary of Terms

"Affiliate(s)" means: (1) any individual, corporation, partnership, limited liability company, limited liability partnership, practice, association, joint stock company, trust, unincorporated organization or other venture or business vehicle (each an "**Entity**") in which a Party owns more than a fifty percent (50%) equity interest; or (2) any Entity which, directly or indirectly, is in control of, is controlled by or is under common control with a Party. For the purpose of this definition, control of an Entity shall include the power, directly or indirectly, whether or not exercised: (1) to vote more than fifty percent (50%) of the securities or other interests having ordinary voting power for the election of directors or other managing authority of such Entity; or (2) to direct or cause the direction of the management or policies of such Entity, whether through ownership of voting securities, partnership interest or equity, by contract or otherwise.

"Authorizations" shall mean any tariffs, permits, certifications, authorizations, licenses, approvals, waivers or similar documentation or consents as may be required by any governmental body or agency having jurisdiction over the business of a Party in connection with this Agreement, including, without limitation, as applicable, in the United States, the Federal Communications Commission and a state Public Utility Commission, Public Service Commission or other various regulatory bodies having authority over telecommunications providers.

"Circuit" means the Services provided between designated end points for the specified Service Term.

"Customer(s)" shall mean those of Reseller's customers to whom Reseller resells Services purchased hereunder or who otherwise are authorized to use such Services under a Customer Contract.

"Customer Contract" shall mean the contract, between the Reseller and its customers for the Services.

"Default" means: (1) in the case of a failure to pay any amount when due under this Agreement, if Reseller fails to pay such amount within ten (10) days following notice specifying such failure; or (2) if Reseller is in material breach of Section 4 of this Agreement; or (3) in the case of any other material breach of this Agreement by either Party, a Party fails to cure such breach within thirty (30) days after notice specifying such breach.

"Direct Damages" means those damages that follow immediately upon the act done and which arise naturally or ordinarily from breach of contract, but as used herein shall expressly exclude any cover-type damages.

"Effective Date" shall mean the date appearing at the beginning of this Agreement.

"Excused Outage: Any outage, unavailability, delay or other degradation of Service related to, associated with or caused by Planned System Maintenance (provided that Reseller was informed at least twenty-four (24) hours in advance); Reseller actions or inactions, Reseller provided power or equipment, any third party other than a third party directly involved in the operation and maintenance of the Carrier network, including, without limitation, Reseller's end users, third party network providers, traffic exchange points controlled by third parties, or any power, equipment or services provided by third parties, or a Force Majeure Event.

"Facilities" shall mean property owned or leased by Carrier or its Affiliates and used to deliver Service, including terminal and other equipment, wires, fiber optic cables, lines, circuits, ports, routers, switches, channel service units, data service units.

"FCC" means the Federal Communications Commission.

"Force Majeure Event" means an event (other than a failure to comply with payment obligations) caused by any of the following conditions: act of God; fire; flood; labor strike; sabotage; fiber cut (not caused by Carrier or its employees, agents, affiliates or subcontractors); material shortages or unavailability or other delay in delivery not resulting from the responsible Party's failure to timely place orders therefor; power blackouts (not caused by non-payment of utility charges); lack of or delay in transportation; government codes, ordinances, laws, rules, regulations, permits or restrictions; failure of a governmental entity or other party to grant or recognize a right of way, war or civil disorder; or any other cause beyond the reasonable control of such Party.

"Government Authority" shall mean any instrumentality, subdivision, court, administrative or law enforcement agency, commission, official or other authority of the United States or any other country or any state, province, prefect, municipality, locality or other government or political subdivision thereof, or any quasi-governmental or private body exercising any regulatory, taxing, importing or other governmental or quasi-governmental authority over telecommunications services within the United States or other country or any state, province, prefect, municipality, locality or other government or political subdivision thereof.

"Off-Net" means a service that originates from or terminates to a location that is not on the Carrier network.

"On-Net" means service that originates from and terminates to locations that are both on the Carrier network.

"Party" means either Carrier or Reseller, and "Parties" means collectively Carrier and Reseller.

"Planned System Maintenance" means maintenance on a network facility that is related to service delivery, either directly (maintenance of transmission equipment, fiber cable, etc.) or indirectly (maintenance of power, environmental systems, etc.).

"Regulatory Activity" means any regulation or ruling (including modifications thereto) by any governmental or quasi-governmental authority, regulatory agency, or court of competent jurisdiction.

"Regulatory Requirement" shall mean any rule, regulation, law or order issued by the Federal Communications Commission, a state Public Utility Commission or Public Service Commission or other various regulatory body having authority over telecommunications providers within its state or a governmental agency or court of competent jurisdiction.

"Service Activation Date" is defined in Section 1.

"Services" shall refer to the Services listed in Service Schedules that Carrier (or one or more of its Affiliates) agrees to provide to Reseller pursuant to this Agreement.

"Universal Service Fund" means the Federal fund administered by the Universal Service Administrative Company or successor entities (and to the extent applicable, equivalent state funds) into which certain providers of telecommunications or other entities must make mandatory contributions.

Exhibit A
To
Reseller Agreement
Services and Pricing

Service Description – Low Latency Optical Wavelength Service

1. **SERVICE DESCRIPTION AND PRICING, CARRIER OPTICAL WAVELENGTH SERVICE**

Reseller Order	MRC per wavelength	NRC per wavelength
One (1) wavelength, One (1) Gbps	$______	$______

The Carrier Optical Wavelength Service provides point-to-point optical connectivity between two locations. The Reseller is provided a private, dedicated optical wavelength capable of transporting one (1) Gbps.

If Reseller exercises the Renewal Option in accordance with Section 1.1, the Service Description and Pricing shall be amended as follows for all orders entered into following the effective date of the Renewal Option:

Reseller Order	MRC per wavelength	NRC per wavelength
One (1) wavelength, One (1) Gbps	$______	$______

Low Latency Optical Wavelength Service:

Carrier's low latency optical wavelength service is available between the termination point at ____________________ _______________:

(c) The premises of a collocation provider at that location;

and the following termination point:

(d) __________________;

Low latency optical wavelength service is available for resale at one (1) Gbps line rate.

The "**Latency Specification**" for this low latency optical wavelength service is a transmission latency of fourteen and ______________________ milliseconds round trip time between the demarcation points of the Service.

Latency is measured between:

(a) The Carrier demarcation point located at ______________________, to
(b) The Carrier Collocation Facility located at ______________________________, or a Carrier demarcation point located at another termination point.

The low latency optical wavelength service is offered as a single path, unprotected service only. Dual entrance facilities are not available for low the latency optical wavelength service.

2. **TECHNICAL SPECIFICATIONS.**

Service Characteristics	Low Latency Optical Wavelength Service Single Path
Line Rate	1 Gbps
Handoff	2 fibers

Availability*	99.9%

*** See Service Level Agreement for applicability and exclusions.**

3. RESTRICTIONS ON USE OF THE LOW LATENCY OPTICAL WAVELENGTH SERVICE:

The Carrier low latency optical wavelength service is subject to restrictions on assignment and use, as set forth in **Section 17** of the Agreement.

Carrier will have a right to inspect Reseller's use of the Service at any time during normal business hours and with at least twenty-four (24) hours prior notice by Carrier, in order to verify Reseller's compliance with Section 17 of the Agreement.

4. PORTABILITY: Reseller may terminate a Service on thirty (30) days' notice without incurring the early termination charges described in **Section 6** of the Agreement, provided that Reseller simultaneously replaces the existing Service with a new Service, and provided that the following conditions are met:

5. The Service that is being terminated must be On-Net, and the replacement Service must be to On-Net locations where Carrier has capacity available.
6. The MRC of the new Service must be equal to or greater than that of the Service that is being replaced.
7. There must not be any amount past due for any Service.
8. The Service Term of the replacement Service must be not less than the longer of (a) the remainder of the Service Term of the Service that is being replaced, or (b) twelve (12) months.

In addition, if Reseller exercises this portability option, Reseller agrees to pay:

9. A nonrecurring charge associated with the new Service in an amount reasonably determined by Carrier.
10. All nonrecurring disconnect and installation charges for the terminated Service, including all termination charges for local access services or any other third party provided services or facilities that are affected by the disconnection and replacement of the Service;
11. All charges for Carrier constructed interconnects, if applicable; and
12. All third party charges, including but not limited to cross-connect charges, early termination charges, cancellation charges, one-time charges, and the new monthly recurring charges, if applicable.

6. DEFINITIONS:

Dual Path Optical Wavelength Services
Dual Path Optical Wavelength Services provide two optical circuit paths between the Carrier demarcation points. These optical circuit paths will be diverse from each other. These Services are delivered with a four-fiber handoff at the Carrier demarcation point.

Single Path Optical Wavelength Services
Single Path Optical Wavelength Services provide a single optical circuit path between the Carrier demarcation points. These Services are delivered with a two-fiber handoff at the Carrier demarcation point.

Exhibit B
To
Reseller Agreement

Service Level Agreement (SLA)

1. Service Provisioning. Subject to availability and Carrier's acceptance of Service orders, Carrier agrees to provide those Services set forth in the Service Order Form and **Exhibit A**. Each Service will be provisioned pursuant to the terms and conditions of the Service Order Form, including but not limited to those service level objectives set forth in this **Exhibit B**.

2. Associated Equipment. Unless otherwise agreed to in writing by the parties, the Reseller is responsible for providing, installing, connecting and maintaining all customer premises equipment ("CPE") associated with any Service provisioned under a Service Order Form. Carrier Service charges will continue to apply regardless of whether the CPE is unavailable or inoperable.

3. SLAs. The Services are subject to the following Service Levels Agreements, as applicable to each particular Service as specified. If Carrier does not achieve a Service Level in a particular month, Carrier will issue a credit to Reseller as set forth below upon Reseller's request. To request a credit, Reseller must contact Carrier Customer Service within thirty (30) days following the end of the month for which a credit is requested. Carrier Customer Service may be contacted by calling toll free in the U.S. (800) 777-3232 or by electronic mail at customerservice@Carriernetworks.com. In no event shall the total amount of all credits issued to Reseller per month exceed the monthly recurring charges ("**MRC**") for the applicable Service.

SERVICE AVAILABILITY

Service availability is measured between the demarcation points at each end of the Carrier Circuit. Availability for dual path, route-diverse service is defined as the relative amount of time that at least one end-to-end Circuit is available for use. Service availability is calculated separately for each individual service connection order.

The Service availability objectives exclude periods of Excused Outages.

Definitions

Service Unavailability

"**Unavailable**" or "**Unavailability**" for wavelength Circuits means the duration of an interruption in transmission measured from the first of ten (10) consecutive severely erred seconds ("**SES**") on the affected Service until the first of ten (10) consecutive non-SES. "**SES**" means the point when the Bit Error Ratio (BER) exceeds 10^{-6} for a period of ten consecutive seconds and ends when the BER drops below 10^{-6} for a period of one second.

For purposes of calculating a credit, Unavailability is to be measured from the time that Carrier receives a verified notification from Reseller that a Circuit is Unavailable or from the time that Carrier detects that a Circuit is Unavailable, whichever is earlier, until the time that Carrier determines that the Service has been restored.

For the Low Latency Optical Wavelength Service only, a Service is also Unavailable if it experiences a Latency Deficiency.

Service Latency

"**Latency**" means the average round trip time period (rounded up to the next full microsecond) for the transmission of an IP packet between the points of the Service, as reported by Carrier, during a month. Latency measurements shall not include latency attributable to outages or disruptions caused by Reseller or the facilities on Reseller's side of the demarcation points.

"**Latency Deficiency**" shall mean transmission Latency in excess of the Latency Specification.

If Carrier discovers or is notified by Reseller that a Service is experiencing a Latency Deficiency, Carrier will determine whether the source of the Latency Deficiency is limited to the Carrier network. If the source of the Latency Deficiency resides outside of the Carrier network, Carrier will use commercially reasonable efforts to identify the cause of the Latency Deficiency and cooperate with the Reseller to resolve such deficiency as soon as possible.

Service Availability Objectives

Carrier's Service Availability Objectives vary depending on the type of service being provided, and for Wavelength services are as follows:

Table B-1 – Service Availability Objectives

Service Type	Service Availability Objectives
Low Latency Optical Wavelength Service (Unprotected - Single Path)	99.9%

Credits Structure

The credit structure for Service that is Unavailable is based on monthly billing calculations. For any billing month in which Carrier fails to meet the Service Availability SLA, the following credit structure will be applied to the net Monthly Recurring Charge for the Circuit(s) affected.

Table B-2 – Service Availability Credit

1. Availability Service Level for Unprotected Low Latency Optical Wavelength Service.

If a Low Latency Optical Wavelength Service becomes Unavailable (as defined above) for reasons other than an Excused Outage, Reseller will be entitled to a service credit. The amount of the credit shall be based on the cumulative Unavailability for the affected Service in a given calendar month, as set forth in the following table:

Cumulative Monthly Unavailability	**Service Level Credit for Unprotected Low Latency Optical Wavelength Service (% of MRC)**
Less than __ hours	No Credit
__ hours or longer but less than 30 hours	__% of the MRC
__ hours or longer but less than 36 hours	__% of the MRC
__ hours or longer but less than 42 hours	__% of the MRC
__ hours or longer but less than 72 hours	__% of the MRC
__ hours or longer but less than 96 hours	__% of the MRC
__ hours or longer but less than 120 hours	__% of the MRC
__ hours or longer	100% of the MRC

The Availability Service Levels and associated credits set forth in this Section shall not apply to Off-Net Service provisioned by Carrier through a third party carrier for the benefit of Reseller. Carrier will pass through to Reseller any availability service level and associated credit (if applicable) provided to Carrier by the third party carrier for such Off-Net Service.

CHRONIC OUTAGE

If Reseller experiences a Chronic Outage with respect to a Service for any reason, including an Excused Outage, Reseller shall have the right to terminate the Service without penalty or further obligation (a "**Chronic Outage Termination**"). In order to exercise the right, Reseller must give notice of termination during the period of Chronic Outage or within thirty (30) days following the end of the period of Chronic Outage. If Reseller fails to give the notice during such 30-day period, then the right to terminate for Chronic Outage shall lapse and shall no longer be exercisable. For purposes of this Section, a Service suffers a "**Chronic Outage**" if the Service is Unavailable for a continuous period of forty-five (45) days or longer.

This Section sets forth the sole and exclusive remedy of Reseller for a Chronic Outage.

TIME TO RESTORE

Time to Restore (TTR) objective is the time required to restore the failure and resume availability of a Service. The time is measured from the moment an outage is reported by Reseller or detected by Carrier, whichever is earlier, until the Service is available.

TTR Objectives:

Carrier's TTR objective is to restore a fiber cut within twelve (12) hours, and to resolve Unavailability from any other cause within four (4) hours.

Part 4, Chapter 12 - Independent Referral Agreement

An Independent Referral Agreement is essentially a broker agreement. A telecommunications carrier engages one or more resellers to market telecommunications services on behalf of the carrier.

A customer of the Reseller may be either an end user customer, a "Subscriber," or may be another telecommunications services provider, a "Service Provider." Neither a Subscriber nor a Service Provider is a customer of the Reseller. Bother are customers of either the Carrier or a Service Provider. (See Section 2.7.) The customer contractual relationship runs directly from the Service Provider to the Subscriber. The Reseller is compensated for bringing the Subscriber to the Service Provider.

In this form, in addition to the terms of the Independent Referral Agreement, the Carrier has adopted a Carrier Referral Program. The terms of the Carrier Referral Program are incorporated into the Independent Referral Agreement. (See Section 1 and Section 2.5.) The Carrier may modify the Carrier Referral Program from time to time, and all of the changes are automatically incorporated into this agreement. If the Representative finds any of the changes unacceptable, the Representative's only recourse would be to terminate the agreement. (See Section 2.5.)

Just as a seller in a master services agreement may accept or reject any request for services from its customers, the Carrier in this agreement reserves the right to reject any request for Services. (See Section 2.3.)

A Carrier may engage more than one Reseller to resell its services. (See Section 3.3 and Section 9.1.) This can give rise to uncertainty over which Representative is entitled to receive a commission in a given instance. It is quite possible that a prospective customer will be approached by more than one Representative or by the Carrier itself. In order to avoid potential disputes, the Carrier reserves the right to determine whether Reseller is entitled to receive a commission. (See Section 2.3 and Section 3.3.)

Furthermore, the Carrier reserves the right to approve all marketing materials that are used to offer Carrier's Services. (See Sections 6.2 and 6.3.)

Commissions are payable to Representative only when the associated revenues are actually received by the Carrier. If a Customer fails to pay for Services, the Representative's commission is reduced accordingly. (See Exhibit B.)

CARRIER INDEPENDENT REFERRAL AGREEMENT

This Independent Referral Agreement ("**Agreement**") is made as of the _____ day of ________, 20__ (the "**Effective Date**"), by and between ____________________, a ________________ having principal offices at ______________________________ ("**Carrier**") and ________________________ a ________ ______________ having principal offices at ______________________________, a __________________ whose federal tax identification number is ______________________________ ("**Representative**").

Recitals:

A. Carrier is engaged in the business of providing certain telecommunications services described in ***Exhibit A*** attached hereto (the "**Services**");
B. Carrier seeks to utilize Representative to provide lead referral services to Carrier for the purpose of generating sales opportunities; and
C. Representative desires to participate in Carrier's lead referral program ("**Referral Program**") and to be appointed as a nonexclusive representative to endorse and promote the sale of the Services to potential customers, pursuant to the terms and conditions of the Referral Program.

Now, therefore, in consideration of the mutual covenants and agreements set forth in this Agreement, the parties agree as follows:

1. Appointment.

Subject to the terms and conditions of this Agreement, and any changes to the Carrier Referral Program described on its website at www.________________.com, Carrier hereby appoints Representative on a nonexclusive basis to endorse and promote the Services. Representative acknowledges that Carrier may provide services directly to end user customers ("**Subscribers**") or to telecommunications carriers ("**Service Providers**") that utilize the Services to sell their services to Subscribers.

2. Responsibilities of Representative.

2.1 Throughout the term of this Agreement, Representative shall use reasonable efforts to endorse and promote the Services and to refer and forward potential customers to Carrier.

2.2 As part of each solicitation, Representative shall complete a "**Representative Lead Form**," a copy of which is attached hereto as ***Exhibit C*** (which may be modified by Carrier from time to time), and submit this form via e-mail to ____________@_______.com. Each Representative Lead Form shall include:

(a) The name of the Service Provider or Subscriber for whom the Service is ordered,
(b) The address of such Service Provider or Subscriber,

(c) The type of service requested, including the "A" and "Z" locations,
(d) The preferred Service Provider (Carrier reserves the right to forward all leads to other service providers if the preferred Service Provider is not selling Carrier services or does not respond to the lead(s) within 10 business days.)

Incomplete Representative Lead Forms will be returned to Representative to be completed.

2.3 Representative acknowledges and agrees that Carrier and Service Providers may accept or reject requests for Services in their sole discretion, without obligation or liability to Representative or any other party.

2.4 Representative shall furnish all supervision, labor, equipment and the premises required for it to perform its obligations hereunder, Representative assumes full responsibility for the actions of persons performing services on its behalf. Persons engaged by Representative to perform services under this Agreement shall be employees or subcontractors of Representative, and shall not for any purpose be considered employees, agents or subcontractors of Carrier. Representative shall assume and pay all of the costs and expenses of conducting its business under this Agreement, without reimbursement by Carrier, including payment of compensation (including withholding of income and social security taxes), workers' compensation, disability benefits, and any insurance requirements. Carrier reserves the right to require Representative to replace immediately unsuitable personnel and replace with personnel acceptable to Carrier, at Representative's sole cost and expense.

2.5 Representative agrees to adhere to and comply with all policies and procedures provided by Carrier to Representative, including those procedures set forth in Carrier Referral Program, as they may be modified from time to time (including modification of the commission schedule and the Services), all of which changes shall be effective when posted on Carrier's website (www.__________.com) or when other notice is provided to Representative. Representative's sole and exclusive remedy for any such changes which it finds unacceptable to Representative shall be to terminate this Agreement.

2.6 Representative represents, warrants and covenants that: (a) it will make only such representations concerning Carrier and the Services as are true and correct, and have been expressly approved in advance by Carrier; and that it will not engage in deceptive or misleading marketing practices; (b) it will not solicit for employment any Carrier employee or representative; and (c) it shall make all of its employees and representatives aware of and ensure their compliance with the provisions of this Agreement.

2.7 When a circuit referral is submitted, Representative shall inform Subscribers that: (a) Carrier or a Service Provider will be contacting them directly to obtain an authorization for service; and (b) Subscriber will be a customer of Carrier or the selected Service Provider, and not the Representative. Accordingly, all Service Provider rules, policies, and operating procedures concerning subscriber orders, subscriber service, and product sales will apply to the Subscriber, as well as Carrier rules, policies, and operating procedures made applicable to Subscriber through its general service agreements with the selected Service Provider.

3. Compensation.

3.1 In consideration of services rendered hereunder, Carrier shall pay to Representative the amounts determined in accordance with, and earned in accordance with, the terms and conditions of the compensation schedule set forth in **Exhibit B** that is attached hereto and incorporated by this reference (the "**Representative Commissions**").

3.2. Representative hereby waives any right to dispute Carrier's calculation of commissions payable hereunder unless Representative provides Carrier with a written notice of such dispute, including the amounts claimed to be due, and justification (including any applicable documentation) for such payment no later than thirty (30) days after the date the Representative Commission has been paid.

3.3. In the event of any dispute between Representative, Carrier personnel or another representative as to who is due a commission, if any, in regard to sale of Services, Carrier shall determine, in its sole discretion, which party shall be entitled to such commission.

3.4. Representative agrees and understands that Carrier is required by law to have the attached ***Exhibit D*** ("**Tax Information**") completed by Representative for payment of Representative Commissions in excess of Six Hundred Dollars ($600).

3.5. Carrier may withhold final payment for a reasonable time to ensure that (a) the correct amount is paid, and (b) Representative has returned all Carrier materials described in Section 6.

4. Term and Termination.

4.1. The term of this Agreement shall commence on the Effective Date and shall terminate one (1) year thereafter. At the end of the term this Agreement shall be automatically renewed on a month-to-month basis, terminable by either party on thirty (30) days' notice to the other party. The foregoing notwithstanding, at any time during the term or a renewal period, either party may terminate this Agreement upon providing the other party thirty (30) days prior written notice. Moreover, Carrier may terminate this Agreement immediately without notice in the following circumstances:

(a) A breach of any covenant, term or condition of this Agreement by Representative which breach continues uncured for a period of ten (10) days after notice to Representative of such breach;

(b) An assignment by Representative for the benefit of creditors or Representative becoming bankrupt or insolvent, or taking benefit of, or becoming subject to, any legislation in force relating to bankruptcy or insolvency, it being understood that the appointment of a receiver or trustee of the property and assets of the Representative is conclusive evidence of insolvency; or

(c) Carrier is unable to offer Services by reason of any law, rule, regulation, or order of any municipal, state or federal authority, including, but not limited to, any regulatory authority having jurisdiction.

4.2. The provisions of Sections 5 and 6 shall survive termination or nonrenewal of this Agreement or any part thereof.

4.4 Effect of Termination. As of the date of any such termination, Representative shall immediately, (a) discontinue soliciting referrals for Services from prospective and existing Subscribers; (b) remove from its website, all links to the Carrier website or other Carrier graphic images, (c) cease using any Carrier name, trademark, service mark, trade dress, Carrier logo, other Carrier materials, and intellectual property; and (d) return to Carrier all materials, including any electronic reproductions or copies thereof, provided by or on behalf of Carrier under this Agreement or in connection with the Referral Program. Representative agrees and understands that termination of this Agreement shall not relieve it of liability for past and future damages and losses incurred by Carrier on account of any act, omission, or breach by Representative hereunder.

5. Confidentiality and Customer Ownership.

It is agreed that all documents, data files, information and other materials made available to Representative in connection with this Agreement, including without limitation all information regarding Services, Service Providers, Subscribers and prospective Subscribers, marketing data, business plans, and technical information and information relating to Carrier's business, as well as information developed by Representative during the term of this Agreement (collectively "**Confidential Information**") is and shall be held by Representative in confidence and shall remain the exclusive property of Carrier during and after the term of this Agreement. Representative shall treat as trade secrets and keep in strict confidence all Confidential Information, and Representative will not at any time during the term of this Agreement or thereafter use such Confidential Information for its own benefit or disclose or permit any of its employees, agents, or representatives to disclose, through any medium, such Confidential Information to any other person or entity without the prior written consent from Carrier.

6. Intellectual Property; Sales and Marketing Materials.

6.1. Representative acknowledges and agrees that Carrier retains ownership rights in and to certain intellectual property of Carrier, including without limitation any Carrier trademark, service mark, or other designation, advertising material and any associated goodwill, whether presently existing or later developed by Carrier (collectively "Intellectual Property"). Unless expressly stated otherwise in this Agreement, nothing contained herein shall give Representative any right to use any Intellectual Property in advertising, publicity or marketing materials.

6.2. If approved in advance and in writing by Carrier, Representative may use advertising or marketing materials prepared by Carrier for purposes of Representative carrying out its obligations under this Agreement. Representative may use such advertising materials only on the terms and conditions stated by Carrier from time to time. Representative may not modify or delete any advertising materials that it is authorized to use without the prior written consent of Carrier.

6.3. Representative agrees to submit to Carrier all advertising, sales, promotion, and other publicity matters wherein the name of any Service Provider, Carrier or any of its affiliates is mentioned, or in which language is used from which the connection of Service Provider, Carrier or its affiliate may, in Carrier's judgment, be inferred or implied; and Representative further agrees not to publish or use such advertising, sales promotion, or publicity matter, or to use the name of any Service Provider, Carrier or any of its affiliates as a reference, without the prior written approval of Carrier. This provision shall be included in all subcontracts of Representative. Further, Representative shall not issue any publicity or press release with respect to this Agreement or Representative's participation in the Referral Program without Carrier's prior written consent. All references to Service Provider in this section shall mean the use of such name as a result of Representative's enrollment in the Referral Program and in performing its duties under this Agreement.

7. Right to Withhold and Offset.

Carrier, without waiver of any rights or remedies which it may at any time have against the Representative, shall be entitled from time to time to deduct from any amounts due or owing by Carrier to Representative in connection with this Agreement, any and all amounts owed by Representative to Carrier in connection with this Agreement or any other agreement. Further, Carrier shall be entitled, from time to time, and without waiver of any rights or remedies which it may at any time have against Representative, to withhold any amount otherwise due, and set off such sums against any liabilities, costs, and expenses that may be incurred arising from: (a) damage to Carrier, its Service Providers, or other parties by Representative; (b) failure of Representative to perform in accordance with this Agreement; and (c) nonpayment for services, labor and/or material supplies.

8. Limitation on Scope of Authority and Relationship Created.

This Agreement does not contemplate, create or constitute a joint venture, partnership, agency, franchise or similar relationship between Representative and Carrier. Representative is, and shall act, operate and hold itself out only as, an independent contractor. Representative has no authority or power to represent, act on behalf of or bind Carrier, including but not limited to making or accepting any offers or representations on Carrier's behalf, except as expressly authorized in this Agreement. Representative shall have no right or authority, nor shall Representative hold itself out as having any right or authority, to create any contract or obligation, express or implied, binding on Carrier or any Service Provider, including accepting orders for Services or agreeing to or offering prices, terms or conditions of sale that have not been pre-approved or otherwise authorized by Carrier and its Service Provider.

9. Nonexclusive Nature of Relationship.

9.1 Nothing set forth herein shall prevent Carrier from contracting with others to perform services of the same or similar type as Representative may be requested to perform.

9.2 Representative agrees and understands that Carrier may at any time (directly or indirectly) solicit customer referrals on terms that may differ from those contained in this Agreement or the Referral Program or operate web sites that are similar to or compete with Representative's web site. Representative hereby waives any right to bring an action or claim for damages against Carrier for any such actions.

10. Representations & Warranties.

10. 1 Representative represents, warrants, and covenants that it has independently evaluated the desirability of participating in the Referral Program and is not relying on any representation, guarantee, or statement other than as set forth in this Agreement.

10.2 Representative represents and warrants that it is under no obligation or restriction, nor will it accept any obligation or restriction, that would in any way interfere with or be in conflict with, the obligations to be undertaken by it under this Agreement.

10.2 Disclaimer of Warranties. CARRIER MAKES NO EXPRESS OR IMPLIED WARRANTIES OR REPRESENTATIONS AS TO THE REFERRAL PROGRAM, ANY MARKETING MATERIALS PROVIDED HEREUNDER, OR THE SERVICE TO BE MARKETED AND SOLD HEREUNDER OR THROUGH THE REFERRAL PROGRAM. CARRIER SPECIFICALLY DISCLAIMS ALL IMPLIED WARRANTIES, INCLUDING WITHOUT LIMITATION, IMPLIED WARRANTIES OF MERCHANTABILITY, FITNESS FOR A PARTICULAR PURPOSES, TITLE, NONINFRINGEMENT OF THIRD PARTY RIGHTS AND PERFORMANCE OR INOPERABILITY OF THE SERVICES WITH ANY END USER CUSTOMER PROVIDED EQUIPMENT OR ANY IMPLIED WARRANTIES ARISING OUT OF A COURSE OF PERFORMANCE, DEALING, OR TRADE USAGE. CARRIER FURTHER MAKES NO REPRESENTATION OR WARRANTY THAT THE OPERATION OF ITS WEBSITE WILL BE UNINTERRUPTED OR ERROR-FREE, AND IT WILL NOT BE LIABLE FOR THE CONSEQUENCES OF ANY INTERRUPTIONS OR ERRORS.

11. Limitation of Liability.

11.1 NEITHER PARTY SHALL BE LIABLE TO THE OTHER FOR ANY INDIRECT, CONSEQUENTIAL, EXEMPLARY, SPECIAL, INCIDENTAL OR PUNITIVE DAMAGES, INCLUDING WITHOUT LIMITATION, LOSS OF USE OR LOST BUSINESS, REVENUE, PROFITS, OR GOODWILL, ARISING IN CONNECTION WITH THIS AGREEMENT, THE REFERRAL PROGRAM AND/OR CARRIER'S OR A SERVICE PROVIDER'S PROVISIONING OF SERVICES (INCLUDING BUT NOT LIMITED TO: (a) ANY SERVICE IMPLEMENTATION DELAYS/FAILURES; (b) LOST, DELAYED OR ALTERED MESSAGES/TRANSMISSIONS; OR (III) UNAUTHORIZED ACCESS TO OR THEFT OF CUSTOMER'S TRANSMITTED DATA), UNDER ANY THEORY OF TORT, CONTRACT, WARRANTY, STRICT LIABILITY OR NEGLIGENCE, EVEN IF THE PARTY HAS BEEN ADVISED, KNEW OR SHOULD HAVE KNOWN OF THE POSSIBILITY OF SUCH DAMAGES. THE TOTAL LIABILITY OF CARRIER TO

REPRESENTATIVE IN CONNECTION WITH THIS AGREEMENT SHALL BE LIMITED TO THE LESSER OF: (a) DIRECT DAMAGES PROVEN BY REPRESENTATIVE; OR (b) THE COMMISSIONS EARNED PURSUANT TO THIS AGREEMENT. THE FOREGOING LIMITATION APPLIES TO ALL CAUSES OF ACTIONS AND CLAIMS, INCLUDING WITHOUT LIMITATION, BREACH OF CONTRACT, BREACH OF WARRANTY, NEGLIGENCE, STRICT LIABILITY, MISREPRESENTATION AND OTHER TORTS. REPRESENATIVE ACKNOWLEDGES AND ACCEPTS THE REASONABLENESS OF THE FOREGOING DISCLAIMER AND LIMITATIONS OF LIABILITY.

11.2 NO CAUSE OF ACTION UNDER ANY THEORY WHICH ACCRUED MORE THAN ONE (1) YEAR PRIOR TO THE INSTITUTION OF A LEGAL PROCEEDING ALLEGING SUCH CAUSE OF ACTION MAY BE ASSERTED BY EITHER PARTY AGAINST THE OTHER.

12. Dispute Resolution.

12.1 Any claim, controversy or dispute, whether sounding in contract, statute, tort, fraud, misrepresentation or other legal theory, related directly or indirectly to this Agreement, whenever brought and whether between the parties to this Agreement or between one of the parties to this Agreement and the employees, agents, affiliated businesses of the other party, anyone purporting to claim through or on behalf of the other party, or anyone purporting to make a claim on the basis of rights conferred by this Agreement, shall be resolved by binding arbitration as prescribed in this section. The Federal Arbitration Act, 9 U.S.C. §§ 1-15, not state law, shall govern the arbitrability of all claims.

12.2 Any arbitration shall be brought in _______, _________, under the auspices of the American Arbitration Association, and in accordance with the Commercial Arbitration Rules of the American Arbitration Association. In addition to such Rules, the arbitration shall be conducted in accordance with the Federal Rules of Civil Procedure, including, without limitation, the applicable rules therein with respect to discovery and the introduction of evidence. The arbitration shall be conducted by a panel of three arbitrators. Each party shall select one arbitrator. The two chosen arbitrators shall then select the third arbitrator. The arbitrators shall have experience in telecommunications matters. The arbitrators shall keep a verbatim transcript of the proceeding, and shall render with the award a detailed statement of the reasons for the award. The arbitration award shall be final, binding and may not be appealed. The parties shall not file any lawsuit or seek judicial review unless in accordance with this Section.

12.3 In such proceeding the parties shall be entitled to pursue all remedies available under such Rules, including equitable relief; provided, however, that in rendering an award no lost revenues or profits, and no indirect, consequential, special or punitive damages shall be payable.

12.4 Any award by the arbitration panel shall be paid within five (5) days following the award of the arbitration panel, and shall bear interest from the date of the award at the annual rate of ten percent (10%) compounded monthly. The parties hereto agree that a judgment of the United States District in ___________, _______, or a court of the State of ________, shall be entered on the award made pursuant to arbitration; and the parties consent to nonexclusive personal jurisdiction of such court to enforce the award so entered. Any court having jurisdiction shall enforce as a binding and final arbitral award any interim measures ordered by the arbitral tribunal.

12.5 If any party files a judicial or administrative action asserting claims subject to arbitration as prescribed herein, and another party successfully stays such action or compels arbitration of said claims, the party filing said action shall pay the other party's costs and expenses incurred in seeking such stay or compelling arbitration, including reasonable attorneys' fees.

13. General.

13.1 Notices. All notices and other communications required or permitted under this Agreement shall be in writing and shall be given by United States first class mail, postage prepaid, registered or certified, return receipt requested, or by hand delivery (including by means of a professional messenger service) addressed as follows:

To Representative at:

____
____
____
Attention: ____________

To Carrier at:

____
____
____
Attention:

With a copy to

____
____
____
Attention: ______________

Any such notice or other communication shall be deemed to be effective when actually received or refused. Either party may, by similar notice given, change the address to which future notices or other communications shall be sent.

13.2 Applicable Law. This Agreement shall be governed by and construed in accordance with the laws of the State of ____________, without regard to the conflict of laws principles of ________.

13.3 Compliance with Laws. Representative agrees to comply with all applicable federal, state, county, and municipal laws, regulations and ordinances applicable to Representative and the services to be performed by it hereunder ("**Laws**").

13.4 Termination of Service. Representative acknowledges Carrier's and the Service Providers' right to terminate Services to Subscribers at any time based on service terms, conditions or other arrangements that may exist between these parties from time to time. Under no circumstances shall Representative be deemed to be a third party beneficiary to any of these arrangements, nor shall Representative be entitled to object to Carrier's or its Service Providers' cancellation or termination of Services for any reason. Representative's sole and exclusive remedy under this Agreement is the right to receive commissions earned through the date of any such cancellation or termination.

13.5 Successors. This Agreement will be binding on, inure to the benefit of, and enforceable against the parties and their respective successors and assigns; provided, however, that this shall not be construed to permit a transfer or assignment except as permitted by the terms of this Agreement.

13.6 Severability. If any provision of this Agreement is held to be invalid by a court with jurisdiction over the parties to this Agreement, such provision shall be deemed to be restated to reflect, as nearly as possible, the original intentions of the parties in accordance with applicable Law, and the remainder of this Agreement shall remain in full force and effect as if the Agreement had been entered into without the invalid portion.

13.7 No Waiver. A failure by Carrier to enforce Representative's strict performance of any provision of this Agreement shall not constitute a waiver of Carrier's right to subsequently enforce such provision or any other provision of this Agreement.

13.8 Entire Agreement. This Agreement, together with Exhibits, constitutes the entire agreement between the parties, and supersedes and replaces any and all contracts, agreements, or understandings, express or implied, between the parties entered into prior to the execution of this Agreement.

13.9 Assignment. Representative shall not: (1) assign this Agreement, whether by operation of law or otherwise; or (2) delegate, subcontract, or otherwise transfer its obligations under this Agreement or any interest herein without Carrier's prior written consent. Carrier may assign this Agreement in whole or part without the consent of Representative or its sureties.

In witness whereof, the parties hereto have executed this Agreement as of the day and year first written above.

Carrier: ____________________	Representative: ____________________
By: ____________________	By: ____________________
Name: ____________________	Name: ____________________
Title: ____________________	Title: ____________________
Date: ____________________	Date: ____________________

EXHIBIT A
TO
CARRIER INDEPENDNET REFERRAL AGREEMENT

CARRIER SERVICES

Representative is authorized to solicit orders only for the Carrier Services listed and described in this Exhibit A.

Carrier offers SONET, Ethernet and Wavelength services.

The Carrier SONET service provides two different options: unprotected or protected SONET connectivity between sites within the applicable Points of Demarcation within a metropolitan area at data rates of DS-3, OC-3, OC-12, OC-48 (individual case basis "ICB") and OC-192.

The Carrier Ethernet service provides Ethernet connectivity between two sites within the applicable Points of Demarcation within a metropolitan area from 10Mbps to 1 Gbps data rates (10 Gbps in near future). Multiple Circuits may be aggregated onto a single Gigabit Ethernet trunk.

The Carrier Optical Wavelength Service provides point-to-point optical connectivity between two sites within the applicable Points of Demarcation within a metropolitan area. The customer is provided a private, dedicated optical wavelength capable of transporting 1.25 Gbps, 2.5 Gbps, or 10 Gbps line rates.

INDEPENDENT REFERRAL AGREEMENT

EXHIBIT B

COMPENSATION: Representative Commission

Carrier shall compensate Representative in accordance with the compensation plans described below for each Service.

Representative Commissions:

Representative shall be entitled to Representative Commissions only on sales of Services:

1. Occurring during the Term, and
2. Ordered pursuant to a Representative Lead Form obtained by Representative, and accepted and installed by Carrier and the designated Service Provider within sixty (60) days following submittal to Carrier
3. Billed and paid for by Subscribers.

Representative Commissions earned through the date of termination of this Agreement will remain payable only if the related orders are not canceled by Carrier or the Service Provider.

Calculation of Representative Commissions:

The amount of the Representative Commissions will be calculated by multiplying the applicable percentage for each Service order entered pursuant to a Representative Lead Form by the monthly recurring charges ***received in cash by Carrier*** for such Services (the "**Qualifying Revenues**"), during the first twelve (12) months of the Service. Qualifying Revenues do not include any nonrecurring charges, discounts, non-cash credits, sales and other taxes, surcharges, charges for customer premises equipment, service credits, pass-through and other third-party charges, and charges paid by Subscribers to Service Providers that may apply to the sale of the Services. In addition, Qualifying Revenues are limited to revenues that are received by Carrier either from Subscriber or from Service Providers, and do not include amounts received by a Service Provider from a Subscriber.

Representative Commissions shall be payable quarterly in arrears ***as and when payment is received*** from Subscribers.

The applicable percentage for Representative Commissions shall be 5% of the Qualifying Revenues.

INDEPENDENT REFERRAL AGREEMENT
EXHIBIT C
Representative Lead Form

Please complete all information before submitting via e-mail to ____________@______.com. Only forms that are signed by an authorized Carrier employee will be considered an Accepted Circuit.

Referring Party Information:

Date Submitted	
Your Company	
Your Name	
Your Address	
Your Telephone Number	
Your E-Mail Address	

For Carrier Use Only

Acceptance Date: __________

Signature: ______________

Name: ________________

Circuit Opportunity Prospect Information:

Prospect Contact Information	
Service Provider or Subscriber for whom Service is ordered	
Contact Name	
Title	
Address	
Telephone Number	
E-Mail	
Relationship with Prospect	

Circuit Opportunity Information			
Circuit Type	TENOS	tenrehtE	h tgnelevaW
Circuit Speed	(spbM 54) 3SD CO -3 (155 Mbps) CO -12 (622 Mbps) CO -48 (2.4 Gbps)	spbM 01 spbM 52 spbM 05 spbM 001 spbM 006 spbM 0001	spbG 52.1 spbG 5.2
Location A			
Location Z			
What type of data application is this circuit to be used for?			
Preferred Service Provider			
What action should Carrier take?	yltcerid pu wolloF	erofeb em tcatnoC following up	:rehtO
Comments:			

Part 4, Chapter 13 - Bandwidth Trading

A bandwidth trading agreement is used by parties that wish to attempt to structure the purchase and sale of telecommunications bandwidth or capacity as though it were a commodity. A bandwidth trading agreement is an attempt to structure bandwidth transactions in a manner similar to commodities transactions. One party wishes to act as market maker or intermediary between the sellers and the purchasers of services. Other parties wish to participate as purchasers or sellers of bandwidth.

However, the payment structure is more complex than a sale of commodities. Trading in commodities involves payment of a purchase price at the time that an order is agreed to or at the time of delivery. Payment for bandwidth usually occurs periodically as the services are used. (See Section 5.1.) Furthermore, credits for interruptions of service are extended periodically as a service experiences an interruption. Therefore, to create a bandwidth trading structure the parties may adopt extraordinary measures to forestall the possibility of a failure to perform under the agreement, including a failure to pay for services. (See Section 5.3, Section 10(c), Exhibit B, and Exhibit C.)

In a typical lit fiber services agreement, a seller will offer outage credits in the event of an interruption of service. This bandwidth trading contract form employs the terms "Cover Price" and "Replacement Price" instead of more common outage credit provisions. The objective is to establish a credit for interruptions of service based on the market value of a substitute service having identical or at least similar characteristics. (See Section 3.2 and Section 3.3(iv)).

MASTER BANDWIDTH PURCHASE AND SALE AGREEMENT
COVER SHEET

This Cover Sheet (the "Cover Sheet") to the attached General Terms and Conditions and all Annexes and Exhibits hereto and thereto and Confirmations in connection therewith, which are hereby incorporated by reference (collectively, the "Agreement") is entered into as of this ___ day of _________, 20__ (the "Effective Date"), by and between:

[INSERT CORPORATE NAME] (Party "A") and [INSERT CORPORATE NAME] (Party "B")

Party "A"	Party "B"
______________________	______________________
Attn: ______________________	Attn: ______________________
Phone: _ Fax: ______________________	Phone: _ Fax: ______________________
Type of Legal Entity: ______________	Type of Legal Entity: ______________
Place of Organization: ______________	Place of Organization: ______________

The parties to this Agreement listed above are each referred to herein as a "Party" and are together referred to herein as the Parties. The Parties' respective contacts for the matters set forth below, as described more fully in the attached General Terms and Conditions, are as follows:

Notices and Correspondence:

______________________	______________________
______________________	______________________
______________________	______________________
Attn.: ______________________	Attn.: ______________________
Fax No.: ______________________	Fax No.: ______________________

Payments:

Bank: ______________________	Bank: ______________________
Attn: ______________________	Attn: ______________________
Phone: ____________ Fax: ____________	Phone: ____________ Fax: ________
Account No. ______________________	Account No. ______________________
ABA Routing No. ______________	ABA Routing No. ________________

Invoices and Accounting Matters:

______________________	______________________
______________________	______________________
______________________	______________________
Attn.: ______________________	Attn.: ______________________
Phone: ____________ Fax: ____________	Phone: ____________ Fax: ________

Technical Matters:

On Site Contact Information:	On Site Contact Information:
______________________	______________________
Attn: ______________________	Attn:: ______________________
Phone: __________ Fax: ____________	Phone: ______________ Fax: ________
Pager: __________ Cell: ____________	Pager: ______________ Cell: ________
E-mail: ______________________	E-mail: ______________________

All references herein to Sections, Annexes and Exhibits are to those set forth in or attached to this Agreement. Reference to any document means such document as amended from time to time and reference to any Party includes any permitted successor or assignee of such Party. The parties hereby make the following elections with respect to the applicability or operation of certain provisions set forth in the General Terms and Conditions and attachments to the Agreement **(select only one from each box):**

Section 1.2 Transaction and Confirmation Procedures	☐ Oral (Section 1.2(b) applies and the Confirming Party is the Party designated below) ☐ Seller ☐ Buyer ☐ Party A ☐ Party B ☐ Either ☐ Written (Section 1.2(c) applies)	**Section 2 Business Day**	☐ Commercial banks in London, England ☐ Commercial banks in Tokyo, Japan ☐ Federal Reserve member banks in New York City ☐ Other
Section 2 Default Rate Source	☐ The Financial Times ☐ Japan Times ☐ The Wall Street Journal ☐ Other____________	**Section 2 GAAP**	☐ Japan ☐ United Kingdom ☐ United States ☐ Other____________
Section 2 Time Zone	☐ Central European Time ☐ Japanese Time ☐ U.S. Eastern Time ☐ Other____________	**Section 5.3 Performance Assurance**	If indicated below, a Party may be required to provide Performance Assurance pursuant to Section 5.3: ☐ Neither Party ☐ Both Parties ☐ Specify Party: ______________
Section 6.2 Included Affiliates List of Party A's Affiliates for purposes of Section 6.2 ____________________ ____________________ ____________________ ____________________ List of Party B's Affiliates for purposes of Section 6.2 ____________________ ____________________		**Section 12.5 Choice of Law**	☐ Japan ☐ United Kingdom ☐ United States ☐ New York ☐ Other____________

________________ ________________	
Section 11 Arbitration ☐ Option A Applies ☐ Option B Applies ☐ Arbitration Does Not Apply	Section 12.4 Confidentiality ☐ Applies ☐ Does Not Apply

Other Terms	List Additional Terms Here or on Attached Separate Page(s)	

CREDIT AND COLLATERAL REQUIREMENTS

Cross Default for Party A: Party A: Cross Default Amount ________ Other Entity: Cross Default Amount ________	**Cross Default for Party B:** Party B: Cross Default Amount ________ Other Entity: Cross Default Amount ________
Party A Credit Protection: Collateral Termination Payment Threshold: ☐ Applies ☐ Does Not Apply If applicable, complete the following: Party B Termination Payment Threshold: ________________ Party B Rounding Amount: ________________	**Party B Credit Protection:** Collateral Termination Payment Threshold: ☐ Applies ☐ Does Not Apply If applicable, complete the following: Party A Termination Payment Threshold: ________________ Party A Rounding Amount: ________________
Guarantor: ☐ Applies ☐ Does Not Apply Party A: If applicable, Party A's Guarantor: ________________	**Guarantor:** ☐ Applies ☐ Does Not Apply Party B: If applicable, Party B's Guarantor: ________________

Minimum Guaranty Amount:	____________	**Minimum Guaranty Amount:**	____________
Material Adverse Change:	☐ Applies ☐ Does Not Apply	**Material Adverse Change:**	☐ Applies ☐ Does Not Apply

IN WITNESS WHEREOF, the Parties hereto have executed this Agreement as of the Effective Date:

____________ *(Party Name)*	____________ *(Party Name)*
By ____________ Title ____________	By ____________ Title ____________

DISCLAIMER: The purpose of this Agreement is to facilitate trade, avoid misunderstandings and make more definite the terms of contracts for purchase and sale of Bandwidth Units. This Agreement is intended for use in connection with transactions to be entered into and performed within the jurisdictions indicated on the Cover Sheet and may not be suitable for transactions to be entered into and/or performed in other jurisdictions. The use of this Agreement by any party is voluntary. This Agreement is not required to be used to purchase or sell Bandwidth Units. Except as expressly modified, each Party to this Agreement agrees to be bound by all of the terms and conditions set forth in the attached General Terms and Conditions, including, without limitation, the Liability and Warranty provisions set forth in Sections 9 and 10 of the General Terms and Conditions.

MASTER BANDWIDTH PURCHASE AND SALE AGREEMENT
GENERAL TERMS AND CONDITIONS

Except to the extent expressly modified by the Parties, these General Terms and Conditions shall apply to, and are incorporated by reference in, the Agreement between the Parties identified on the Cover Sheet.

1. Agreement for Transactions.

1.1. Scope of Agreement. From time to time, the Parties may, but shall not be obligated to, enter into binding Transactions pursuant to this Agreement for the purchase or sale of Bandwidth Units, as hereinafter defined. Each Transaction shall be effectuated, evidenced and governed in accordance with this Agreement and the documents and other confirming evidence, whether a Transaction Agreement (as defined below) or any other form of Confirmation (as defined below). All Transactions are entered into in reliance on the fact that this Agreement, constitutes a single agreement between the Parties and the Parties would not otherwise enter into such Transactions. Any capitalized term used herein and not defined in the Section in which it appears shall have the meaning set forth in Section 2 hereof.

1.2. Transaction Procedures.

(a) General. It is the intent of the Parties to facilitate Transactions in accordance with the procedures set forth in this Section 1 and to assure that such Transactions are valid and enforceable for the mutual benefit of the Parties. The Parties shall elect one of the following transaction procedures on the Cover Sheet. Except as otherwise provided in this Agreement, in the event of any irreconcilable, express conflict between the provisions of this Agreement and any term or condition set forth in a Confirmation (as further described below), the term or condition set forth in the Confirmation shall control for the relevant Transaction.

(b) Oral Transaction Procedure. If the Parties have elected on the Cover Sheet to use the oral transaction procedure, any Transaction may be formed and effectuated and shall be legally binding from the moment the Parties agree to such Transaction by a telephone conversation that may be recorded (each Party thereby and hereby consenting to the recording of its representatives' telephone conversations without any further notice) whereby an offer and acceptance shall constitute the agreement of the Parties to a Transaction. As a material part of the consideration for entering into an oral Transaction, each Party hereby waives and agrees not to contest or assert any defense relating to (i) the validity or enforceability of Transactions under laws relating to whether certain agreements are to be in writing or signed by such Party to be thereby bound, or (ii) the authority of any employee or representative of such Party to enter into a Transaction. All telephone recordings may be introduced into evidence and shall be the controlling and original evidence used to prove the existence of oral agreements as to Transactions between the Parties in the event a Confirmation (as hereinafter defined) is not fully executed (or deemed accepted) by both Parties. Any Transaction formed between the Parties pursuant to the procedures set forth in this Section 1.2(b) shall be considered to be a "writing" or "in writing" and to have been "signed" by the Parties, and any tape recording of a Transaction shall be considered to constitute an "original" document evidencing the Transaction. The Parties agree that the party designated as the "Confirming Party" on the Cover Sheet may confirm a Transaction entered into pursuant to the provisions of this Section 1.2 (b) by giving to the other Party, within three (3) Business Days after the Trade Date, a written or electronic notice confirming the specific terms of such Transaction. The failure to send such Confirmation or to send such Confirmation within such three (3) Business Day period shall not invalidate any Transaction previously entered into in accordance with this Section 1.2(b). If such written or electronic notice is executed or accepted by both Parties, it shall constitute a Confirmation hereunder. If such written or electronic notice is not executed by the other Party and returned (or is not otherwise objected to by written notice) within three (3) Business Days after a Party's receipt thereof, it will be deemed correct and accepted as given and shall constitute a Confirmation and shall be conclusive and binding evidence of such Transaction and the final expression of the terms thereof. If such written or electronic notice is objected to by the receiving Party in writing within such three (3) Business Days, such objection and the existence of any agreement with respect to such Transaction shall be determined in accordance with Section 11 of these General Terms and Conditions.

(c) Written Transaction Procedure. If the Parties have elected on the Cover Sheet to use the written transaction procedure, any Transaction may be formed and effectuated only by written Confirmation executed by both Parties whereby execution of such Confirmation shall constitute the agreement of the Parties with respect to a Transaction.

Within the time period set forth in the Confirmation, if any, applicable to a Transaction, the Seller agrees to provide information related to the Demarcation Point and any other information reasonably required to make available and test the Bandwidth Unit(s) relating to such Transaction. By way of example but not limitation, the information to be provided by Seller may include, where applicable, LOA, LOA/CFA, bay panel, jack, and other tie down information of Seller or the common cross connect location. If no time period is set forth in a Confirmation, the Seller shall provide such information related to the Demarcation Point as set forth above within thirty (30) Days prior to the start of the Term for the Transaction or if the start of the Term is scheduled to begin in less than thirty (30) Days from the Trade Date, then within five (5) Business Days after the Trade Date. In the event that Seller fails to so provide the Demarcation Point information or such other information, and such failure continues for a period of five (5) Business Days following receipt of written notice thereof, Buyer may at its option cancel the Transaction with neither Party having any further liability to the other Party. If Buyer elects not to cancel the Transaction, Seller's failure to deliver the required information as set forth above shall constitute an Outage for the first Period of the Transaction under Section 3.3 (a) hereof and Buyer shall be entitled to the applicable remedies in respect thereof. In the event Seller fails to provide the information as set forth above within thirty (30) Days prior to the start of the second Period, Seller's failure to deliver the required information as set forth above shall constitute a Critical Outage, and Buyer shall be entitled to the applicable remedies in respect thereof. Providing inaccurate or invalid information related to the Demarcation Point shall be treated as a failure to provide such information. An event giving rise to any payment obligation under this Section 1.3 or Section 3.3 shall not constitute an "Event of Default" to the extent that such payment obligation is satisfied on a timely basis.

1.3. Term of Agreement. This Agreement shall be effective as of the Effective Date and shall remain in effect until terminated by either Party upon thirty (30) days prior written notice to the other Party; provided, however, that this Agreement shall continue to govern any Transaction(s) entered into prior to the effective date of such termination.

2. Definitions. Terms used, but not expressly defined in this Section or elsewhere in this Agreement, shall have the meanings attributed to such terms in the Confirmation. The following definitions and any terms defined elsewhere in this Agreement shall apply to this Agreement and all notices and communications made pursuant to this Agreement.

"Affiliate" means, with respect to any Person, any other Person that directly or indirectly, through one or more intermediaries, controls or is controlled by, or is under common control with, such Person. For these purposes, "control" of any Person shall mean the ownership of, or otherwise having the power to direct the voting of, more than fifty percent (50%) of the common stock or other equity interests having ordinary voting power for the election of directors (or persons performing comparable functions) of such Person.

"Agreement" is defined on the Cover Sheet.

"Bandwidth" means the capacity to carry or transmit data and other information between two Demarcation Points.

"Bandwidth Unit" means the "Bandwidth Unit" specified by the Parties in connection with a particular Transaction.

"Bankruptcy Proceeding" means with respect to a Party or any other Person, such Party or Person (a) makes an assignment or general arrangement for the benefit of creditors, (b) files a petition or otherwise commences, authorizes or acquiesces in the commencement of a proceeding or cause of action under any bankruptcy, reorganization, debt restructuring, insolvency, liquidation or other law for the protection of debtors or creditors (or analogous proceedings in the jurisdiction of such Party or Person), (c) has such a petition filed against it or (d) otherwise becomes bankrupt or insolvent (howsoever evidenced).

"Business Day" means (i) with respect to payments, a day, other than a Saturday or Sunday, on which the banks designated under the "Payment" section on the first page of the Cover Sheet are open for business; and (ii) with respect to notices or other communications, any day, other than a Saturday or Sunday, on which commercial banks in the city designated by the recipient party under the "Notices and Correspondence" section on the first page of the Cover Sheet are open for business.

"Buyer" means the Party to a Transaction who is obligated to purchase one or more Bandwidth Units.

"Claims" means all third party claims, demands or actions in connection with this Agreement, threatened or filed and whether groundless, false or fraudulent, that directly or indirectly relate to the subject matter of an indemnity or remedy hereunder, and the resulting losses, liabilities, obligations, damages, expenses, attorneys' fees and court costs, whether incurred by or in connection with a settlement or otherwise, and whether such claims, demands or actions are threatened or filed prior to or after the termination of this Agreement.

"*Confirmation*" means a writing substantially in the form of Exhibit A hereto (or in such other form as may be mutually agreed upon by the Parties) which has either been fully executed (or deemed accepted) by both Parties pursuant to Section 1.2(b) or (c).

"*Contract Price*" means the price per Period to be paid by Buyer to Seller for the purchase of Bandwidth Units, calculated as the Unit Price multiplied by the number of Bandwidth Unit(s).

"*Costs*" means brokerage fees, commissions, installation charges or similar nonrecurring charges and other transactional costs and expenses reasonably incurred by a Party as a result of entering into new arrangements to replace any terminated Transaction(s), and Legal Costs.

"*Cover Price*" means, with respect to any Period during the Term of a Transaction in which an Outage occurs, (a) the Replacement Price applicable to the Bandwidth Unit to be provided during such Period under such Transaction, minus (b) the Contract Price applicable to such Bandwidth Unit; provided, that the Cover Price shall be deemed to be zero in the event that the aforementioned Contract Price exceeds the Replacement Price.

"*Cover Sheet*" is defined in the preamble to this Agreement.

"*Credit Annex*" means the Credit Annex attached hereto as Exhibit B.

"*Critical Outage*" means either: (I) an Outage which occurs at the commencement of the first Period of the Term of the Transaction as set forth on the Confirmation and continues in duration for fifty percent (50%) of such first Period or (ii) any other Outage which occurs after the Seller has commenced providing Bandwidth Units and occurs for 10% of the Period in the aggregate.

"*Cross Default Amount*" means the cross default amount, if any, set forth in the Cover Sheet for a Party.

"*Day*" means a 24-hour period commencing at 12:00 midnight (in the Time Zone) on a calendar day and ending at 12:00 midnight (in the Time Zone) on the next calendar day.

"*Defaulting Party*" is defined in Section 6.1.

"*Default Rate*" means, for any day, unless otherwise agreed by the Parties, that rate of interest from time to time published in the source designated on the Cover Sheet as the prime commercial lending rate, as such rate may change when and as such prime commercial lending rate (or such comparable rate, if such source does not so designate a "prime commercial lending rate") changes, plus two (2) percent per annum; provided that the Default Rate shall never exceed the maximum rate permitted by applicable law.

"*Demarcation Point*" means, with respect to a Bandwidth Unit, the single point of interconnection of Seller and Buyer at each end point applicable to a Segment in a particular physical location designated by Buyer and Seller with respect to such Bandwidth Unit, provided that if the Parties fail to specify a Demarcation Point with respect to any end point of a Segment, the Demarcation Point shall be deemed to be Seller's fiber distribution panel at such end point.

"*Early Termination Date*" is defined in Section 6.1(h)(i).

"*Event of Default*" is defined in Section 6.1.

"*Expiration Date*" means the specified date on which an option or other financial instrument expires.

"*Force Majeure*" means an event that is not within the reasonable control or the result of the negligence of the Party claiming suspension, and which is of such nature and scope as to prevent both the performance of obligations under the Transaction (or part(s) thereof) by the party claiming suspension and similar performance at the same time by other similarly situated third parties involved in the market for Bandwidth (notwithstanding the exercise of due diligence by such Party or third parties to overcome or avoid such events). The term Force Majeure shall include, without limitation, those events that meet the definition in the immediately preceding sentence, and that involve Acts of God, labor disputes, floods, earthquakes, natural disasters, wars, civil disturbances, sabotage, power failures (other than with respect to a failure to obtain or maintain any required power facilities), cable cuts, any law, order, regulation, ordinance or requirement of any governmental or legal authority, or a failure of a Demarcation Point or related facilities within such Demarcation Point. Without limiting the foregoing, the term Force Majeure shall not include (i) the loss of Buyer's markets or Buyer's inability to economically use or resell Bandwidth Units purchased hereunder, (ii) Seller's ability to sell Bandwidth Units in the market at a more advantageous price, (iii) any mechanical or equipment failure (other than any such failure that would otherwise constitute an event of Force Majeure as described above), or (iv) any change in law or regulation after the Trade Date of the affected Transaction unless such change in law or regulation affects the market for Bandwidth generally by preventing performance by all similarly situated parties under similar transactions.

"*GAAP*" means accounting principles that are generally accepted in the jurisdiction specified on the Cover Sheet.

"*Gains*" means, with respect to a Party, an amount equal to the economic benefit (exclusive of Costs), if any, to such Party directly resulting, or which would result directly, from the termination of its rights and obligations under terminated Transactions. The Nondefaulting Party shall calculate Gains in a commercially reasonable manner.

"*Guarantor*" means, in respect of a Party, any guarantor of such Party's obligations hereunder identified on the Cover Sheet.

"*Guaranty*" means the Guaranty, if any, executed by a Party's Guarantor in accordance with the Credit Annex.

"*Legal Costs*" means the reasonable out-of-pocket expenses incurred by a Party, including legal fees and costs of collection, in connection with the enforcement and protection of its rights and remedies under this Agreement.

"*LOA*" means letter of agency.

"*LOA/CFA*" means letter of agency/carrier facilities assignment.

"*Local Access Service*" means, in connection with a Transaction, the facilities connecting Buyer-designated termination points to the Demarcation Points at each end of the Segment.

"*Losses*" means, with respect to a Party, an amount equal to the economic loss (exclusive of Costs), if any, to such Party directly resulting, or which would result directly, from the termination of its rights and obligations under a terminated Transaction (or the applicable part thereof); provided, however, that in the case of an exercised Option, the "Losses" with respect to such Option shall mean: (a) with respect to a Transaction in which the Nondefaulting Party is the Option Buyer, the Market Value of the Option determined in a commercially reasonable manner as of the Early Termination Date; and (b) with respect to a Transaction where the Nondefaulting Party is the Option Seller, zero. The Nondefaulting Party shall calculate Losses in a commercially reasonable manner.

"*Moody's*" means Moody's Investor Services, Inc. or its successor.

"*Nondefaulting Party*" is defined in Section 6.1.

"*Nonrecurring Charges*" means any charges assessed in connection with the use of Bandwidth Units for a Transaction, including, by way of example and not by way of limitation, installation charges or hook-up fees.

"*Option*" means a Transaction granting the right but not the obligation to purchase or sell Bandwidth Units.

"*Option Buyer*" means the Party identified as the Option Buyer in connection with an Option.

"*Option Seller*" means the Party identified as the Option Seller in connection with an Option.

"*Other Support*" means credit support in the form of Letter(s) of Credit or other agreed form of credit support.

"*Outage*" means the unavailability of a Bandwidth Unit during any period of the Term.

"*Outage Credit*" means in respect of an Outage, where Buyer notifies Seller in writing of such Outage pursuant to Section 3.3(a) and such Outage continues, after a trouble ticket has been opened for such Outage, for a period that is equal to or greater than (a) one (1) consecutive hour, or (b) two (2) hours and fifty-two (52) minutes in the aggregate during a Period, a credit that is equal to: the sum of (1) the pro rata portion of the Cover Price applicable to the portion of the Bandwidth Unit affected by such Outage, and (2) the pro rata portion of the Contract Price applicable to the portion of the Bandwidth Unit affected by such Outage.

"*Payment Date*" means, with respect to a Transaction, the last Business Day of the calendar month or if such Day is not a Business Day, the next following Business Day, unless such next following Business Day falls in the next calendar month, in which event "Payment Date" shall mean the immediately preceding Business Day.

"*Performance Assurance*" means credit support in the form of cash or such other form as may be agreed to by the Parties and which may be as set forth in the Credit Annex.

"*Period*" means the period of time agreed to between the Parties in connection with a Transaction. If the Parties fail to agree to a Period applicable to the Transaction, the Period for such Transaction shall be deemed to be thirty (30) Days.

"*Person*" means an individual, partnership, corporation, limited liability company, business trust, joint stock company, trust, unincorporated association, joint venture, firm or other entity, or a government or any political subdivision or agency, department or instrumentality thereof.

"*Planned Network Maintenance*" is defined in Section 4.1.

"*Premium*" means the premium, if any, related to an option that is specified by the Parties.

"*Qualified Institution*" means a commercial bank or trust company (a)(i) organized under the law of the United States or a political subdivision thereof, or (ii) organized under the law of a jurisdiction other than the United States and having one or more branch offices in the United States, and (b) having a Credit Rating of at least "A-" in the case of S&P or "A3" in the case of Moody's.

"*Replacement Price*" means, upon the occurrence of an Outage with respect to any Bandwidth Unit(s) to be provided by Seller to Buyer during any Period, the price at which Buyer, acting in a commercially reasonable manner, purchases or could purchase Bandwidth Unit(s) ("Replacement Bandwidth Unit(s)") substantially similar to such Outage affected Bandwidth Unit(s) (plus Costs reasonably incurred by Buyer in purchasing Replacement Bandwidth Unit(s), including additional provisioning and connection charges, if any, incurred by Buyer) or, absent any such purchase, the market price for such Replacement Bandwidth Unit(s) as determined by Buyer in a commercially reasonable manner. In determining the market price, Buyer may consider, among other valuations, any or all of the settlement prices of any Bandwidth futures contracts, quotations from recognized industry dealers in Bandwidth capacity contracts, and other bona fide third party offers, all adjusted for differences in provisioning and connection charges. For purposes of this definition, Replacement Bandwidth Units shall be deemed to be Bandwidth units having substantially the same characteristics (*e.g.*, location of Segment end points and capacity) as the Outage affected Bandwidth Units and having a term equal to the applicable Outage; provided that if no Replacement Bandwidth Unit is available for such a term, the Replacement Bandwidth Units shall have a term closest to but greater than the term of the applicable Outage. No actual replacement transaction shall be required in order to determine the Replacement Price. In the event that either a Replacement Bandwidth Unit is not available or the market price thereof cannot be determined, the Replacement Price shall be equal to one hundred twenty five percent (125%) of the Contract Price of the Outage affected portion of the Bandwidth Unit.

"*S&P*" means the Standard & Poor's Rating Group (a division of McGraw-Hill, Inc.) or its successor.

"*Secured Party*" is defined in the Credit Annex.

"*Segment*" means a continuous path for Transmission between two Demarcation Points.

"*Seller*" means the Party to a Transaction who is obligated to sell and make available Bandwidth Units for the Term.

"*Strike Price*" means the price that is specified or otherwise determined in the Transaction related to the applicable Option.

"*Taxes*" means any or all ad valorem, Bandwidth, consumption, electronic commerce, excise, fiber optic, gross receipts, privilege, property, occupation, sales, telecommunication, (including without limitation, any tax or charge levied pursuant to the Telecommunications Act of 1996, or any state equivalent) transaction, transport, use, utility, and other taxes, levies, duties, imports, governmental charges, licenses, fees, permits and assessments, or increases therein, other than those based on net income or net worth.

"*Term*" means the time period described in Periods and agreed to by the Parties in connection with a particular Transaction.

"*Time Zone*" means the time zone designated on the Cover Sheet.

"*Trade Date*" means the date on which the Parties agree to enter into a Transaction.

"*Transaction*" means a transaction agreed to between the Parties in accordance with Section 1 hereof, relating to the purchase or sale of Bandwidth Units or any Option and any amendment or modification of such transaction in accordance herewith.

"*Transaction Contract Number*" means contract identifying information assigned to a particular Transaction by a Party in a Confirmation.

"Transfer" means, with respect to any Performance Assurance, payment or interest amount, and in accordance with the instructions of the payee or its Custodian, as applicable:

(a) in the case of cash, payment or delivery by wire transfer into one or more bank accounts specified by the recipient;
(b) in the case of certificated securities that cannot be paid or delivered by book-entry, payment or delivery in appropriate physical form to the recipient or its account accompanied by any duly executed instruments of transfer, assignments in blank, transfer tax stamps and any other documents necessary to constitute a legally valid transfer to the recipient;
(c) in the case of securities that can be paid or delivered by book-entry, the giving of written instructions to the relevant depository institution or other entity specified by the recipient, together with a written copy thereof to the recipient, sufficient if complied with to result in a legally effective transfer of the relevant interest to the recipient;
(d) in the case of other forms of Performance Assurance, as otherwise specified by the payee or its Custodian; and
(e) in the case of Letters of Credit, by delivery of such duly executed instruments as are necessary to evidence the creation of such Letters of Credit by the Qualified Institution in favor of the recipient.

"*Transmission*" means the simultaneous two-way transfer of a signal, message or other form of data between two Demarcation Points.

"*Unit Price*" means the price per Bandwidth Unit per Period agreed to in respect of a relevant Transaction.

"*U.S. Dollar,*" "*Dollar,*" "*US$,*" "*$*" *and* "*USD*" means the lawful currency of the United States of America.

"*Value*" means, for any Day for which Value is calculated, with respect to (a) Performance Assurance that is (i) cash, the amount thereof, or (ii) a security, the bid price obtained by the Party determining the Value, or (b) a Letter of Credit, the lesser of the available face amount and the stated amount thereof.

3. Obligations.

3.1. Purchase and Sale. With respect to each Transaction, Seller agrees to sell and make available, and Buyer agrees to purchase for the relevant Contract Price, Bandwidth Unit(s) for the Term, provided, however, that with respect to Options, the obligations set forth in the preceding sentence shall only arise if Option Buyer exercises an Option in accordance with its terms. Any

Local Access Service charges are the sole responsibility and expense of Buyer. Taxes in connection with any Transaction shall be allocated in accordance with Section 8.

3.2. Availability. As agreed between the Parties in accordance with Section 1 hereof in connection with any Transaction, Seller agrees to make available to Buyer Bandwidth Unit(s) over the Segment that are available for use 99.6% of the time during each Period of the Term. Such availability standard excludes events of Force Majeure or Planned Network Maintenance from its calculation.

3.3. Damages. The remedies set forth in this Section 3.3 shall constitute the sole and exclusive remedies for Buyer, in respect of an Outage, and for Seller, in respect of Buyer's failure to accept Bandwidth Units pursuant to this Agreement prior to the Early Termination Date.

(a) Failure of Seller to Provide Bandwidth Units.

(1) Upon the occurrence of an Outage during any Period within the Term, Seller shall apply an Outage Credit to the payment due from Buyer on the next following Payment Date, as liquidated damages and not as a penalty. Seller shall have no obligation to apply and Outage Credit to the payment due from Buyer for any Outage to the extent such Outage is the result of: (I) Planned Network Maintenance, (ii) Force Majeure, or (iii) the actions of Buyer or Buyer's agents.

(2) To the extent that the Outage Credits due to Buyer pursuant to this Section 3.3(a) exceed the payment due from Buyer on the next following Payment Date, Seller shall pay to Buyer such excess amount, as liquidated damages and not as a penalty, on such next following Payment Date or, in the event there is no such Payment Date, the last Business Day of the Term applicable thereto.

(3) Seller expressly limits its liability to Buyer for any Outage to the damages specified in this Section 3.3(a); provided, however, that in the event that (A) Seller fails to timely credit (or pay, as the case may be) any damage amount, and such failure is not remedied within three (3) Business Days after written notice thereof from Buyer, or (B) in any Period, an Event of Default Triggering Event occurs, such event shall constitute an Event of Default with respect to Seller under Section 6.1(f).

(4) Buyer shall notify Seller of the occurrence of an Outage as soon as practicable after Buyer becomes aware of such Outage. Seller shall open a trouble ticket for the Outage upon receipt of notice of an Outage from Buyer. To receive the applicable Outage Credit, Buyer must request the Outage Credit in writing within five (5) Business Days after the occurrence of such Outage. Failure by Buyer to provide such notice within such time period shall result in a forfeiture of Buyer's right to claim such damages.

(b) Failure of Buyer to Accept Bandwidth Units. Unless excused by Force Majeure or Seller's failure to perform, if Buyer fails to accept any Bandwidth Unit(s) on and as of the start of any Period within the Term, Buyer shall pay Seller, as liquidated damages and not as a penalty, the Contract Price applicable to such Bandwidth Unit in accordance with Section 5.

(c) Satisfaction of Payment Obligation. An event giving rise to a payment obligation by a Party pursuant to this Section 3.3 shall not constitute an Event of Default to the extent that such payment obligation is satisfied on a timely basis.

(d) Dispute on Payment of Liquidated Damages. If Seller disputes Buyer's calculation of the Replacement Price for Replacement Bandwidth Units used to determine the Outage Credit, in whole or in part, the Seller shall within five (5) Business Days of receipt of Buyer's explanation of the calculation of the Outage Credit, pay any undisputed amount and provide to the Buyer a detailed written explanation of the basis for such dispute.

4. Bandwidth Interconnection and Maintenance.

4.1 Interconnection. Seller shall be solely responsible for the provision of all local distribution facilities, ports, interconnection facilities, network equipment, testing equipment and procedures, and the maintenance and repair of the foregoing, up to each relevant Demarcation Point, and for other facilities or actions necessary for it to sell and provide the Bandwidth Units being sold hereunder. Buyer shall be solely responsible for the provision of all local distribution facilities, ports, interconnection facilities, network equipment, testing equipment and procedures, and the maintenance and repair of the foregoing, up to each relevant Demarcation Point, and for other facilities or actions necessary for it to purchase, accept and use any Bandwidth Unit(s) being purchased hereunder. In addition, Buyer shall be responsible for any costs and/or expenses associated with its interconnection to Seller at the Demarcation Point. The Parties agree to cooperate and to provide reasonable assistance to each other in initial testing of each Bandwidth Unit, in accordance with industry standard testing procedures and protocols. Except as otherwise provided herein, each Party shall conduct its operations and use Bandwidth Units in a manner that does not interrupt, impair or interfere with the operations of the other Party's system.

4.2 Planned Network Maintenance.

(a) Timing. Seller shall avoid performing network maintenance between 0600 to 2200 p.m. in the applicable Time Zone, Monday through Friday, inclusive, that will have a disruptive impact on the continuity or performance level of Customer's Service. However, the preceding sentence does not apply to restoration of continuity to a severed or partially severed fiber optic cable, restoration of dysfunctional power and ancillary support equipment, or correction of any potential jeopardy conditions. Seller will use commercially reasonable efforts to notify Buyer prior to emergency maintenance.

(b) Notice. Seller shall provide Buyer with electronic mail, telephone, facsimile, or written notice of all non-emergency, planned network maintenance (i) not less than three (3) Days prior to performing maintenance that, in its reasonable opinion, has a substantial likelihood of affecting Buyer's traffic for up to fifty (50) milliseconds, and (ii) not less than ten (10) Business Days prior to performing maintenance that, in its reasonable opinion, has a substantial likelihood of affecting Buyer's traffic for more than fifty (50) milliseconds ("*Planned Network Maintenance*"). If Seller's planned activity is canceled or delayed, Seller shall promptly notify Buyer and shall comply with the provisions of this Section in rescheduling any delayed activity.

5. Payments/Settlements.

5.1. Payments. On or about the first Business Day of each calendar month after the start of the Term, Seller shall provide a billing statement to Buyer setting forth the amount owed by Buyer in respect of (a) the Bandwidth Units to be provided pursuant to Transactions during the next following calendar month, (b) the Bandwidth Units sold pursuant to Transactions during the then current calendar month and any prior calendar month, to the extent that any payment is due and owing by Buyer in respect of such Bandwidth Unit(s). Such billing statement shall also include any applicable Outage Credits.. Buyer shall pay for all Bandwidth Units made available whether used or not. All amounts payable hereunder are due on the Payment Date of the calendar month in which the billing statement is issued. All payments hereunder shall be Transferred in immediately available funds to payee on or before the due date in respect thereof. Payments shall be sent and made in accordance with the Payments information as set forth on the Cover Sheet. Interest on past due amounts shall accrue at the Default Rate.

5.2. Netting and Setoff. If the Parties are required to pay each other any amount on the same Day for Transactions under this Agreement, then the Parties shall discharge their obligations to pay through netting, in which case, the Party, if any, owing the greater amount shall pay to the other Party the difference between the amounts owed. Each Party reserves to itself all rights, setoffs, counterclaims, combination of accounts, liens and other remedies and defenses which such Party has or may be entitled to (whether by operation of law or otherwise).

5.3. Performance Assurance. If the Parties have elected on the Cover Sheet that this Section 5.3 is applicable, and if a Party that is eligible to request Performance Assurance from the other Party pursuant to this Section 5.3 has reasonable grounds to believe that such other Party's creditworthiness or performance under this Agreement has become unsatisfactory, the requesting Party may provide the other Party with written notice requesting Performance Assurance in an amount determined by the requesting party to be commercially reasonable. Upon receipt of such notice, the other Party shall within five (5) Business Days provide such Performance Assurance to the requesting Party. Any Performance Assurance furnished pursuant to this Agreement shall be deemed to be a "margin payment" within the meaning of Section 101(38) of the United States Bankruptcy Code.

5.4. Payment for Options. Premium shall be paid by Option Buyer within two (2) Business Days of receipt of an invoice from Option Seller. Seller may issue an invoice for the Option Premium on the applicable Trade Date. Upon the exercise of an Option, Buyer shall pay the Strike Price to Seller in accordance with Section 5.1.

6. Events of Default and Remedies.

6.1. Events of Default. An event of default (an "*Event of Default*") with respect to a Party (the "*Defaulting Party*") shall mean any of the following:

(a) The failure of the Defaulting Party to pay any amount when due under this Agreement (other than an amount described in Section 5.3, if applicable), if such failure is not remedied within five (5) Days after receipt of written notice thereof;

(b) The failure of the Defaulting Party to comply with any of its material obligations under this Agreement (other than defaults that are otherwise specifically set forth in this Section 6.1, the failure to provide or accept Bandwidth Unit(s) for which exclusive remedies are provided in Section 3.3, and such failure is not excused by Force Majeure or cured within ten (10) Days after receipt of written notice thereof to the Defaulting Party;

(c) The Defaulting Party shall be subject to a Bankruptcy Proceeding or shall fail to provide or the Defaulting Party's agents shall fail to provide Performance Assurance when required to do so under Section 5.3;

(d) The failure of a Party's Guarantor, if any, to execute its Guaranty, to perform any covenant in its Guaranty, the expiration, termination or cessation of such Guaranty, or the occurrence of a Bankruptcy Proceeding with respect to such Guarantor;

(e) Any representation or warranty made by a Party hereunder or by its Guarantor under its applicable Guaranty shall prove to have been untrue in any material respect when made or deemed made;

(f) The occurrence of a Critical Outage;

(g) With respect to Party A, at any time it or its Guarantor shall have defaulted on its indebtedness to third parties, resulting in its or its Guarantor's obligations in excess of Party A's Cross Default Amount being accelerated or capable of being accelerated, or with respect to Party B, at any time it or its Guarantor shall have defaulted on its indebtedness to third parties, resulting in its or its Guarantor's obligations in excess of Party B's Cross Default Amount becoming accelerated or capable of becoming accelerated;

(h) The failure of a Party to Transfer Performance Assurance or Other Support to the Secured Party when due under this Agreement or to otherwise comply with any provision of any applicable Credit Annex, if such failure is not remedied within three (3) Business Days after receipt of written notice thereof.

(i) *[The occurrence of a Material Adverse Change with respect to the Defaulting Party; provided that such Material Adverse Change shall not be considered an Event of Default if the Defaulting Party Transfers and maintains, for so long as such Material Adverse Change is continuing, Performance Assurance or Other Support in favor of the Nondefaulting Party in form and amount acceptable to the Nondefaulting Party.]*

6.2. Nondefaulting Party's Rights for Events of Default. Upon the occurrence and during the continuation of an Event of Default as to the Defaulting Party (other than an Event of Default as described in clause (f), above), the other party (*"Nondefaulting Party"*) may, in its sole discretion,

(a) Accelerate and liquidate the Parties' respective obligations in respect of all Transactions outstanding under this Agreement by giving not more than twenty (20) Days' notice to the Defaulting Party declaring a date (*"Early Termination Date"*) (which shall be no earlier than the date of such notice) on which date such Transactions shall terminate; and

(b) Suspend performance of its obligations (including the making of any payment) in respect of all Transactions under this Agreement.

6.3. Nondefaulting Party's Rights for Section 6.1(f) Event of Default. If the Event of Default is one described in clause 6.1(f) above, Buyer may, in its sole discretion,

(a) Accelerate and liquidate the Parties' respective obligations with respect to only those Transactions affected by the Critical Outage by giving not more than twenty (20) Days' notice to the Defaulting Party declaring an Early Termination Date (which shall be no earlier than the date of such notice) in respect of such Bandwidth Unit(s) on which date such obligations with respect to such Bandwidth Unit(s) shall terminate; and

(b) Suspend its performance of obligations (including the making of any payment) with respect to only those Transactions affected by the Critical Outage for which the Buyer has declared an Early Termination Date.

(c) If the Buyer has elected to declare an Early Termination Date for the Transaction applicable to those Bandwidth Units affected by the Critical Outage, this Agreement shall remain in effect without prejudice to the Buyer's rights under this Section 6 to declare an Early Termination Date as to such remaining Transactions, Bandwidth Units (or part(s) thereof) upon a subsequent Event of Default.

6.4. Early Termination Date. If an Early Termination Date is declared under this Section 6, the Early Termination Date will occur on the designated date, whether or not the relevant Event(s) of Default is then continuing. Any rights of a Nondefaulting Party under this Section 6 shall be in addition to such Party's other rights under this Agreement and at law, to the extent not waived hereunder.

6.5. Early Termination Payment.

(a) If an Early Termination Date is declared pursuant to Sections 6, the Nondefaulting Party shall in good faith calculate its Gains, Losses and Costs, resulting from the terminated Transactions or part(s) thereof. Such Gains, Losses and Costs with respect to all terminated Transactions (or part(s) thereof) shall be aggregated with all damages payable under Section 3.3 into a single net amount, and the Non Defaulting Party shall notify the other Party of such net amount owed (such notice being the "Damage Notice"). The Gains, Losses and Costs shall be determined by the Non Defaulting Party by comparing the value (as determined by it) of each terminated Transaction (or part(s) thereof), had it not been terminated to the Replacement Price for each such terminated Transaction (or part(s) thereof). If the Non Defaulting Party's aggregate Losses and Costs and amounts due to it under Sections 3.3, if any, exceed its aggregate Gains, the other Party shall, within five (5) Days of its receipt of the Damage Notice, pay the amount of such excess owed to the Calculating Non Defaulting Party as an early termination payment ("Early Termination Payment"). Such Early Termination Payment shall be paid together with interest thereon (before as well as after judgement), from (and including) the relevant Early Termination Date to (but excluding) the date such amount is paid at the Default Rate, plus any other unpaid amounts (but, as to the terminated Transactions or part(s) thereof, excluding any unpaid amounts which are taken into account in the definition of "Losses") owing by the other Party under this Agreement. Such interest will be calculated on the basis of daily compounding and the actual number of days elapsed. If the Non Defaulting Party's aggregate Gains exceed its aggregate Losses and Costs and amounts due to it under Sections 3.3, if any, resulting from such early termination, the Party shall, after giving effect to any setoff rights, pay the amount of such excess as an Early Termination Payment without interest to the other Party on or before the date twenty (20) Days after the Early Termination Date. The Non Defaulting Party shall determine its Gains, Losses and Costs as of the Early Termination Date, or, if that is not possible, at the earliest date thereafter that is reasonably possible. This Section shall not be construed as an exception or limitation to the Parties rights and agreement in Section 9 which limits both Parties' liability for consequential, incidental, special, punitive, exemplary or indirect damages, loss profits or other business interruption damages.

(b) If an Event of Default occurs and/or an Early Termination Date is designated and occurs, the Non Defaulting Party may (at its election) set off any or all amounts which the other Party owes to the Non Defaulting Party or its Affiliates (under this Agreement or otherwise) against any or all amounts which the Non Defaulting Party owes to the other Party (whether under this Agreement or otherwise). Notwithstanding any provision to the contrary contained in a Confirmation, the Non Defaulting Party shall not be required to pay to the other Party any amount under the Agreement until the Non Defaulting Party receives confirmation satisfactory to it in its reasonable discretion that (i) all amounts due and payable as of the Early Termination Date by the other Party under all Transactions with the Non Defaulting Party or any of its

Affiliates listed on the Cover Sheet have been fully and finally paid, and (ii) all other obligations of any kind whatsoever of the other Party to make any payments to the Non Defaulting Party or any of its Affiliates listed on the Cover Sheet under this Agreement or otherwise have been fully and finally performed.

(c) The Parties acknowledge and agree that any amount payable by one Party to the other under this Section 6.2 shall constitute liquidated damages, that actual damages are difficult to ascertain and that such liquidated damages are a reasonable estimation of actual damages and are not a penalty.

(d) Dispute on Payment of Early Termination Payment. If the Defaulting Party disputes the Nondefaulting Party's calculation of the Early Termination Payment, in whole or in part, the Defaulting Party shall, within five (5) Business Days following receipt of Nondefaulting Party's explanation of the calculation of the Early Termination Payment, provide to the Nondefaulting Party a detailed written explanation of the basis for such dispute; provided, however, that if the Early Termination Payment is due from the Defaulting Party, the Defaulting Party shall pay any undisputed amount and transfer into an escrow account with a Qualified Institution a payment Nondefaulting in an amount equal to the disputed amount of the Termination Payment.

6.6. Expenses. The Defaulting Party shall, on demand, indemnify and hold harmless the Nondefaulting Party for any and all direct Costs, incurred by the Nondefaulting Party in connection with the enforcement of its rights under this Agreement; provided, that no amount paid by the Defaulting Party shall be duplicative of any amount previously paid by the Defaulting Party to the Nondefaulting Party.

6.7. Early Termination. If an Early Termination Date is designated or deemed to occur, an amount equal to the Value of the Performance Assurance, if any, held by a Party, determined as of the Early Termination Date, will be deemed to be a Gain of such Party for purposes of Section 6.2.

7. Force Majeure. In the event that either Party is rendered unable, wholly or in part, by an event of Force Majeure to carry out its obligations to provide or accept any Bandwidth Unit under a Transaction, such Party shall give oral notice and full details of such event of Force Majeure to the other Party as soon as practicable after the occurrence thereof and shall provide to the other Party written details of such event of Force Majeure within five Business Days after the date of such oral notice. The obligations of the Parties with respect to such Bandwidth Unit(s) under such Transaction shall be suspended to the extent required by such event and during its continuance. The Party claiming Force Majeure shall make reasonable efforts to mitigate the effects of such event of Force Majeure with reasonable dispatch. If an event of Force Majeure persists for a continuous period of thirty (30) Days after the date the event of Force Majeure is declared, the Party not claiming the Force Majeure shall have the option, upon three (3) Days' notice, to terminate the affected Transaction(s) (or part(s) thereof) and the obligations of the Parties thereunder (other than payment obligations for prior performance thereunder). Upon the occurrence of an event of Force Majeure, Bandwidth Units not scheduled during the applicable Term by reason of such event shall not be rescheduled except by mutual agreement of the Parties.

8. Taxes. Allocation of and Indemnity for Taxes. Buyer is liable for and shall pay, cause to be paid (or reimburse Seller if Seller has paid) all Taxes applicable to the Transaction, including any Taxes imposed or collected by a taxing authority with jurisdiction over Buyer, unless Buyer

has presented Seller with a valid tax exemption certificate. Buyer agrees to pay any such applicable Taxes and to indemnify and hold Seller harmless from any Claims for such Taxes.

9. Limited Liability for Damages; Exclusion of Certain Warranties. TO THE EXTENT THAT AN EXPRESS REMEDY OR MEASURE OF DAMAGES IS PROVIDED IN THIS AGREEMENT FOR ANY BREACH, SUCH EXPRESS REMEDY OR MEASURE OF DAMAGES SHALL BE THE SOLE AND EXCLUSIVE REMEDY OF A PARTY. THE LIABLE PARTY'S LIABILITY SHALL BE LIMITED AS SET FORTH IN SUCH PROVISION AND ALL OTHER REMEDIES OR DAMAGES AT LAW OR IN EQUITY ARE WAIVED UNLESS OTHERWISE EXPRESSLY PROVIDED IN THIS AGREEMENT. IF NO REMEDY OR MEASURE OF DAMAGES IS EXPRESSLY HEREIN PROVIDED, THE LIABLE PARTY'S LIABILITY SHALL BE LIMITED TO DIRECT, ACTUAL DAMAGES ONLY, SUCH DIRECT, ACTUAL DAMAGES SHALL BE THE SOLE AND EXCLUSIVE REMEDY HEREUNDER AND ALL OTHER REMEDIES OR DAMAGES AT LAW OR IN EQUITY ARE WAIVED. NOTWITHSTANDING ANYTHING HEREIN TO THE CONTRARY, NEITHER PARTY SHALL BE LIABLE FOR CONSEQUENTIAL, INCIDENTAL, SPECIAL, PUNITIVE, EXEMPLARY OR INDIRECT DAMAGES, LOST PROFITS OR OTHER BUSINESS INTERRUPTION DAMAGES, IN TORT, CONTRACT, UNDER ANY INDEMNITY PROVISION OR OTHERWISE. THE PARTIES ACKNOWLEDGE THE DUTY TO MITIGATE DAMAGES. Without limiting any of the foregoing, Seller assumes no responsibility for the security and confidentiality of any data or other information Transmitted over any Bandwidth Unit. Seller shall have no responsibility for the security or confidentiality of Buyer's data or information, and Seller makes no representation, warranty or assurance as to information or its security or confidentiality. Seller shall make no effort to validate Buyer's data and/or information for content, correctness or usability. Seller makes no representation, warranty or assurance that Buyer's equipment and data will be compatible with any Bandwidth Unit being sold or with any connection or interconnection with the Demarcation Points at each end of a Segment.

10. Representations, Warranties and Covenants. On the date of entering into this Agreement and on the date of entering into each Transaction,

(a) Each Party represents and warrants to the other Party that:

(1) It is (A) a commercial participant (provider, commercial user or merchant dealing with Bandwidth Units) who in connection with its activities, incurs risk, in addition to price risk, related to Bandwidth Units, has a demonstrated capacity or ability directly or through separate bona fide contractual arrangements, to provide or accept Bandwidth Units under the terms of a Transaction and (B) is either (1) a corporation, partnership or proprietorship, organization, trust or other business entity with a net worth exceeding $1,000,000 or has total assets exceeding $5,000,000 or (2) an entity, the obligations of which under this Agreement are guaranteed or otherwise supported by a letter of credit or keepwell support, or other agreement by any such entity or by an entity referred to in subsection (B)(1) above,

(2) It is duly organized, validly existing and in good standing under the laws of the jurisdiction of its formation and is qualified to conduct its business,

(3) It has all authorizations, licenses and consents necessary to legally perform its obligations under this Agreement and each Transaction,

(4) The execution, delivery and performance of this Agreement and each Transaction are within its power, have been duly authorized by all necessary action and do not violate its governing documents or any law applicable to it,

(5) This Agreement constitutes, and each Transaction when entered into in accordance with this Agreement will constitute, a legally valid and binding obligation enforceable against it in accordance with its terms, subject to any equitable defenses,

(6) With respect to each Transaction involving the purchase or sale of an Option, it is entering into such Transaction for purposes related to its line of business,

(7) There are no Bankruptcy Proceedings pending or being contemplated by it or to its knowledge, threatened against it,

(8) There are no legal proceedings that materially adversely affect its ability to perform under this Agreement or any Transaction,

(9) It is a forward contract merchant within the meaning of the United States Bankruptcy Code,

(10) It has entered into this Agreement and each Transaction in connection with the conduct of its business and it has the capacity or ability to make or take delivery of all Bandwidth Units referred to in any Transaction to which it is a Party,

(11) It is a producer, processor, commercial user or merchant, dealing with Bandwidth Units, and it is entering into such Transaction for purposes related to its business as such,

(12) The material economic terms are subject to individual negotiation between the Parties,
(xiii) it has entered into this Agreement and each Transaction as principal (and not as agent, advisor or fiduciary) and with a full understanding of the material risks and terms thereof, and is capable of assuming those risks, and

(13) It has not received from the other Party any promise, advice or assurance as to the expected performance or result of this Agreement and any and all Transactions.

Each Party covenants that it will cause the representations and warranties set forth in this Section 10(a) to be true and correct on the date hereof, on the date any Transaction is entered into and throughout the term of each Transaction.

(b) Each Party represents and warrants to the other Party (which representation and warranty will be deemed repeated as of each date on which it Transfers Performance Assurance or Other Support (or any other amounts Transferred pursuant to the Credit Annex, if any)) that it is the sole owner of or otherwise has the right to transfer such Performance Assurance, Other Support or other amounts it Transfers to the other Party pursuant to this Agreement, free and clear of any security interest, lien, encumbrance or other restriction (other than a lien routinely imposed on all securities in a relevant clearance system).

(c) If requested by a Party, the other Party shall deliver (i) within 120 Days following the end of each fiscal year, a copy of its annual report and/or the annual report of its Guarantor containing audited consolidated financial statements for such fiscal year certified by independent certified public accountants, and (ii) within 60 Days after the end of each of its first three fiscal quarters of each fiscal year, its unaudited consolidated financial statements and/or the unaudited consolidated financial statements of its Guarantor for such fiscal quarter. In all cases, such financial statements shall be for the most recent accounting period and shall be prepared in accordance with GAAP or such other accounting principles then in effect in the relevant jurisdiction and on a basis consistent with prior financial statements. In the event that any requested financial statement is not available due to a delay in preparation or certification, such delay shall not be considered a default with respect to a Party so long as such Party diligently pursues the preparation, certification and delivery of such statements.

(d) Buyer represents and warrants to Seller that it shall not use any Bandwidth Unit for any illegal purpose or in any other unlawful manner.

11. Arbitration.

Option A: If the Parties have elected on the Cover Sheet to have this Option A apply, any dispute, controversy or claim arising out of, connected with, or relating in any way to this Agreement shall be resolved by binding arbitration governed by the U.S. Federal Arbitration Act ("FAA") and conducted in accordance with the American Arbitration Association ("AAA") Commercial Arbitration Rules. Nothing herein shall, however, prohibit a Party from seeking temporary or preliminary injunctive relief in a court of competent jurisdiction. The number of arbitrators shall be three, each Party having the right to appoint one arbitrator, who shall together appoint a third neutral arbitrator with at least five years professional experience in either the telecommunications backbone network industry or the commodity trading industry, within 30 Days after the appointment of the last party-designated arbitrator. The Parties expressly waive any right of appeal to any court. All arbitration proceedings shall take place in New York, New York (or another location mutually agreed to by the Parties) and shall be conducted in the English language. Only damages allowed pursuant to this Agreement may be awarded and the arbitrators shall have no authority to award treble, exemplary, consequential, indirect or punitive damages of any kind under any circumstances regardless of whether such damages may be available at law for the relevant Transaction or under the FAA or AAA. All substantive issues arising in an arbitration brought pursuant to this provision shall be resolved by application of the provisions of the Agreement and the governing law selected by the Parties on the Cover Sheet, and the arbitrators shall have no authority to add to, modify or change any provision of the Agreement. Any arbitration brought pursuant to this provision and any related award shall be confidential and not disclosed, except to a Party's Affiliates, accountants and counsel. Judgment upon any award granted in a proceeding brought pursuant to this provision may be entered in any court of competent jurisdiction.

Option B: If the Parties have elected on the Cover Sheet to have this Option B apply, any dispute, controversy or claim arising out of, connected with, or relating in any way to this Agreement shall be resolved by binding arbitration governed by the Rules of Arbitration of the International Chamber of Commerce ("ICC"). Nothing herein shall, however, prohibit a Party from seeking temporary or preliminary injunctive relief in a court of competent jurisdiction. The number of arbitrators shall be three, each Party having the right to appoint one arbitrator, who shall together appoint a third neutral arbitrator with at least five years professional experience in either the telecommunications backbone network industry or the commodity trading industry, within 30 Days after the appointment of the last party-designated arbitrator. The Parties expressly waive any right of appeal to any court. All arbitration proceedings shall take place in the location selected on the Cover Sheet and shall be conducted in the English language. Only damages allowed pursuant to this Agreement may be awarded and the arbitrators shall have no authority to award treble, exemplary, consequential, indirect or punitive damages of any kind under any circumstances regardless of whether such damages may be available at law for the relevant Transaction or under the FAA or AAA. All substantive issues arising in an arbitration brought pursuant to this provision shall be resolved by application of the provisions of the Agreement and the governing law selected by the Parties on the Cover Sheet, and the arbitrators shall have no authority to add to, modify or change any provision of the Agreement. Any arbitration brought pursuant to this provision and any related award shall be confidential and not disclosed, except to a Party's Affiliates, accountants and counsel. Judgment upon any award granted in a proceeding brought pursuant to this provision may be entered in any court of competent jurisdiction.

12. General.

12.1. Successors and Assigns; Assignment. Transactions arising under this Agreement shall be binding on and inure to the benefit of, and may be performed by, the respective successors and assigns of the Parties, except that no assignment, pledge, or other transfer by either Party (the "Assigning Party") shall operate to release the Assigning Party from any of its obligations under this Agreement unless: (i) consent to such release is given in writing by the Non-Assigning Party, which consent shall not be unreasonably withheld or delayed; (ii) such assignment, pledge or transfer is made to an Affiliate of the Assigning Party and such Affiliate is at least as creditworthy as the Assigning Party, or (iii) such assignment, pledge or transfer is incident to a merger, reorganization, consolidation or other transaction in which substantially all of the assets of the Assigning Party are transferred to another Person who assumes all of the obligations of the Assigning Party under this Agreement and such Person is at least as creditworthy as the Assigning Party.

12.2. Warranties. OTHER THAN AS EXPRESSLY PROVIDED IN SECTION 10 OR A CONFIRMATION, SELLER MAKES NO OTHER REPRESENTATION OR WARRANTY, WRITTEN OR ORAL, EXPRESS OR IMPLIED, INCLUDING BUT NOT LIMITED TO, ANY REPRESENTATION OR WARRANTY THAT BANDWIDTH UNIT(S) SOLD WILL BE FIT FOR A PARTICULAR PURPOSE OR WILL BE MERCHANTABLE.

12.3. Notices. All notices required or permitted to be given hereunder in writing shall, unless expressly provided otherwise, be in writing, properly addressed, postage pre-paid and delivered by hand, facsimile, overnight mail or courier. A notice will be deemed effective as indicated: (a) if in writing and delivered in person or by courier, on the date it is delivered, (b) if sent by facsimile transmission, on the date that transmission is received, (c) if sent by registered or certified mail (airmail if overseas) or the equivalent (return receipt requested) on the date that delivery is attempted, or (d) if sent by electronic messaging system, on the date that the electronic message is received, unless the date of delivery (or attempted delivery) or receipt, as applicable, is not a Business Day, or such electronic message is delivered (or delivery is attempted) or received, as applicable, after the close of business in the location of the recipient on a Business Day, in which case it shall be deemed given and effective on the next following Day that is a Business Day. A Party may change its address by providing notice thereof in accordance with this Section.

12.4. Confidentiality. If the Parties have elected on the Cover Sheet to make this Section 12.4 applicable, neither Party shall disclose the terms of any Transaction to a third party (other than the employees, lenders, counsel or accountants of the Party and its Affiliates or prospective purchasers, directly or indirectly, of a Party or all or substantially all of a Party's assets or of any rights under this Agreement, provided such Persons shall have agreed to keep such terms confidential) except (i) in order to comply with any applicable law, order, regulation or exchange rule, or (ii) to the extent necessary to implement or enforce any Transaction. Each Party shall notify the other Party of any proceeding of which it is aware which may result in disclosure of the terms of any Transaction (other than as permitted hereunder) and use reasonable efforts to prevent or limit the disclosure. The provisions of this Agreement, other than the terms of any Transaction, are not subject to this confidentiality obligation. The Parties shall be entitled to all remedies available at law or in equity (and the Parties acknowledge for these purposes that monetary damages may not provide an adequate remedy) to enforce, or seek relief in connection with this confidentiality obligation; provided that all monetary damages shall be limited in accordance with Section 9.

12.5. Governing Law. THIS AGREEMENT AND EACH TRANSACTION AND THE RIGHTS AND DUTIES OF THE PARTIES ARISING HEREFROM AND THEREFROM SHALL BE GOVERNED BY AND CONSTRUED, ENFORCED AND PERFORMED IN ACCORDANCE WITH THE LAWS OF THE JURISDICTION DESIGNATED ON THE COVER SHEET, WITHOUT GIVING EFFECT TO THE CONFLICTS OR CHOICE OF LAWS RULES OF SUCH JURISDICTION. IF THE LAWS OF THE STATE OF NEW YORK ARE DESIGNATED AS APPLICABLE ON THE COVER SHEET, IT IS AGREED THAT EACH TRANSACTION SHALL BE ENFORCEABLE AS A "QUALIFIED FINANCIAL CONTRACT" WITHIN THE MEANING OF NEW YORK GENERAL OBLIGATIONS LAW § 5-701(b).

12.6. Entire Agreement; Amendments; Joint Work Product. This Agreement and any Annexes and Exhibits hereto are a part hereof, and each Transaction, including any Confirmation, the written text of any recorded telephonic communications or electronically exchanged communications forming a binding agreement in respect of a Transaction, constitute the entire agreement between the Parties relating to the subject matter contemplated by this Agreement, and supersede any prior or contemporaneous agreements or representations affecting the same subject matter. Except for any matters which, in accordance with the express provisions of this Agreement, may be resolved by verbal agreement between the Parties, no amendment, modification or change to this Agreement or any Transaction shall be enforceable unless reduced to a writing and executed by the Party against whom such amendment, modification or change is sought to be enforced and unless specifically referencing this Agreement and the relevant Transaction. This Agreement is the joint work product of the Parties and has been negotiated by the Parties and their respective counsel and shall be fairly interpreted in accordance with its terms and, in the event of any ambiguities, no inferences shall be drawn against either Party.

12.7. Counterparts; Severability. This Master Agreement, any Credit Annex and each Confirmation (if executed) may be executed in several counterparts, each of which is an original and all of which constitutes one and the same instrument. Except as may otherwise be stated herein, any provision or Section hereof that is declared or rendered unlawful by any applicable court of law or regulatory agency, or deemed unlawful because of a statutory change, will not otherwise affect the lawful obligations that arise under this Agreement or a Transaction. In the event any provision of this Agreement or a Transaction is declared unlawful, the remainder of this Agreement and each Transaction shall survive and remain in full force and effect and the Parties will promptly renegotiate to restore this Agreement including all Transactions, as near as possible to its original intent and economic effect.

12.8. Non-Waiver; No Partnership or Third Party Beneficiaries. No waiver by any Party of any of its rights with respect to the other Party or with respect to this Agreement, any Transaction, or any matter or default arising in connection with this Agreement, shall be construed as a waiver of any other right, matter or default. Any waiver shall be in writing signed by the waiving Party. Neither Party shall be deemed to be the employee, agent, partner, joint venturer or contractor of any other Party under or in connection with this Agreement. This Agreement and each Transaction is made and entered into for the sole benefit of the Parties, and their permitted successors and assigns, and no other Person shall be a direct or indirect legal beneficiary of, have any rights under or have any direct or indirect cause of action or claim in connection with this Agreement.

12.9. Nature of Rights. All rights related to Bandwidth Units purchased and sold under this Agreement and all obligations incurred under this Agreement are purely contractual in nature. Nothing contained in this Agreement shall have the effect of granting to Buyer any leasehold, ownership, proprietary or possessory rights in any physical telecommunications system or facility. In the event of a dispute involving both Parties with a customer of one Party, both Parties shall assert the applicability of any limitations on liability to customers that may be contained in either Party's applicable contracts or tariff(s). Nothing in this Agreement shall be construed as the grant of a license, either express or implied, with respect to any patent, copyright, trademark, trade name, trade secret or any other proprietary or intellectual property now or hereafter owned, controlled or licensable by either Party. Neither Party may use any patent, copyrightable materials, trademark, trade name, trade secret or other intellectual property right of the other Party except in accordance with the terms of a separate license agreement between the Parties granting such rights.

12.10. Survival. The Parties' obligations under Sections 6, 8, 9, 11, 12.4, 12.5, 12.11 and 12.12 of this Agreement shall survive the termination or expiration of this Agreement.

12.11. Indemnity.

(a) Each Party shall defend, indemnify and hold harmless the other against and from any and all Claims for physical property damage, personal injury or wrongful death to the extent that such Claims arise out of the negligence or willful misconduct of the respective indemnifying party or either of their respective employees, agents, or contractors in connection with the provision of Bandwidth Units or other performance hereunder.

(b) With respect to parties that use the Bandwidth Units provided by the Seller hereunder, Buyer shall defend, indemnify and hold harmless Seller against any Claims by such parties arising or resulting from any defect in or failure to provide the Bandwidth Units sold hereunder.

(c) The indemnifying party agrees to defend the other Party against Claims and to pay all reasonable Legal Costs, settlement payments, and any damages awarded or resulting from such Claims. A Party entitled to indemnification pursuant to this Agreement shall, with respect to any claim made against such indemnified Party for which indemnification is available, notify the other Party in writing of the nature of the claim as soon as practicable but not more than ten (10) days after the indemnified Party receives notice of the assertion of the claim. (The failure by an indemnified Party to give notice as provided, above, shall not relieve the indemnifying Party of its obligations under this Section 12.11, except to the extent that the failure results in the failure of actual notice and the indemnifying Party is damaged as a result of the failure to give notice.) Upon receipt of notice of the assertion of a claim, the indemnifying Party shall employ counsel reasonably acceptable to the indemnified Party and shall assume the defense of the claim. The indemnified Party shall have the right to employ separate counsel and to participate in (but not control) any such action, but the fees and expenses of such counsel shall be at the expense of the indemnified Party unless (a) the employment of counsel by the indemnified counsel has been authorized by the indemnifying Party, (b) the indemnified Party has been advised by its counsel in writing that there is a conflict of interest between the indemnifying Party and the indemnified Party in the conduct of the defense of the action (in which case the indemnifying Party shall not have the right to direct the defense of the action on behalf of the indemnified Party), or (c) the indemnifying Party has not in fact employed counsel to assume the defense of the action within a reasonable time following receipt of the notice given pursuant to this Section 12.11, in each of which cases the fees and expenses of such counsel shall be at the expense of the indemnifying Party. An indemnifying Party shall not be liable for any settlement of an action effected without its written consent (which consent shall not be unreasonably withheld), nor shall an indemnifying Party settle any such action without the written consent of the indemnified Party (which consent shall not be unreasonably withheld). No indemnifying Party will consent to the entry of any judgment or enter into any settlement that does not include as an unconditional term thereof the giving by the claimant or plaintiff to the indemnified Party a release from all liability with respect to the claim. Each Party shall cooperate in the defense of any claim for which indemnification is available and shall furnish such records, information, testimony and attend such conferences, discovery proceedings, hearings, trials and appeals as may reasonably be requested by the other Party.

12.12. Forward Contract. This Agreement shall constitute a "forward contract" under and in all Bankruptcy Proceedings and will be treated similarly under and in all Bankruptcy Proceedings (regardless of the jurisdiction of application or competence of such law), rulings, orders, directives or pronouncements, made pursuant thereto.

IN WITNESS WHEREOF, the Parties hereto have made and executed this Agreement, signed by their respective duly authorized officers or individuals, as of the Effective Date.

[COUNTERPARTY]	[COMPANY]
By: ______________________________	By: ______________________________
Name: ______________________________	Name: ______________________________
Title: ______________________________	Title: ______________________________

EXHIBIT A
TO THE
MASTER BANDWIDTH PURCHASE AND
SALE AGREEMENT
(FORM OF CONFIRMATION) Model Form of Confirmation
(Bandwidth Purchase and Sale)

Date: ________________
To: ______________________________

Fax No. _____________
[This Transaction is subject to (and this Confirmation is provided pursuant to and in accordance with) the Master Bandwidth Purchase and Sale Agreement dated ____________, (which, as it may be amended from time to time, is herein called the "Master Agreement"). All terms used herein and not otherwise defined shall have the meanings set forth in the Master Agreement. This Confirmation shall confirm the Transaction ("Transaction") agreed to pursuant to [a telephone conversation] [an exchange of electronic communications between the Parties' representatives] on the Trade Date (defined below) by and between ________________ ("Company") and ____________________ ("Counterparty") (collectively, the "Parties")][2]. [This confirmation (this "Confirmation") evidences a complete and binding agreement between ________________________ ("Company") and _________ ("Counterparty") (collectively, the "Parties") as to the terms of the transaction to which this Confirmation relates. In addition, Company and Counterparty agree to use all reasonable efforts promptly to negotiate, execute and deliver an agreement in the form of the Master Bandwidth Purchase and Sale Agreement [attached hereto][previously delivered by Company to Counterparty] (the "Master Agreement"), with such modifications as Company and Counterparty will in good faith agree. Upon execution by the Parties of such an agreement, this Confirmation will supplement, form a part of, and be subject to that agreement. All provisions contained in or incorporated by reference in that agreement upon its execution will govern this Confirmation. Until we execute and deliver that agreement, this Confirmation, together with all other documents referring to the Master Agreement and confirming transactions (each a "Transaction") entered into between the Parties (notwithstanding anything to the contrary in a Confirmation), shall supplement, form a part of, and be subject to an agreement in the form of the Master Agreement as if we had executed an agreement in such form on the Trade Date of the first such Transaction between the Parties. In the event of any inconsistency between the provisions of that agreement and this Confirmation, this Confirmation will prevail for the purpose of this Transaction.][3] The terms of the particular Transaction to which this Confirmation relates are as follows:

TRANSACTION CONTRACT NUMBER: [_________]

Trade Date: [mm/dd/yy]

Seller: [Company][Counterparty]

Buyer: [Company][Counterparty]

Term: [ ] Month[s]
Commencing [12:00:00 a.m. (midnight)] EST on [INSERT DATE] and
Terminating [11:59:59 p.m.] EST on [INSERT DATE]

[2] Insert this provision if a Master Agreement is in place between Company and Counterparty.

[3] Insert this provision if there is NO Master Agreement in place between Company and Counterparty.

Segment: A Location :
Suite:
Floor:
City:
State:
Zip Code:
NPA-NXX:

Z Location:
Suite:
Floor:
City:
State:
Zip Code:
NPA-NXX:

Bandwidth Unit: __________ to be made available over the Term

[DS-Os per
Bandwidth Unit: [ ]

Number of Bandwidth
Units: [______]

[Vertical & Horizontal (DS-0)
Miles for Segment: [ ]]

Quoted Price: US$_______________ (price per DS-0 mile per Month)

Unit Price: US$ _____________ per Month (calculated as follows: Quoted Price x DS-0 per Bandwidth Unit x the number of Vertical & Horizontal (DS-0) Miles for Segment).

Contract Price: Unit Price multiplied by the number of Bandwidth Units

[[this Confirmation])]

Demarcation Information:

Payment: Buyer to pay Seller the Contract Price on or before each Payment Date during the Term.

Governing Law: New York.

Conditions & General Terms: This Transaction (and Confirmation) constitutes part of and is subject to the provisions of the Master Agreement.

Additional Provisions:
Please confirm that the foregoing correctly reflects the agreement between the Parties as to this Transaction by returning an executed copy of this Confirmation by facsimile to Company at the Company's fax number specified below for Confirmations. If Counterparty does not return this

Confirmation or otherwise object to the accuracy of this Confirmation by written notice to Company within three Days of receipt of this Confirmation, this Confirmation shall be deemed correct as given and shall constitute, conclusive and binding evidence of this Transaction, and the final expression of the terms of this Transaction. This Confirmation supersedes any broker confirmation concerning this Transaction.

[NOTE: This provision should be consistent with the Master Agreement.]

Sincerely,

[COMPANY]

By:______________________________
Name:____________________________
Title:_____________________________

CONFIRMED:
[COUNTER PARTY NAME]

By:______________________________
Name:____________________________
Title:_____________________________

<u>Technical Contacts:</u>

Company: Counterparty:
Full Name: Full Name:
Phone No.: Phone No.:
Fax No.: Fax No.:
Pager: Pager:
Cellular (Optional): Cellular (Optional):
Email Address: Email Address:

EXHIBIT B
to the
MASTER BANDWIDTH PURCHASE AND SALE AGREEMENT

FORM OF CREDIT ANNEX

THIS CREDIT ANNEX ("Credit Annex") is attached to and made a part of the Master Bandwidth Purchase and Sale Agreement ("Master Agreement"), dated as of ________________, 200_, by and between ______________________ ("Party A") and ________________ "("Party B").

1. Definitions. The following terms, when used in this Credit Annex, shall have the meanings set forth in this Section 1. Capitalized terms used in this Credit Annex and not defined in this Section 1 or elsewhere in this Credit Annex shall have the meanings ascribed to them in the Master Agreement.

"*Credit Rating*" means with respect to a Party (or its Guarantor or the issuer of a Letter of Credit, as the case may be) or entity, on any date of determination, the respective ratings then assigned to the unsecured, senior long-term debt or deposit obligations (not supported by third party credit enhancement) of such Party (or its Guarantor or the issuer of a Letter of Credit, as the case may be) or entity by S&P, Moody's or the other specified rating agency or agencies.
"*Letter of Credit*" means one or more irrevocable, standby letters of credit issued by a Qualified Institution.
"*Material Adverse Change*" means, with respect to a Party, such Party having a credit rating below the credit rating applicable to such Party, if any, indicated on the Cover Sheet of the Master Agreement to which this Credit Annex is attached.
"Performance Assurance" means credit support in the form of (i) cash, (ii) a Letter of Credit from a Qualified Institution in a form acceptable by the Beneficiary Party (as defined below), or (iii) such other form of credit support as may be reasonably acceptable to the Beneficiary Party.
"Qualified Institution" means a commercial bank or trust company (a)(i) organized under the law of the United States or a political subdivision thereof, or (ii) organized under the law of a jurisdiction other than the United States and having one or more branch offices in the United States, and (b) having a Credit Rating of at least "A-" in the case of S&P or "A3" in the case of Moody's.

"Termination Payment Threshold" means the amount with respect to each Party, if applicable as set forth on the Cover Sheet of the Master Agreement to which this Credit Annex is attached.

2. Performance Assurance Requirement. If at any time and from time-to-time during the term of the Master Agreement (notwithstanding whether an Event of Default has occurred), the Early Termination Payment that would be owed to a Party in respect of all Transactions, if then terminated, exceeds the Termination Payment Threshold, then such Party (the "Beneficiary Party") on any Business Day, may request the other Party (the "Posting Party") to provide Performance Assurance in such form as the Beneficiary Party shall reasonably request. Such Performance Assurance shall have a value equal to or in excess of the difference between (a) such Early Termination Payment, and (b) the Threshold Amount (rounding upwards for any fractional amount to the next nearest [Dollar]), if positive, subject to the Minimum Delivery Amount. In determining the value of any Performance Assurance, the Beneficiary Party may, in its sole discretion, discount the fair market value of any such Performance Assurance in order to reflect the actual proceeds that the Beneficiary Party could reasonably expect to receive upon liquidation of, or the exercise of its rights under and in respect of, such Performance Assurance. The Posting Party shall deliver performance assurance within two (2) Business Days after the date of such

request. On any Business Day (but no more frequently than weekly with respect to Letters of Credit and daily with respect to cash), the Posting Party, at its sole cost and expense, may request that Performance Assurance be returned or reduced to the extent of any reduction of the amount of the Early Termination Payment (calculated as set forth above in this Section) (rounding upwards for any fractional amount to the next nearest [Dollar]).

3. Holding and Use of Performance Assurance. (a) *Care of Performance Assurance.* Without limiting the Beneficiary Party's rights under Section 6.6(c), the Beneficiary Party will exercise reasonable care to assure the safe custody of all Performance Assurance to the extent required by applicable law, and in any event, the Beneficiary Party will be deemed to have exercised reasonable care if it exercises at least the same degree of care as it would exercise with respect to its own property. Except as specified in the preceding sentence, the Beneficiary Party will have no duty with respect to Performance Assurance, including, without limitation, any duty to collect any dividends or other distributions, or enforce or preserve any rights pertaining thereto.

(b)(i) *General.* The Beneficiary Party will be entitled to hold Performance Assurance or to appoint an agent (a "Custodian") to hold Performance Assurance for the Beneficiary Party. Upon notice by the Beneficiary Party to the Posting Party of the appointment of a Custodian, the Posting Party's obligations to make any Transfer will be discharged by making the Transfer to that Custodian. The holding of Performance Assurance by a Custodian will be deemed to be the holding of that Performance Assurance for which the Custodian is acting.

(c) *Use of Performance Assurance.* (i) Without limiting the rights and obligations of the Parties under this Agreement, if the Beneficiary Party is not a Defaulting Party and no Early Termination Date has occurred or been designated as a result of an Event of Default with respect to the Beneficiary Party, then the Beneficiary Party will, notwithstanding Section 9-207 of the New York Uniform Commercial Code, have the right to:

(i) sell, pledge, re-hypothecate, assign, invest, use, commingle or otherwise dispose of, or otherwise use in its business any Performance Assurance it holds, free from any claim or right of any nature whatsoever of the Posting Party, including any equity or right of redemption by the Posting Party; and

(ii) register any Performance Assurance in the name of the Beneficiary Party, its Custodian or a nominee of either.

For purposes of the obligation to transfer Performance Assurance pursuant to this Section 3 and any rights or remedies authorized under this Agreement, the Beneficiary Party will be deemed to continue to hold all Performance Assurance and to receive dividends and other distributions made thereon, regardless of whether the Beneficiary Party has exercised any rights with respect to any Performance Assurance pursuant to (A) or (B) above.

If a Party is an Ineligible Party (the event that causing it to be ineligible to hold Performance Assurance being referred to as a "Credit Rating Event"), then:

the provisions of Section 3(c)(i) will not apply with respect to the Ineligible Party as the Beneficiary Party; and

(b) the Ineligible Party shall be required to deliver (or cause to be delivered) not later that the close of business on the second Business Day following such Credit Rating Event all Performance Assurance in its possession or held on its behalf to a Qualified Institution approved by the non-Ineligible Party (which approval shall not be unreasonably withheld) to a segregated, safekeeping or custody account ("Collateral Account") within such Qualified Institution with the title of the Collateral Account indicating that the property contained therein is being held as Performance Assurance for the Beneficiary Party. The Qualified Institution shall serve as Custodian with respect to the Performance Assurance in the Collateral Account, and shall hold such Performance Assurance in accordance with the terms of this Section 1.2 and for the security interest of the Ineligible Party and, subject to such security interest, for the ownership of the non-Ineligible Party.

4. Interest, Dividends and Other Amounts. Unless otherwise specified by the Beneficiary Party, interest shall accrue on any Performance Assurance in the form of cash at the Federal Funds Effective Rate. So long as no Event of Default with respect to the Posting Party has occurred and is continuing, and to the extent that an obligation to deliver Performance Assurance would not be created or increased, the Beneficiary Party shall Transfer to the Posting Party, any interest, dividends or other amounts paid with respect to the Performance Assurance on the last Business Day of the calendar month in which such interest, dividends or other amounts were received by the Beneficiary Party. On or after the occurrence of an Event of Default with respect to the Posting Party, the Beneficiary Party shall retain any such interest, dividends or other amounts received by the Beneficiary Party in respect of the Performance Assurance until all obligations of the Posting Party under this Agreement have been satisfied.

5. Events of Default under the Credit Annex. An Event of Default will include: (a) with respect to a Party , if at any time such Party or its Guarantor shall have defaulted on its indebtedness to third parties, resulting in obligations on the part of the such Party or its Guarantor in excess of the amount specified on the first pages of the Master Agreement to which this Credit Annex is attached being accelerated or capable of being accelerated,; (b) the failure of the Posting Party to deliver Performance Assurance to the Beneficiary Party when due under this Agreement or to otherwise comply with Section 3 hereof; (c) the occurrence of a Material Adverse Change with respect to the Defaulting Party; provided, that such Material Adverse Change shall not be considered an Event of Default if the Defaulting Party establishes and maintains for so long as such Material Adverse Change is continuing, Performance Assurance in favor of the Nondefaulting Party in form and amount acceptable to the Nondefaulting Party.

6. Guaranty Agreement. If indicated as applicable on the first pages of the Master Agreement to which this Credit Annex is attached, the Parties may, in order to secure all payment obligations of a Party or the Parties, require such other Party or Parties to cause their respective Guarantor(s) to execute and deliver to the other Party a guaranty agreement in the amount of specified on such pages, which guaranty agreement shall be substantially in the form of Exhibit C-1 attached hereto or in such other form as may be mutually agreed upon by the Parties.

EXHIBIT C-1
to the
MASTER BANDWIDTH PURCHASE AND SALE AGREEMENT

FORM OF GUARANTY AGREEMENT

This Guaranty Agreement (this "Guaranty"), dated as of ________________, is made and entered into by ______________, a corporation ("Guarantor").

W I T N E S S E T H:

WHEREAS, ________________ (the Company will enter into a Master Bandwidth Purchase and Sale Agreement (the "Agreement") effective as of the date of this Guaranty with ____________________________ ("Counterparty") pursuant to which Company and Counterparty may enter into Transactions related to the purchase and sale of Bandwidth Units (as defined in the Agreement); and

WHEREAS, Guarantor will directly or indirectly benefit from the Agreement.

NOW THEREFORE, in consideration of Counterparty entering into the Agreement, Guarantor hereby covenants and agrees as follows:

1. GUARANTY. Subject to the provisions hereof, Guarantor hereby irrevocably and unconditionally guarantees the timely payment when due of the obligations under the Agreement of Company and any Affiliate (as defined in the Agreement) of Company or any other entity to which Company shall assign, pledge or transfer such obligations pursuant to Section 12.1(2) or (3) and (4) of the Agreement (the "Obligations") to Counterparty in accordance with the Agreement. Upon an assignment, pledge or transfer of the Obligations by Company pursuant to Section 12.1(2) or (3) and (4) of the Agreement, all references herein to "Company" shall be deemed to be references to the Affiliate (as defined in the Agreement) of Company or the other entity to which Company has so assigned, pledged or transferred the Obligations. To the extent that Company shall fail to pay any Obligations, Guarantor shall promptly pay to Counterparty the amount due. This Guaranty shall constitute a guaranty of payment and not of collection or performance. The liability of Guarantor under the Guaranty shall be subject to the following:

(a) Guarantor's liability hereunder shall be and is specifically limited to payments expressly required to be made in accordance with the Agreement (even if such payments are deemed to be damages) and, except to the extent specifically provided in the Agreement, in no event shall Guarantor be subject hereunder to consequential, exemplary, loss of profits, punitive, tort, or any other damages, costs, equitable relief or attorney's fees.

(b) The aggregate amount payable under this Guaranty shall not exceed _________ U.S. Dollars ($___________).

2. DEMANDS AND NOTICE. If Company fails or refuses to pay any Obligations, Counterparty shall notify Company in writing of such failure or refusal and demand that payment be made by Company. If Company's failure or refusal to pay continues for a period of fifteen (15) days after the date of Counterparty's notice to Company, and Counterparty has elected to exercise its rights under this Guaranty, Counterparty shall make a demand upon Guarantor (hereinafter referred to as a "Payment Demand"). A Payment Demand shall be in writing and shall reasonably and briefly specify in what manner and what amount Company has failed to pay and shall include an explanation of why such payment is due, with a specific statement that Counterparty is calling upon Guarantor to pay under this Guaranty. A Payment Demand satisfying the foregoing requirements shall be deemed sufficient notice to Guarantor that it must pay the Obligations. A single written Payment Demand shall be effective as to any specific default during the continuance of such default, until Company or Guarantor has cured such default, and additional written demands concerning such default shall not be required until such default is cured.

3. REPRESENTATIONS AND WARRANTIES. Guarantor represents and warrants that:

(a) It is a corporation duly organized and validly existing under the laws of the State of _______ and has the corporate power and authority to execute, deliver and carry out the terms and provisions of the Guaranty;

(b) No authorization, approval, consent or order of, or registration or filing with, any court or other governmental body having jurisdiction over Guarantor is required on the part of Guarantor for the execution and delivery of this Guaranty; and

(c) This Guaranty constitutes a valid and legally binding agreement of Guarantor, except as the enforceability of this Guaranty may be limited by the effect of any applicable bankruptcy, insolvency, reorganization, moratorium or similar laws affecting creditors' rights generally and by general principles of equity.

4. SETOFFS AND COUNTERCLAIMS. Without limiting Guarantor's own defenses and rights hereunder, Guarantor reserves to itself all rights, setoffs, counterclaims and other defenses to which Company or any other affiliate of Guarantor is or may be entitled to arising from or out of the Agreement or otherwise, except for defenses arising out of the bankruptcy, insolvency, dissolution or liquidation of Company.

5. AMENDMENT OF GUARANTY. No term or provision of this Guaranty shall be amended, modified, altered, waived or supplemented except by a writing signed by the parties hereto.

6. WAIVERS. Guarantor hereby waives (a) notice of acceptance of this Guaranty; (b) presentment and demand concerning the liabilities of Guarantor, except as expressly hereinabove set forth; (c) any right to require that any action or proceeding be brought against Company or any other person, and (d) except as expressly hereinabove set forth, any right to require that Counterparty seek enforcement of any performance against Company or any other person, prior to any action against Guarantor under the terms hereof.

Except as to applicable statutes of limitation, no delay of Counterparty in the exercise of, or failure to exercise, any rights hereunder shall operate as a waiver of such rights, a waiver of any other rights or a release of Guarantor from any obligations hereunder.

Guarantor consents to the renewal, compromise, extension, acceleration and any other changes in the time of payment of or other changes in the terms of the Obligations, or any part thereof and any changes or modifications to the terms of the Agreement.

Guarantor may terminate this Guaranty by providing written notice of such termination to Counterparty and upon the effectiveness of such termination, Guarantor shall have no further liability hereunder, except as provided in the last sentence of this paragraph. No such termination shall be effective until one (1) business day after receipt by Counterparty of such termination notice. No such termination shall affect Guarantor's liability with respect to any Transaction (as defined in the Agreement) entered into prior to the time the termination is effective, which Transaction shall remain guaranteed pursuant to the terms of this Guaranty.

7. NOTICE. Any Payment Demand, notice, request, instruction, correspondence or other document to be given hereunder by any party to another (herein collectively called "Notice") shall be in writing and delivered personally or mailed by certified mail, postage prepaid and return receipt requested, or by telecopier, as follows:

To Counterparty:

____

____

- Attn.: ________________________

- Fax No.: ________________________

To Guarantor: ____

Attn.: ________________________
Fax No.: ________________________

Notice given by personal delivery or mail shall be effective upon actual receipt. Notice given by telecopier shall be effective upon actual receipt if received during the recipient's normal business hours, or at the beginning of the recipient's next business day after receipt if not received during the recipient's normal business hours. All Notices by telecopier shall be confirmed promptly after transmission in writing by certified mail or personal delivery. Any party may change any address to which Notice is to be given to it by giving notice of such change of address as provided above.

8. MISCELLANEOUS. THIS GUARANTY SHALL IN ALL RESPECTS BE GOVERNED BY, AND CONSTRUED IN ACCORDANCE WITH, THE LAWS OF THE STATE OF [NEW YORK], WITHOUT REGARD TO PRINCIPLES OF CONFLICTS OF LAWS. This Guaranty shall be binding upon Guarantor, its successors and assigns and inure to the benefit of and be enforceable by Counterparty, its successors and assigns. The Guaranty embodies the entire agreement and understanding between Guarantor and Counterparty and supersedes all prior agreements and understandings relating to the subject matter hereof. The headings in this Guaranty are for purposes of reference only, and shall not affect the meaning hereof. This Guaranty may be executed in any number of counterparts, each of which shall be an original, but all of which together shall constitute one instrument.

EXECUTED as of the day and year first above written.

- **[GUARANTOR]**
-
-
- By:____________________
- Name:________________________
- Title:_________________________

Printed in Great Britain
by Amazon

62165012R00222